Computational Context

The Value, Theory and Application of Context with AI

Editors

William F. Lawless
Professor, Math & Psychology
Paine College
Augusta, Georgia, USA

Ranjeev Mittu
Branch Head
Naval Research Laboratory
Washington, D.C., USA

Donald A. Sofge
Computer Scientist
Navy Center for Applied
Research in Artificial Intelligence
Naval Research Laboratory
Washington, D.C., USA

CRC Press is an imprint of the
Taylor & Francis Group, an **informa** business
A SCIENCE PUBLISHERS BOOK

MATLAB® and Simulink® are trademarks of The MathWorks, Inc. and are used with permission. The MathWorks does not warrant the accuracy of the text or exercises in this book. This book's use or discussion of MATLAB® and Simulink® software or related products does not constitute endorsement or sponsorship by The MathWorks of a particular pedagogical approach or particular use of the MATLAB® and Simulink® software.

CRC Press
Taylor & Francis Group
6000 Broken Sound Parkway NW, Suite 300
Boca Raton, FL 33487-2742

Printed on acid-free paper
Version Date: 20180810

International Standard Book Number-13: 978-1-138-32064-2 (Hardback)

Library of Congress Cataloging-in-Publication Data

Names: Lawless, William F. (William Frere), 1942- editor. | Mittu, Ranjeev, editor. | Sofge, Donald A., editor.
Title: Computational context : the value, theory and application of context with AI / editors, William F. Lawless, Professor, Math & Psychology, Paine College, Augusta, Georgia, USA, Ranjeev Mittu, Branch Head, Naval Research Laboratory, Washington, D.C., USA, Donald A. Sofge, Computer Scientist, Navy Center for Applied, Research in Artificial Intelligence, Naval Research Laboratory, Washington, D.C., USA.
Description: Boca Raton, FL : Taylor & Francis Group, [2018] | "A science publishers book." | Includes bibliographical references and index.
Identifiers: LCCN 2018033205 | ISBN 9781138320642 (hardback : acid-free paper)
Subjects: LCSH: Computational intelligence. | Artificial intelligence. | Context-aware computing.
Classification: LCC Q342 .C597 2018 | DDC 006.3--dc23
LC record available at https://lccn.loc.gov/2018033205

Visit the Taylor & Francis Web site at
http://www.taylorandfrancis.com

and the CRC Press Web site at
http://www.crcpress.com

Preface

This book was inspired by an Association for the Advancement of Artificial Intelligence (AAAI) Symposium, March 2017 on "Computational context: Why it's important, what it means, and can it be computed?" This book is not a proceedings of the conference; instead, it is based on revisions of some of the presentations at the symposium, and it includes chapters by authors who did not attend the event. Inspired by the symposium, this book has a revised title *"Computational context: The value, theory and application of context with AI."* The goal of this book is to address the current state of the art of determining context computationally by examining the gap in the existing research that must be addressed to better integrate human systems and autonomous systems (both machines and robots). The research offered to the public in this book helps to advance the next generation of systems that are already planned or might be planned in the near future, ranging from autonomous machines and robots to teams of autonomous systems that support or provide for better support to human operators of these systems, decision makers and even society at large.

The book explores how context figures into social explanations about the humans are affected by the environment and its influences on human perception, cognition and action. Context can be clear, uncertain or an illusion. Examples of clear context could be a car with a flat tire; a toothache; a marriage license. It could be the word sequence in a sentence that allows humans to interpret an unknown word. Biomedical research uses context-specific language. The ever-changing context in an organization context is affected by market factors from day-to-day, but also by the management of the organization, its culture and its location and geography.

Examples of uncertain contexts spring readily to mind; e.g., the fog of war; a judge's reaction to unplanned counter-arguments; a shout of "Fire!" on a ship or in a theater.

But illusions and deceptions are common in social reality. Awareness may factor in to determining context. This book explores whether those who respond to a given context must be aware, as is required in the case of deception or an illusion. It is noteworthy that individuals are often not aware of why they act differently as the context shifts; e.g., they act differently whether alone, with another person or in a team.

Can computational context with AI adapt to clear, uncertain and unaware contexts, to contexts that change over time, and to whether context is affected by arbitrary combinations of individuals, machines and robots as part of a team?

The Introduction in this book describes the current state of the art for AI research in computational context. We have introduced various themes in this book and the contributions from world-class researchers, scientists and practitioners. The chapters that follow are elaborations on key research and engineering topics at the heart of AI for effective and efficient integration of human-machine-robot systems. These topics include, for example, the analyses of computational context in intelligence communities; the challenge of verifying the situation for today's and future autonomous systems; comparisons between humans and machines today and their evolution; the impact of human information on AI and context to avoid errors that may impede missions or lead to human tragedy (e.g., avoiding a train accident); systems able to reason to computationally determine context; context needed to establish autonomy; and hybrid teams, where the term hybrid represents arbitrary combinations of humans, machines and robots.

The contributions to this book are written by many of the leading scientists and practitioners across the field of autonomous agent research, ranging from academia to industry and government. Given the wide diversity represented by the research in this volume, we strove to thoroughly examine the challenges and trends of systems that implement and exhibit Artificial Intelligence (AI); the implications for society of systems made autonomous with AI; systems with AI seeking to build on established and new relationships trusted by humans, machines, and robots; and the effective human systems integration that must result for trust in these new systems and their applications to be diminished, sustained or increased.

William F. Lawless
Ranjeev Mittu
Donald A. Sofge

Content

1

Introduction

W.F. Lawless,[1,*] *Ranjeev Mittu*[2] and *Donald Sofge*[3]

1.1 Introduction

An Association for the Advancement of Artificial Intelligence (AAAI) symposium organized and held at Stanford in Spring 2017 inspired this book (details about this symposium, including the agenda, are in the Preface of this book). After the symposium was completed, the conference organizers solicited book chapters from those who participated in the symposium and in the broader AI community. While most of the chapters included in this book were contributed by authors who attended the symposium, other chapters were contributed by authors who did not attend. In this introduction we review briefly the symposium and then introduce the individual contributed chapters that follow.

1.2 Background of the 2017 Symposium

Our symposium was held at Stanford in 2017 and titled "Computational context: Why it's important, what it means, and can it be computed?" The title for this book evolved: *"Computational context: The value, theory and application of context with AI"*. The symposium was organized by Ranjeev Mittu, Branch Head, Information Management & Decision Architectures

[1] Paine College, Departments of Math & Psychology, Augusta, GA.
[2] Branch Head, Information Management & Decision Architectures Branch (CODE 5580), Information Technology Division, U.S. Naval Research Laboratory, 4555 Overlook Avenue, SW, Washington, DC 20375; ranjeev.mittu@nrl.navy.mil
[3] Computer Scientist, Distributed Autonomous Systems Group, Code 5514, Navy Center for Applied Research in Artificial Intelligence, Naval Research Laboratory, 4555 Overlook Avenue S.W., Washington D.C. 20375; don.sofge@nrl.navy.mil
* Corresponding author: w.lawless@icloud.com

Branch, Information Technology Division, US Naval Research Laboratory, Washington, DC; Donald Sofge, Computer Scientist, Distributed Autonomous Systems Group, Navy Center for Applied Research in Artificial Intelligence, US Naval Research Laboratory, DC; and W.F. Lawless, Paine College, Departments of Mathematics & Psychology, Augusta, GA.

The 2017 symposium on computational context addressed the increasing use of artificial intelligence (AI) to manage and reduce the threats to complex systems composed of individual machines and robots as well as to teams, including hybrid teams composed arbitrary numbers of humans, machines, and robots. AI has been useful in modeling the defenses of individuals, organizations, and institutions, as well as the management of social systems. However, foundational problems remain for continuing the development of AI with autonomy for individual agents and teams, especially with objective measures able to optimize their function, performance and composition. We dedicate this book to addressing one of these problems, namely "context".

Supposedly, context accounts for how the environment influences human thinking and behavior. It can be clear, uncertain or illusory. Clear contexts, like giving money to a street beggar, are universally understood. Uncertain contexts, like those surrounding a loud argument in a strange language, impede human behavior and choices. Illusory contexts are apparently clear contexts that may surround the use of deception by one human against another to hide corruption or to participate in an illicit behavior, e.g., a cyber-attack (see Chapter 13).

Alternatively, some contexts are difficult for humans to perceive, especially when change is occurring or when joining or exiting a group. Bragging to a friend is less risky than bragging to a boss, an authority figure (a judge) or a charismatic celebrity. But bragging to, or intimidating, an opponent may be necessary to impede an undesired, opposing action. These complications are often subtle shades that may greatly affect behavior, as a consequence, increasing the difficulty of deriving a computational context.

But if a software program can automatically "know" the context that improves performance, it may not matter whether the context is clear, uncertain or illusory. In preparing for combat, the Department of Defense needs individual agents and hybrid teams that are capable of "having a common perception of the surrounding world and able to place it into context" (RCTA, 2016). DoD's needs present a computational challenge that we believe can be achieved by integrating systems to work for the members of individual agents and teams, an opportunity to advance the science of context for individual agents and teams, one of our research interests.

Knowing the context is critical for the establishment of an autonomous agent operating in a social setting. AI approaches often attempt to address autonomy by modeling aspects of individual human decision-making or behavior. Behavioral theory is either based on modeling the individual, such as through cognitive architectures or, more rarely, and with more difficulty, through group dynamics and interdependence theory. Approaches focusing on the individual assume that individuals are more stable than the social interactions in which they engage. Interdependence theory assumes the opposite, that a state of mutual dependence among participants in an interaction affects the individual and group beliefs and behaviors of participants whether these behaviors are perceived, recognized or not. The latter is conceptually more complex, but both approaches must satisfy the demand for manageable outcomes as autonomous agents, teams or systems grow in importance and number. Prediction in social systems is presently considered a human skill that can be enhanced (Tetlock and Gardner, 2015). But the skill of prediction in social affairs has been found to be wanting, whether in political polling, economics, stock markets or government policies (reviewed in Lawless, 2017, e.g., Tetlock and Gardiner's first superforecast failed to predict Brexit and their second failed to predict Trump would become President; see: http://goodjudgment.com/superforecasting/index.php/2016/11/03/is-donald-trump-mr-brexit/).

Despite its theoretical complexity, including the inherent uncertainty and nonlinearity wrought by social interdependence, we argue that complex autonomous systems must consider multi-agent interactions in order to develop manageable, effective and efficient individual agents and hybrid teams. Important examples include cases of supervised autonomy, where a single human oversees several interdependent autonomous systems; where an autonomous agent is working with a team of humans, such as the cyber defense of a network; or where an autonomous agent is intended to replace effective, but traditionally worker-intensive team tasks, such as warehousing and shipping. Autonomous agents that seek to fill these roles, but do not consider the interplay between the participating entities, will likely disappoint.

This book offers readers with or without an AI background the opportunity to address these and other fundamental issues about the need for autonomous systems to be able to determine the "computational contexts" within which they operate, whether clear, uncertain, or illusory, and whether static or changing.

1.3 Relevance of the 2016 Symposium

Human or machine errors, accidents or failures can dramatically change a context, whether unexpected or intended (e.g., a terror attack or an "active shooter" event). We introduced aspects of this problem earlier at our symposium at Stanford in 2016, titled "AI and the mitigation of human error: Anomalies, team metrics and thermodynamics"; it was organized by the same four individuals as was the 2016 symposium.

Human errors can dramatically affect context. AI has the potential to mitigate human error by reducing car accidents, airplane accidents, and other mistakes made mindfully or inadvertently by individual humans or by teams. One worry about the bright future with AI is that jobs may be lost. Another is from the perceived and actual loss of human control with common carriers like trains, ships and airplanes. For example, despite the loss of all aboard several commercial airliners in recent years, commercial airline pilots reject being replaced by AI (e.g., Markoff, 2015).

Posed by AI is the context of an existential threat that AI might somehow conspire to end the existence of humans, raised by physicist Stephen Hawking, entrepreneur Elon Musk and computer billionaire Bill Gates. While recognizing what these leaders have said, Etzioni (Etzioni and Etzioni, 2016), CEO of the Allen Institute for Artificial Intelligence and Professor of Computer Science at the University of Washington, has disagreed. Regardless, we must know the contexts that our autonomous agents possess of us as humans.

Across a wide range of occupations and industries, human error and human performance is a primary cause of accidents (Hollnagel, 2009). In general aviation, the FAA attributed accidents primarily to skill-based errors and poor decisions (e.g., Wiegmann et al., 2005).

Making a context more stable to dampen the sources of human error, safety is an important area but one that organizations often skimp on to save money. The diminution of safety coupled with human error led to the explosion in 2010 that destroyed the Deepwater Horizon in the Gulf of Mexico (USDC, 2012). Human error emerges as a top safety risk in the management of civilian air traffic control (Moon et al., 2011). Human error was the cause attributed to the recent sinking of Taiwan's Ocean Researcher V in the fall of 2014 (Showstack, 2014). Human behavior is a leading cause of cyber breaches (Howarth, 2014).

Human-caused accidents often lack situational awareness (context), caused by a convergence to one-sided or incomplete beliefs until it is too late ("the engineer made a comment regarding an over speed condition ... the application of the locomotive's brakes just before [the accident]"; in NTSB, 2017), or by emotional decision-making (for example, the Iranian Airbus flight erroneously downed by the USS Vincennes in 1988; in Craig et al., 2004). Other contexts where human errors occur include poor problem diagnoses; poor planning, communication and execution; and poor organizational functioning.

By better determining the context, AI might be able to mitigate the human role in the cause of accidents, e.g., reducing a problem unsuspected by a team, like an impending suicide that involves the team (the German copilot, Libutz, who killed 150 aboard his Germanwings commercial aircraft; in Levs et al., 2015); and by mitigating mistakes in operational contexts by military commanders who act without knowing the dangers their troops face (for example, the 2001 sinking of the Japanese tour boat by the USS Greeneville; in NTSB, 2001).

AI can mitigate human error by finding anomalies in human operations before an accident or a terrorist attack, and by discovering, for example, when teams have gone awry, whether AI should intercede in the affairs of humans, e.g., by taking control of a train speeding in a zone when it should not be speeding.

1.4 Contributed Chapters

The second chapter, "Learning Context through Cognitive Priming", was written by Laura Hiatt, Wallace Lawson and Mark Roberts, all with the U.S. Naval Research Laboratory in Washington DC; they have written that the context of the larger environment affects the way humans think. They believe that this "priming" by the context can be learned or copied by computational systems (i.e., with machine perception). In this view, context is the glue of shared understanding. Then, if attention is focused at any one instant on the items or objects in memory, context connects the related items together that humans experience. Theirs is a well-established model of priming. With it, priming memory illuminates the associated items that have been learned to recognize and identify an object in a scene. But it can do more by connecting newly related elements. New objects can be recognized, aberrance detected, sequences of decisions recognized, and new actions expressed. With the priming of items as a focus of attention in memory, the authors argue that the features of a context can be parsed, re-constructed and learned. Priming connects elements of context jointly experienced, but it also avoids connections among independent items encountered separately. Priming means that when focused on elements in the external world, context can be refined with internal states and thoughts. The authors describe a computational approach to cognitive context with priming that demonstrates its value for learning, applying primed context to object recognition, anomaly detection, decision-making, tool selection and action choices. Overall, with extensive work in their chosen area, this project based on computational priming for machines is becoming a mature and well-established program of experimental study with a bright future. In the authors' pursuit of high-level reasoning to improve the performance of autonomous machines, one of the areas of future research might be with the context shared between agents, an agent and an operator, or an agent in a team.

The third chapter, "The Use of Contextual Knowledge in a Digital Society", was written by Shu-Heng Chen in Taipei, Taiwan and Ragupathy Venkatachalam in London, UK. Shu-Heng Chen is a Distinguished Professor, Department of Economics, Director of the AI-ECON Research Center and Vice President of the National Chengchi University in Taipei; he is also the Economics Editor-in-Chief of the New Mathematics and Natural Computation, the Journal of Economic Interaction and Coordination; the Editor of Economia Politica, the Global and Local Economic Review, and the International Journal of Financial Engineering and Risk Management; and he is the Associate Editor of Computational Economics and Evolutionary and Institutional Economics Review. Ragupathy Venkatachalam is a lecturer in economics at the Institute of Management Studies, Goldsmiths, University of London. In their chapter, the authors state that the advancements in Information and Communication Technology (ICT), the Internet and big data tools more and more determine the context where humans generate, store, communicate or exchange (price) information. The authors examine the effect that this somewhat seeming aggregation by "centralized" ICT-enabled platforms has on contexts. These new platforms, the authors argue, contrast with knowledge generation by the widely dispersed stock market. To make their points, with their model, they reprise today the old debate that took place among economists in the past over the Socialist Calculation Debate for establishing the context where the value of goods was calculated centrally in the absence of money setting the price of goods; however, this central aggregation performed poorly. Given this debate, ICT-enabled platforms are attempting to centralize the same calculations but, among other things, when they take over these calculations, they face many of the same intractable problems as before, i.e., the aggregation of contextual, tacit knowledge; the lack of effective coordination devices; as well as the lack of criteria to discipline the data generated. Even if ICT is not effective as a centralised aggregation mechanism, because the context of the stock-market is complex, it continues to motivate decentralised, non-market platforms (markets without prices) to perform the same function not only to better understand the stock market, but also to at least disperse information. Overall, the authors offer a cautionary study of the complexity of centralizing the computation of context to mindfully construct knowledge as well as the value of learning how the stock market performs the same function mindlessly by aggregating and dispersing information and constructing knowledge. This area of research is very rich, but its complexity calls for more and more details at the micro-level to better understand how the market determines the context shared at the market level.

Kristin E. Schaefer, Derya Aksaray and Julia Wright are with the Robotics Collaborative Technology Alliance (RCTA) in the Human Research and Engineering Directorate, US Army Research Laboratory, Aberdeen Proving Ground, MD; and Nicholas Roy is with the Massachusetts Institute of Technology (MIT) in the Computer Science and Artificial Intelligence Laboratory; they have written Chapter four on "Challenges with addressing the issue of context within AI and human-robot teaming". The authors address context for human-robot teams. They write that context-driven AI is critical for robotic systems of the future faced by dynamic and uncertain environments when making complex decisions. The authors describe two main difficulties with developing context-driven artificial intelligence. First, there is not an agreed upon representation or model for the subtleties of context that can be directly integrated into computational reasoning. As a result, it is difficult to have a unified understanding of research needs with advanced algorithms and intelligence architectures; in fact, slight changes in a context can produce significant changes in its interpretation. The authors consider a series of models to tackle this challenge. But they conclude, however, that if these models and representations can be developed, they expect that the cues for a context can be incorporated into robots to enhance their capability to know the world, advancing human-agent teams with the shared communication of intent in a given context. Interpreting events in the world depend on understanding the shared context held by individual agents. Thus, developing context-driven AI without considering the human interaction will not necessarily improve future teaming initiatives unless all involved agree on the shared context. Another problem directly impact shared communication is the development of team cognition with humans and robots. The authors review the current terminology for computational contexts with an overview of the open context-dependent models and representations and research challenges. Their long-term challenge is to establish shared human-robot contexts; their immediate goal is to better understand the impact of context-driven AI on effective human-robot teaming. Overall, the authors report from the field with actual robots. Although their past research was focused on individual robots, they are shifting their focus to teams involving a context shared between humans and robots. With limited results to date, establishing the context for a robot with rules is proving to be difficult for a well-constrained setting, but hard for a social setting. While there are no breakthoughs yet evident, we agree with the authors that the subtlety in a shared context is a difficult challenge for the future, but one with a large payoff.

Chapter five was written by Luke Marsh of Defence Science and Technology Group in Edinburgh, Australia; and Iryna Dzieciuch and Douglas S. Lange with the Space and Naval Warfare Systems Center-Pacific (SPAWAR) in San Diego, CA. The title of their chapter is "Machine Learning Approach for Task Generation in Uncertain Contexts". The authors note that the command and control of unmanned vehicles is mentally difficult and intensive for human operators, especially those operators who are working with uncertainty where the contexts are unknown. They state that efficient and successful operator performance depends on many parameters, like training, human abilities, human factors, timing and situational awareness (these form the context in an environment that includes patterns that change over time). Humans can multitask in uncertain environments, process (context) situational data and use autonomous agents across different contexts. But the process can easily lead to information overloads that adversely impact missions even with computational task managers. The authors conclude that while the contexts that change are a challenge that remains unsolved, these contexts need to be addressed computationally so that the optimization and control of unmanned vehicles can shift to become autonomous. Today's models with machine learning are providing a number of benefits by running continuously, collecting data and providing state-space searches across decision domains to plan and execute missions. However, they require real time task optimization and coordination by establishing context from the temporal-spatial patterns in an environment. In addition, the rate of events in an uncertain environment is always changing, making these models inefficient when tasks proliferate. In the pursuit of autonomy, the authors propose a model for decision-making that can be used in uncertain contexts under different levels of complexity through the optimization of assignment coordination. Overall, the authors define a suite of models that can be used in uncertain contexts for different complexities to work with the optimization and coordination of task assignments. The authors lay out a long and difficult path ahead for the research that needs to be conducted to achieve autonomy with agents operating under uncertainty.

Chapter six, "Creating and Maintaining a World Model for Automated Decision Making", was written by Hope Allen and Donald Steiner, who are both with Northrop Grumman Corporation. Allen is a Product Manager, Autonomous Vehicles, with Nvidia Corporation in Santa Clara, CA. Her coauthor, Steiner, formerly with Northrop Grumman Corporation in McLean, VA, was the Product Area Architect for Autonomy and Cognition as well as a Technical Fellow. (Donald Steiner passed away shortly after revising the manuscript; the editors would like to honor him by recognizing his wit, his intellect and his superb contribution to our book in the chapter he co-authored with Hope Allen.). Allen and Steiner present a computational approach to context by modeling knowledge in the world; if successful, their "World Model" will establish the context for decision-making by autonomous agents along with aids for making decisions. The authors begin by defining terms from the Defense Science Board, including autonomy and automation; they use these terms as they focus on sensors and models as a prelude to decisions and then actions. Their techniques allow them to pursue an in-depth study of the contexts in war across the domains of air, land, maritime, space and cyberspace. The authors focus on satisfying missions instead of controlling vehicles; they address how autonomous systems can work in human and machine teams to fulfill a commanding officer's goal in a war zone. By autonomy, the authors mean the ability to complete missions without control by humans or other machines all the while sensing and adapting to a changing environment. They use 'cognitive' to mean the ability to learn from experience or from others to detect the unforeseen changes in the context and the ability to reason about these changes and other unplanned events while still achieving the mission. Most importantly, in their view, a core agent model must be able to achieve its objectives even when the context becomes uncertain. With this in mind, based primarily on individual agents but with the goal of teams working on a single task, they have designed and implemented a framework to automate the military's OODA (Observe, Orient, Decide and Act) loop for systems of individual autonomous agents to enable the seamless transition from automation to autonomy. Overall, the authors provide a well-defined and promising approach to the development of autonomy for a system based primarily on individual heterogeneous agents but with the goal of an autonomous team dedicated to perform tasks. Overall, the authors are following a broad research path to the development of context for application to systems of autonomous agents and autonomous teams. They anticipate problems with developing and sharing complex domain- and mission-specific world models across agents in environments with poor communications.

Chapter seven, "Probabilistic Scene Parsing", was written by Michael Walton and Douglas S. Lange at the Space and Naval Warfare Systems Center-Pacific; and by Song-Chun Zhu, at the University of California in Los Angeles, CA. Their chapter reviews research on the interpretation of complex, real-world image-based contextual scenes dependent on non-evident abstractions. For this context, the authors represent decompositions of visual scenes using structured probabilistic graphical models (PGMs). They discuss the application of And-Or and parse graphs to build PGM decompositions of visual scenes and the visual entities in their appropriate context; this parsing is used temporally, spatially, causally and textually. For the authors, they write that the context might be determined from a small subset of state variables that other context-sensitive state variables depend. The models the authors discuss allow context to be uncertain and dependent on partial observations. The authors devote a section to applications like answering queries, text summaries and modeling behavior, all with an aim to the modelling of contexts for Naval applications in the field and at sea. Several visual scenes are introduced as examples that they parse with PGMs. Afterwards, they review their algorithms of joint distributions and factor graphs that they use to infer statistical dependences. Overall, a computational model for determining context derived from visual scenes that can be deployed in the field and at sea in real time during operations would be a significant advance. There is much research for the authors to complete at this stage of the wide-ranging research being conducted by the authors, including complex scenes; future complex scenes that might be addressed include partially and fully obscured (camouflaged) images or those containing illusions.

Chapter eight, "Using Computational Context Models to Generate Robot Adaptive Interactions with Humans", was written by Wayne

Zachary, Taylor J. Carpenter and Thomas Santarelli of CHI Systems in Plymouth Meeting, PA. In their chapter, the authors write that compatible context understanding is the key to effective human-to-human interaction, and that humans rely on implicitly shared representations of context to understand each other and to communicate, negotiate, collaborate, and perform teamwork. However, from the perspective of technology, people find coordination and teamwork difficult with interactive non-human agents (e.g., robots and smart information systems) because those agents are currently not able to represent context in a form compatible with the way humans do. But computational context, the authors argue, can be engineered to represent context in a human-like way, providing non-humans and information systems with context-reasoning that can enable and simplify teamwork interactions with humans. The challenge with context is that it is embedded in the social interaction and often not obvious. The authors label the framework they are developing as the Narratively-Integrated Multilevel (NIM) framework. Their goal is applying NIM to robots to allow them to engage in teamwork with astronauts. The authors use three lower levels of human cognition in their model (perception, comprehension, projection) and four higher levels (expectation, action, stories, narratives). From these levels, they build a NIM computational architecture to replicate the process humans use to construct context. NIM is thus a computational model based on the theory of context inspired by human mental models. They use NIM to define a specific context model for robots that need to team with humans in a space habitat. The authors implement the model with a toolset labeled as the Integrated Context Engine (ICE), which they have developed to build, execute and embed NIM context models in non-human agents such as robots. They close by considering future research directions for NIM and ICE. Overall, the authors recognize that context is a shared understanding important for humans to understand the choices they make and the actions they execute for the interactions within which they participate. At the same time, the authors are preparing for future robots by describing in significant detail how context is constructed and shared. Already, theirs is a complicated, detailed model designed to be applied to a robot that will team with astronauts as a member of their teams. Making space habitats with mixes of human and robots safe for the humans is the key driver of this application.

The next chapter, chapter nine, was written by Manisha Mishra (with Aptiv, Inc., in Kokomo, Indiana), David Sidoti (University of Connecticut), Gopi V. Avvari (with Aptiv, Inc., in Kokomo, Indiana), Pujitha Mannaru (University of Connecticut), Diego F. M. Ayala (the former Technical Staff at Qualtech, now at Columbia University), and Krishna R. Pattipati (University of Connecticut, Storrs, Connecticut, CT). The title of their chapter nine, «Context-Driven Proactive Decision Support: Challenges and Applications», covers the challenges of rapid mission planning, re-planning and execution in a highly dynamic, asymmetric, and unpredictable mission environment. The authors begin with a canonical mission planning abstraction that they use to describe the key queries and elements of a mission. The result is a very high-level computational proactive decision support (PDS) system that anticipates new contexts yet is adaptable to changes in missions or to threats to the Naval Fleet. Existing decision support systems are overloaded with data unprocessed into information, in turn causing an overload with decision makers (DMs), increasing the likelihood of failure. To overcome this problem by detecting the context, the authors deploy a process that they label as "6R": the "right" data, information or knowledge must be delivered from the "right" sources in the "right" mission context to the "right" DM at the "right" time for the "right" purpose. The authors define context as a multi-dimensional event space that evolves with features that consist of the goals for a mission, its environment and assets, the threats or tasks that might be entailed, and the mental state of the DMs. But they also want to determine the dynamic changes in the context and how to disseminate that information to operators and DMs. The authors propose a PDS framework for: dynamically defining the knowledge relevant for a mission's context; detecting a mission's context changes; analyzing and predicting the context to develop and then disseminate "what-if" scenario analyses for missions; and providing courses of recommended actions, all the while considering workload, time pressure, risk styles and the expertise of DMs. The authors provide three applications to illustrate and discuss their PDS framework (e.g., anti-submarine warfare; counter-smuggling operations; unmanned aerial systems, or UASs). Overall, this system is a very strong application of a sophisticated computational context already operational and deployed by the Fleet at sea and in the field (e.g., unmanned aerial systems). It is used to track and allocate assets, to counter threats and to determine courses of action (COAs); it sets plans, for example, to counter pirates and smugglers on the open sea; it plans interdictions in dynamic environments in good and bad weather; it is deployed with UASs. And its dynamic scheduler can solve some NP-Hard problems in a reasonable time. While meta-aspects of the system are not represented mathematically interdependent, the elements of the system are fully interdependent by being mutually responsive to interdependent operational changes (e.g., with a Bayesian server that models mutual changes in assets, threats, DMs and the environment). For example, the graphical model includes dependencies among MEAT-H elements (viz., Mission, Environment, Assets, Tasks/Threats-Humans); e.g., approaching bad weather may impact the threats to the Fleet more than the blue assets available to the Fleet.

Chapter ten was written by Beth Cardier who is at Sirius-Beta in Virginia Beach, VA. Her chapter is titled "The Shared Story–Narrative

Principles for Innovative Collaboration". The author uses the contexts constructed by human narratives as an everyday method to reason about how contextual change occurs in an open world. She argues that to formally model this, general reference frameworks need to be adjusted towards specific circumstances using mechanisms similar to those found in storytelling, such as analogy and nesting. In this way, narrative reveals the special process by which humans integrate information from different contexts to produce new frames of reference that can be shared. In her chapter, she demonstrates these principles by extending ontology diagrams to model how the coherence of a narrative becomes, and can be represented as, a shared context (examples come from a television show, a movie and literature—in this introduction, we refer to her use of the well-known story of Red Riding Hood). She then builds on the Red Riding Hood example to demonstrate how those inferential processes can also be applied to the formation of a human team. In her view, a core notion is that the convergence needed for collaboration depends on identifying and matching structure at the level of systems using devices such as an analogy, nesting, governing influence and retroactive reinterpretation. This enables her to model changes from the beginning to the end of a scene or story. Her model can also identify individual and system levels. It allows her to model zooming through scales, the derivation of local structure from general structure, multi-system complexity, governing influence, analogy, nesting and the interaction and relations among groups. The rate of changes in structures between systems can be modeled as can the communication signals across a system. She provides and discusses visuals for using these principles to capture the interactions between cognitive and biological systems in neurobiology. Overall, the author provides an excellent model of narration for building contexts and drawing inferences by humans. Her visuals are useful, illustrative and impressive; she is beginning to connect her model to actions in the field that will make it more and more useful to other applications and research scientists.

Chapter eleven, "Algebraic Modeling of the Causal Break and Representation of the Decision Process in Contextual Structures", was written by Olivier Bartheye at CREC-St-Cyr, Military School of Coëtquidan at Guer Cedex in France; and Laurent Chaudron, Director, ONERA-Provence Research Center (the French Air Force Academy), and also with the Polytechnique in Provence, France. The authors write that a due to a logic break in causality between action and perception, determining the context with computation is likely the most difficult problem in AI to represent decision process with logic. They conclude that a context is the interface that agrees with universal human interpretations of the environment by the intuitions of one or more intelligent agents. Making context computable with logic, however, is unsolved. Yet for autonomous agents, computable contexts are necessary to generate the behaviors expected from and by other individual agents. To address this problem with logic, they require a categorical contextual algebraic structure that identifies stable contexts and context changes. For example, while the (social) environment influences human perception and the actions taken, actions influence changes in perception. The authors begin by classifying the epistemology of contexts into four knowledge bases (empirical, conceptual, formal, methodological) with a focus on conceptual and methodological bases. With these, the authors believe that contexts defined conceptually can be understood and are universal. However, for computable contexts, human behavior is implemented methodologically, breaking causality, preventing the knowledge from these two bases to be unified per the context validity theorem: action and perception remain in a logical phase shift, never existing as simultaneous signals. For action with AI, both acceptable operators or models and unacceptable operators or counter-models while equally important are incompatible. Completeness can be managed with the incompatibility theorem, but it eliminates proofs for precise contexts, a paradox, not unlike what humans find in their attempts to determine and define the unknown contexts they confront. Due to the context validity theorem, since concepts are based on signals, this means that perception precedes action, requiring the context validity theorem to be reversed, i.e., action must precede perception to deal with the causal break and to manage context change. Overall, and very importantly, the authors recognize that mathematical proofs to establish the context for action and perception are problematic and in some cases unpredictable. The authors offer a path forward with non-commutative research. The ideas posed by the authors should continue to be developed. Of particular interest is whether the signals for action and perception can be dealt with independently as with interconnected Bayesian systems (e.g., Chapter nine).

Chapter twelve, "A Contextual Decision-Making Framework", was written by Eugene Santos Jr., Hien Nguyen, Keum Joo Kim, Jacob A. Russell, Gregory M. Hyde, Luke J. Veenhuis, Ramnjit S. Boparai, Luke T. De Guelle and Hung Vu Mac; Nguyen is an Associate Professor of Computer Science at the University of Wisconsin, in Whitewater, WI; the other authors are at Dartmouth College in Hanover, NH; in addition, Eugene Santos is a Professor of Engineering at the Thayer School of Engineering at Dartmouth. The research by these authors is directed at inferring relevant context to provide military Commanders with the correct information needed to make the best time-critical decision possible, leading to their development of a Proactive Decision Support system. But after they had reconstructed the decision environment, they found that for a given context, different Commanders respond uniquely when making decisions. Mindful of these unique styles, the authors built on their previous research with dynamic models of Markov Decision Processes, Inverse Reinforcement Learning and Double Transition Models to identify context separately from styles to explain how Commanders make decisions based on styles that impact subsequent decisions. They define styles as "the learned habitual response pattern exhibited by an individual confronted with a decision situation". Then, for example, they use their models in State-Action-State tuples for the decisions a Commander might make with the sequences of prior tuples helping to define the context. In three different testbeds, with synthetic commanders and later with human subjects, the authors compared these Commanders and decision styles with a new algorithm they discuss in detail. In these testbeds, the three styles tested are the rational planner who attempts to maximize rewards over longer time horizons; the novice who explores the decision space suboptimally; and a combination of these two styles. They conclude that these styles can be optimized for a problem alone, but also optimized for an individual commander, context, and problem together within a framework that allows the authors to compare different decision styles. They conclude that they are off to a good start but with many challenges remaining. Overall, the authors model decision-makers with computational tools long-established in their laboratory to address their model of commanders with different decision-making styles as they check for both good and bad decisions. In the testbeds that they have used, they are able to test synthetic decision makers cross-checked with human subjects, too. The authors should generalize their model for future computational agents (robots) and teams and, as they already plan, to strengthen their model with data collected in the field from human commanders during operational settings.

Chapter thirteen was written by W.F. Lawless, Departments of Mathematics and Psychology, Paine College, Augusta, GA; Ranjeev Mittu, Ira Moskowitz and Don Sofge at the Naval Research Laboratory, Washington, DC; and Steve Russell at the Army Research Laboratory, Adelphi, MD. Mittu is the Branch Chief of the Information Management and Decision Architectures Branch in NRL's Information Technology Division; Moskowitz is a mathematician in the same Branch; Sofge is a computer scientist in the Distributed Autonomous Systems Group at the Navy Center for Applied Research in Artificial Intelligence; and Russell is ARL's Branch Chief of the Battlefield Information Processing Branch. Their chapter is on the context of proactive cyber-security. From their perspective of context, by overlooking the value of teams in the past to place more emphasis on what proactive individuals must do to improve individual cybersecurity practices (psychological identity; self-reported individual perceptions; individual education), the social sciences have undervalued the physical importance of cybersecurity (physical facility barriers; red-blue team training; an organization's security software updates). Instead, the authors argue, context based on psychology does not generalize to the property of teams, organizations and systems. In contrast to a focus on individual approaches to context, a theory of teams opens context to considering the multiple interpretations of physical (social) reality; the uncertainty produced by a focus on only one view (e.g., management's or a worker's); and the social situations where individual perceptions of context cannot simply be summed or aggregated to determine a team's or organization's context (e.g., individuals are poor at multitasking, but a team's function is to multitask). The authors do not favor either a psychological or physical approach alone, but instead they argue that the shifts between individual and team contexts are subtleties that if combined and mastered, should generalize to future hybrid teams (arbitrary combinations of humans, machines and robots). To improve cyber-defenses and to advance theory, in the context of proactive cyber-defenses, they address cyber-security from the perspective of cognitive, multitasking, and social-physical (team) approaches to context. The authors conclude that in today's context, cyber-risks and cyber-defenses are both increasing, but that a combined new theory of context can point the way to improve both cognitive and physical cyber-defenses. Overall, the authors have taken on the difficult task of creating theory for a context of autonomous teams where none yet exists, with their research suggesting that it will generalize to autonomous hybrid teams. The authors have begun what looks to be a long but hopeful journey.

Questions raised during the Symposium for speakers and attendees at AAAI-2017 and for readers of this book

The Spring AAAI-2017 symposium offered speakers opportunities with AI to address the intractable, fundamental and difficult questions about computational context for humans, machines and robots; how context affected autonomy and its management; the malleability of context based on preferences and beliefs in social settings; or how context affected autonomy for hybrids at the individual, group and system levels.

A list of unanswered but fundamental questions included:

- How has context affected why we have not yet determined the theory and principles underlying individual, team and system behaviors; is awareness a factor?
- Can autonomous systems be controlled without managing context to solve the problems faced by teams while maintaining boundaries and defenses against threats while minimizing mistakes in competitive environments (e.g., how context changes in response to human error, cyber attacks, system failure)?
- Do individuals seek to self-organize into autonomous groups like teams to better understand context to defend against attacks (e.g., cyber

attacks, hostile merger offers, competition over scarce resources) or for other reasons (e.g., least entropy production, or LEP; and maximum entropy production, or MEP)?

- What is it about context that is needed by an autonomous organization to predict its path forward and govern itself? What are the AI tools available to help an organization to be more adept, more adaptive and more creative?
- What signifies an adaptation to a context? For AI, when an adaptation occurs at an earlier time, does that prevent or moderate responses to changes in context, such as with the environment or a competitive opponent?
- What do superordinate goals have to do with MEP? Can MEP be achieved without a superordinate goal? Is the stability state of a hybrid team its ground state, or the state that where the team is able to generate the MEP rate?
- If maintaining social order requires MEP, if context is built around social order, and if the bistable perspectives present in debate (courtrooms; politics; science) lead to a stable superordinate goal, is the chosen decision an LEP or MEP state?
- Considering the evolution of superordinate goals for social systems (e.g., in general, Cuba, North Korea, Venezuela, Palestine, and many others have devolved), are the systems that adjust to MEP more effective and efficient?

New threats to context may emerge due to the nature of the technology of autonomy (as well as the breakdown in traditional Verification and Validation, or V&V; and Test & Evaluation, or T&E, due to the expanded development and application of AI). This nature of technology leads to other key AI questions for consideration now and in the future:

Fault Modes

- Are there new types of fault modes that can be exploited by outsiders that change context (with and without deception)?

Detection

- How can we detect that the context for an intelligent, autonomous system has been or is being subverted?

Isolation

- Is there a "fail safe" or "fail operational" mode for an autonomous system, can it be implemented, and can users be made aware of this context change?
- Implication of cascading faults on context (AI; system; cyber).

Resilience and repair

- What are the underlying causes of the symptoms of faults on context (e.g., nature of the algorithms, patterns of data, etc.)?

Consequences of physical and system cyber and context vulnerabilities

- How fault modes are induced?
- How deception occurs (including false flags)?
- How Subversion occurs?

The human element alone, or in a social context (reliance, trust, and performance).

References

Craig, D., Morales, D. and Oliver, M. 2004. USS Vincennes Incident, MIT Aeronautics & Astronautics, Slide presentation, from https://ocw.mit.edu/courses/aeronautics-and-astronautics/16-422-human-supervisory-control-of-automated-systems-spring-2004/projects/vincennes.pdf.

Etzioni, A. and Etzioni, O. 2016. Designing AI systems that obey our laws and values. Communications of the ACM 59(9): 29–31. doi:10.1145/2955091.

Hollnagel, E. 2009. The ETTO Principle: Efficiency-Thoroughness Trade-off. Ashgate.

Howarth, F. 2014, 9/2. The Role of Human Error in Successful Security Attacks. SecurityIntelligence, from https://securityintelligence.com/the-role-of-human-error-in-successful-security-attacks/.

Lawless, W.F. 2017. The entangled nature of interdependence. Bistability, irreproducibility and uncertainty. Journal of Mathematical Psychology 78: 51–64.

Levs, J., Smith-Spark, L. and Yan, H. 2015, 3/26. Germanwings Flight 9525 co-pilot deliberately crashed plane, officials say. CNN, from http://www.cnn.com/2015/03/26/europe/france-germanwings-plane-crash-main/.

Markoff, J. 2015, 4/6. Planes Without Pilots. New York Times, from https://www.nytimes.com/2015/04/07/science/planes-without-pilots.html.

Moon, W., Yoo, K. and Choi, Y. 2011. Air traffic volume and air traffic control human errors. Journal of Transportation Technologies 1(3): 47–53, doi: 10.4236/jtts.2011.13007.

NTSB. 2001, 3/2. News Release: USS Greeneville/Ehime Maru Collision Update, National Transportation Safety Board, from https://www.ntsb.gov/news/press-releases/Pages/USS_GreenevilleEhime_Maru_Collision_Update.aspx.

NTSB. 2017, 12/22. NTSB News Release. National Transportation Safety Board Office of Public Affairs, from https://www.ntsb.gov/news/press-releases/Pages/PR20171222.aspx.

RCTA. 2016. Robotics Collaborative Technology Initiative (RCTA), ARL-HBCU/MI Partnered Research Initiative (PRI), from http://www.arl.army.mil/www/default.cfm?page=288.

Showstack, R. 2014, 10/21. Taiwan Shipwreck Is Major Loss for Ocean Research, Scientists Say. The 10 October shipwreck of Taiwan's R/V Ocean Researcher V, which resulted in two deaths, is a major setback for ocean research in Taiwan, according to scientists. EOS, from https://eos.org/articles/taiwan-shipwreck-major-loss-ocean-research-scientists-say.

Tetlock, P.E. and Gardner, D. 2015. Superforecasting: The Art and Science of Prediction, Crown.

USDC. 2012, 2/22. United States District Court Eastern District of Louisiana, MDL No. 2179 "Deepwater Horizon" in the Gulf. In re: Oil Spill by the Oil Rig of Mexico, on April 20, 2010. Applies to: 10-4536. Case 2:10-md-02179, Document 5809.

Wiegmann, D., Shappell, S., Boquet, A., Detwiler, C., Holcomb, K. and Faaborg, T. 2005. Human error and general aviation accidents: A comprehensive, fine-grained analysis using HFACS. Federal Aviation Administration, Office of Aerospace Medicine Technical Report No. DOT/FAA/AM-05/24. Office of Aerospace Medicine: Washington, DC.

2

Learning Context through Cognitive Priming

Laura M. Hiatt, Wallace E. Lawson* and *Mark Roberts*

2.1 Introduction

Context is a way in which one's setting, environment or mindset affects one's thought. Words change meanings depending on what other words, if any, surround them; social cues and gestures can be differently interpreted in separate contexts and cultures. While people are able to understand contextual cues fairly easily, such reasoning does not come automatically for machines and autonomous systems.

One major challenge of capturing context in a computational setting is determining the appropriate context for a given situation. Approaches to computational context need to ensure that they are including the appropriate features or states in the context's representation so that it provides useful information; on the other hand, such approaches also need to consider that larger representations of context can quickly become more computationally costly to compute.

We approach this problem by taking inspiration from how people reason about and represent context. In human cognition, cognitive priming is the framework upon which context rests. In theories of cognitive priming, related items in the mind are connected to one another. As the mind thinks and interacts with the world, priming spreads along these connections, such that items that are currently being thought about prime related items

Naval Research Laboratory, 4555 Overlook Avenue S.W., Washington D.C. 20375.
* Corresponding author

in memory. At any given time, the primed items serve as a representation of context that affects forthcoming thought and behavior.

A key part of this theory of context is that the items that are currently being thought about determine not only what is being primed, but also the existence and strength of these connections between items. If two concepts are thought about together, they become connected; if one is then thought of later, the other is primed. This thought-driven approach serves to limit contextual connections to those that are relevant or related to the task or situation at hand. For example, in the kitchen, tools like knife, cutting board and pan might strongly prime one another, because those are used for cooking tasks and so are thought about while cooking. On the other hand, they will not as strongly prime other kitchen items like dish soap because, although it is also a kitchen item, it is used for cleaning, and so is not typically thought of while cooking.

This natural limiting of the connections between concepts in the mind provides useful guidelines for approaches to computational context. In our view, it indicates that context, even when focused on elements in the *external* world, can be learned and refined using one's *internal* state and thought processes. It guides context towards connecting together elements that have been experienced together, meaningfully, in the past, and avoids creating connections between items that were encountered independently or separately. Overall, this means that cognitive priming creates enough connections to provide rich and meaningful context, without creating so many as to become computationally burdensome.

In the rest of this chapter, we further our argument that cognitive priming is a critical component of learning computational context. We begin by discussion more details on our theory of cognitive priming, how it provides context in human cognition, and our implementation of it on a computational platform. Then, we discuss how contextual information can be leveraged for autonomous systems where context has the potential to help performance. To date, we have shown the benefit of this context on object recognition, significantly improving the precision of difficult recognition problems (Lawson et al., 2014). We have also shown that it can effectively identify out-of-context objects, robustly identifying anomalous features in an automated surveillance task (Lawson et al., 2016a; Lawson et al., 2016b). In these works, contextual information was a blend of categorization information (objects are primed by the rooms or areas in which they appear, since they are thought about at the same time), as well as spatial/co-occurrence information (objects are often primed by other nearby objects, again because they are thought about at roughly the same time). These sources of information are automatically blended and combined by the priming mechanisms inherent in our theory.

We also demonstrate how context can be used to help machines make more effective decisions (Roberts and Hiatt, 2017). We apply priming to two sequential decision problems in the game of Minecraft: choosing the next action in a maze, and selecting the best tool to mine blocks. Our work shows that using priming to make these decisions performs comparably to decision trees, but can also reduce regret. Here, context included the features of the environment, such as the surrounding terrain (the terrain primes subgoals that have been used in that terrain in the past), as well as connections between materials and the tools used to mine them. Collectively, these results suggest that cognitive priming is an effective mechanism for learning and representing computational context.

2.2 Cognitive Priming

Because our overarching approach takes inspiration from how people reason about and represent context, our theory of cognitive priming is situated in a broader, integrated theory of human cognition that is implemented in a computational framework. The overarching framework models, in part, human working memory, which includes what a person (or a computational model) is thinking of, looking at, or has as their goal at any given time (Trafton et al., 2013). Using this framework to ground our theory of priming allows us to develop our theory and test it against human data to determine its fidelity to human cognition.

At a high level, connections, or *associations*, are formed between items that are in working memory at the same time (Thomson et al., 2015; Thomson et al., 2017); then, the more often items are thought about together, the stronger their association becomes. The contents of working memory also serve as the source of cognitive priming: at any given time, items in memory are primed according to the strength of their associations with the current contents of working memory. Importantly, using working memory as the basis for creating and strengthening associations, as well as for spreading priming, creates an integrated view of priming that is based on the model's entire internal state and all of its modalities (vision, aural, its goals and reasoning, etc.), and can incorporate both semantic and statistical correlation information.

As part of our work developing this theory, we have shown cognitive priming to be a fundamental component in higher-level cognitive processes. Similarity, for example, is a complex mental construct that is critical to tasks such as object categorization (Nosofsky, 1992), problem solving (Novick, 1990) and decision-making (Medin et al., 1995). We have shown that similarity has strong roots in priming; items which more strongly prime one another are typically considered more similar (Hiatt and Trafton, 2017), such as items that share semantic features, or are commonly seen together.

Priming also has a key role in providing context to category-based feature inferences, where people infer missing features of objects in different categories (such as an object's color or shape) (Hiatt, 2017). Additionally, priming can affect how people perceive and act in the world in other ways, such as explaining errors that people make on routine procedural tasks (Hiatt and Trafton, 2015a), and explaining induced cases of mind wandering (Hiatt and Trafton, 2015b). We next provide a more detailed look at cognitive priming.

2.2.1 A Primer on Priming

The computational framework we use for priming is the cognitive architecture ACT-R/E (Trafton et al., 2013), an embodied version of the ACT-R cognitive architecture (Anderson, 2007). At a high level, ACT-R/E is an integrated, production-based system, and models in ACT-R/E capture the root cognitive processes that people go through as they undergo tasks. At its core are the contents of its working memory. Working memory is represented as a set of limited-capacity buffers that can contain thoughts or memories, and both indicates what the model is currently thinking about, as well as drives future behavior.

At any given time, there is a set of *productions* (if-then rules) that may fire because their preconditions are satisfied by the current contents of working memory. From this set, the production with the highest predicted usefulness is selected to fire. The fired production can either change the model's internal state (e.g., by adding something to working memory) or its physical one (e.g., by pressing a key on a keyboard). Here, we abstract over these productions and instead describe processes at a higher level (i.e., we say that we look at an object, instead of discussing the 3–4 productions that must fire to find the object, attend to it, etc.).

In addition to the symbolic information (i.e., factual information) represented as part of memories, memories have activation values that represent their relevance to the current situation, and guide what memories are retrieved from long-term memory and added to working memory at any given time. Activations has three components—activation strengthening, spreading activation, and activation noise—that together are excellent predictors of human declarative memory (Anderson et al., 1998; Anderson, 1983; Schneider and Anderson, 2011; Thomson et al., 2017).

Activation strengthening is learned over time and is a function of the frequency and recency with which the memory has been in working memory in the past. It is designed to represent the activation of a memory over longer periods of time (e.g., hours, days). Noise is a random component that models the noise of the human brain; its presence, in part, models the randomness of the human mind. We seek to understand the underlying principles of the role of spreading activation (i.e., priming) in computational context, and so generally focus on noiseless versions of models without activation strengthening.

Priming, in contrast to activation strengthening and activation noise, is a short-term activation that originates from working memory, distributing activation along associations between the contents of working memory and other memories (Hiatt and Trafton, 2017; Harrison and Trafton, 2010). As we stated above, memories become associated when they are in working memory at the same time. This means that associations are created between objects that are seen in close temporal proximity to one another, such as between each object and the setting in which they appear. For example, if the model is in a kitchen and sees a refrigerator and a toaster, refrigerator and toaster will each be associated with kitchen, as well as with each other.

Once created, associations have a corresponding strength value which affects how much activation is spread along them. Mathematically, the strength of an association from item j to item i (S_{ji}) is:

$$S_{ji} = mas \cdot e^{\frac{-1}{al \cdot R_{ji}}} \tag{2.1}$$

$$R_{ji} = \frac{f(N_iC_j)}{f(C_j) - f(N_iC_j) + 1} \tag{2.2}$$

These equations reflect two parameters: *mas*, the maximum associative strength; and *al*, the associative learning rate. The function f tallies the number of times that item j has been in working memory, either independently (C_j) or at similar times to when i has been in working memory (N_iC_j). The strengths are thus a function of how often the two items are in working memory at roughly the same time, versus how often each one is in working memory without the other. Because the count functions are easy to update, the strengths also are learned, iteratively, over time, allowing cognitive priming to quickly learn and adapt to new situations. These equations are explained further in Hiatt and Trafton (2017).

Intuitively, an associative strength is a Bayesian-inspired value that reflects how strongly an item in working memory (whether an object the model is looking at or a setting the model knows it is in) predicts that an item it is associated with will be seen next. This allows priming to capture correspondences between memories that are expected to be relevant at the same time (such as items used for the same task), as well as memories that are semantically related (such as an object and its corresponding color and shape). This formulation of the association strengths is a non-standard adaptation of ACT-R's Bayesian-based priming mechanisms. We use this adaptation to account for the large numbers of associations and objects seen and represented by people in their daily lives, which ACT-R's original formulation is unable to do, as well as to capitalize upon its theory that priming stems from working memory; see Hiatt and Trafton (2017) for more information.

2.3 Improving Machine Perception with Cognitive Priming

We now turn to discussing the role of cognitive priming in machine perception. Context is used heavily in the human visual system to bias both where we look and what objects we expect to see (Oliva and Torralba, 2007). This idea has been incorporated into many approaches to machine perception problems, and a variety of contextual cues have been used to improve performance (Li et al., 2010; Divvala et al., 2009; Lawson et al., 2014).

In the past, context has been shown to help improve object recognition performance by providing suggestions about what objects are likely to be present in several different forms:

- *Pixel level context* uses the position of an object within an image as well as surrounding regions to provide cues as to what an object may be. Objects toward the bottom of an image are likely to be those found near the ground or floor; similarly, objects near the top of the image are likely to be those typically found near the ceiling or in the sky. Pixel level context also includes the use of the pixels immediately surrounding an object. For example, if an object were surrounded by sky, it would greatly limit the number of possible object classes to those things we would expect in such a context (e.g., birds, airplanes).
- *Semantic context* uses neighboring objects to suggest what may be present. If a keyboard and mouse are present, semantic context would strongly suggest that a computer monitor may also be present. So, despite the fact that television sets and computer monitors look similar, context provides cues to distinguish between the two classes.
- *Scene context* uses the setting of an image to determine what objects may be present. In a kitchen, we might expect to see things like a refrigerator and coffee maker. In an office, we may expect things such telephones, computers, desks, and chairs.

A lot of success has come from combining multiple sources of context. For example, combining scene level context with pixel level context can help to identify objects. Divvala et al. (2009) demonstrated this by using scene information along with a grid to predict which objects are present. Similarly, Li et al. (2010) illustrated combined location, depth, and saliency to improve object recognition.

Typically, these approaches are implemented *statically*, where the context is learned during training and then remains static during testing (Divvala et al., 2009; Galleguillos et al., 2008; Felzenszwalb et al., 2010). Static context is limited in the sense that it does not allow context to revise its knowledge as it interacts with either new objects, or known objects in new contexts. As we describe above, one challenge to continually learning is deciding which information is useful, and which is not. Here, we use cognitive priming to provide that guidance. As context interacts with multiple objects in a meaningful way, relevant *external* information impacts its *internal* model of priming and guides it towards learning useful, appropriate, *dynamic* context.

We demonstrate the use of cognitive priming as computational context for machine perception in two different problems. In the first, we describe a novel approach where we use cognitive priming to provide dynamic context to improve object recognition in real-world environments. In the second, we use cognitive priming as the basis of context to locate things that are *out* of context in real-world environments.

2.3.1 Cognitive Priming in Object Recognition

The first task we consider is object recognition. Despite significant progress in recent years (Redmon and Farhadi, 2017; He et al., 2016), perception in real world environments, such as those encountered by mobile, autonomous systems, is still challenging. Poor resolution, non-iconic viewpoints, specular reflections, and image blur due to motion and poor focus can all adversely affect many perception algorithms.

Our approach breaks new ground on the role of context in object recognition by combining a machine learning algorithm for computer vision with three types of context: *static* local pixel context, as introduced above, and *dynamic* scene and semantic context from cognitive priming. Static local pixel context is incorporated by augmenting an object's feature vector with the pixel size and pixel centroid of the object within the image (Felzenswalb et al., 2010). This then allows its location in the image to bias its classification. Cognitive priming, on the other hand, undergoes no explicit training period. Instead, during testing, the model learns to associate a scene and nearby objects whenever an oracle provides ground truth. This is analogous to an autonomous system performing online object recognition and being corrected by a human partner when necessary. These associations allow cognitive priming to represent both scene context, where knowing a scene primes objects that commonly appear in that scene, and semantic context, where neighboring objects suggest what might be seen next. As cognitive priming learns these associations and encounters new situations, the items currently being primed serve as a prior probability for recognizing objects in that context.

We evaluate our approach on the NYU Depth V2 dataset (Silberman et al., 2012), which is a dataset that particularly embodies the difficulty of recognizing objects *in situ*. The dataset provides both depth and RGB images; here, we use only the RGB images. It also includes ground-truth labels for objects, as well as for the scene depicted by each image (e.g., kitchen, office, etc.). Although not commonly used for object recognition, it provides an excellent baseline for this task, and we suggest this as a new recognition challenge.

To permit a comparison against other approaches, we also evaluate our approach on the well-known Pascal VOC 2007 dataset (Everingham et al., 2007). This dataset includes ground-truth labels for objects, but does not include any scene information. It should be noted that although it is widely used, Pascal VOC is less ideal for evaluating context since images contain an average of only about 2.5 objects in each image.

2.3.1.1 Object Recognition Components

Our baseline object recognition approach uses a linear support vector machine (SVM) to recognize objects using features from a pre-trained convolutional neural network (CNN). These features, sometimes referred to as *off-the-shelf* features or *deep* features (Razavian et al., 2014; Oquab et al., 2014), are from a CNN within the AlexNet architecture, pre-trained on the ImageNet database (Krizhevsky et al., 2012). The AlexNet architecture is a deep convolutional neural network featuring 7 layers: 5 layers of convolutional features and 2 fully connected layers (each containing 4096 neurons). To extract deep features, we propagate an image (sized 256 × 256 × 3) forward through AlexNet. However, rather than using the predicted class, instead we extract the objects' features from the last fully convolutional layer (denoted *fc7*) and provide those, as well as the objects' labels, to an SVM to train on. As part of this approach, we assume that the classifier has been provided with detection regions in the form of bounding boxes; this information is available for both of the image datasets we consider here.

In order to combine with cognitive context, the SVM needs to output more detailed information than its standard output of a predicted class for each test image: we also need a probabilistic estimate, p_i, that the image comes from each available class. Probabilities are thus found by minimizing the following equation:

$$\min_{p} \frac{1}{2} \sum_{i=1}^{k} \sum_{j:j \neq i} (r_{ji} p_i - r_{ij} p_j)^2 \tag{2.3}$$

where $p_i \geq 0$, $\sum_{i=1}^{k} p_i = 1$, and r_{ij} is the probability that the sample comes from either class i or class j, estimated using 5-fold cross validation.

When incorporating static pixel level context, we also augment the SVM's feature vector with three variables representing the center (c_{xi}, c_{yi}) and area α_i of object i (Felzenszwalb et al., 2010). It is thus learned (and utilized) directly as part of training the SVM classifier.

2.3.1.1.1 The Priming Model

The priming model is not trained ahead of time; it learns during testing. With respect to semantic context, as it is sequentially viewing objects, this means that a priming model is: (1) creating or strengthening associations between objects that appear in close temporal (and therefore, we assume, spatial) proximity to one another; and (2) spreading priming along existing associations based on what it is currently looking at.

Similarly, for scene context, while viewing a series of objects in a labeled scene, the model performs two analogous functions: (1) creating or strengthening associations between objects and the scene they appear in; and (2) spreading priming along existing associations based on what scene it is in.

Thus, at any given time, when viewing an object in a particular scene, objects that have been experienced in meaningful way with the current object as well as objects associated with the scene in general receive priming. In other words, the priming model's internal state determines which of these external concepts become associated. Objects appearing in common situations (both with other objects they typically appear with, and in a scene they typically are part of) receive the highest amounts of priming, facilitating their recognition: they receive strong priming both from other objects they commonly appear with, as well as from the scene they usually appear in. Objects that have not been connected in a meaningful way or that appear out of place typically receive much lower priming, hindering their recognition. This process occurs incrementally and online, allowing the model to predict objects it may see even after a very low number of previously-viewed objects.

2.3.1.1.2 Combining Context with Object Recognition to Recognize Objects

To combine priming with the probabilities output by the SVM classifier, we first estimate the probability that an object will appear in the current context. We do this by translating the amount of priming an item receives to the probability that item will appear in working memory next. To calculate this probability for an item i, we use the "softmax" equation provided as part of the architecture (Anderson et al., 1998):

$$P(i) = \frac{e^{\frac{s_i}{t}}}{\sum_k e^{\frac{s_k}{t}}} \tag{2.4}$$

where the variable S_i is the priming of item i, $\sum_k e^{\frac{S_k}{t}}$ iterates over the set of all objects (including i), and $t = 0.5 \cdot \frac{\sqrt{6}}{\pi}$. Intuitively, this equation represents each item's proportional share of priming, translated to a probability value.

The probability of each object from the SVM classifier (Eq. 2.3) and cognitive context (Eq. 2.4) are then combined multiplicatively. In a sense, cognitive context provides a prior probability for recognizing each object. This has several implications. First, in cases where there may be several equally valid objects, cognitive context can boost the probabilities based on what it expects to see. In cases where an object appears out of context, it can still be recognized but the object recognition system must be very sure of what it sees.

2.3.1.2 Experiment and Results

In our evaluation, both of the two datasets we consider, NYU Depth V2 and Pascal VOC 2007, are split into images used for training and sequestered testing images. The NYU Depth V2 database is randomly split into training and testing scenes (with approximately half in each category). In the case of Pascal VOC 2007, we use the training/testing split as set up by the challenge.

We begin our evaluation by training our SVM/deep feature classifier to recognize objects both with and without static local pixel context. For each of these classifiers, we then evaluate them both with and without cognitive priming providing scene and semantic context. This creates four different evaluation conditions in all. When utilizing cognitive priming, as each object is viewed for evaluation, the priming model also views each object, and uses what it has learned so far to suggest what it is likely to see next. For semantic context, this means that the objects it has seen recently prime other objects associated with it. With respect to scene context, for the NYU Depth V2 dataset, where a scene label is provided as part of the data, the cognitive model is given that scene label, which serves as an additional source of context. This is similar to an autonomous system localizing itself on a map to know what type of room it is currently located in. Since the Pascal VOC 2007 database does not include such labels, we only use dynamic semantic context with that database.

After cognitive priming has made its suggestion of what it expects to see next given the current context, the priming model is provided with the ground truth of the object it just saw (i.e., each image is sequestered, but then used to learn in an online manner). Given its own classification label and input from cognitive priming, the overall classifier then outputs its overall classification given object recognition and cognitive context using the combination method described above.

Because our dynamic semantic context is sensitive to the order in which objects are seen (it affects what objects are seen in close temporal proximity to one another), during evaluation we varied this order for each scene. The first "seen" object in an image was selected randomly; subsequent objects were sorted greedily by their proximity to the last seen object.

2.3.1.2.1 Results on NYU Depth V2

We begin with the NYU Depth v2 dataset. The dataset includes 464 objects over a wide variety of classes. Unfortunately, many of the object classes appear only once or twice, so we limit our evaluation to those classes with at least 20 examples. We also exclude common classes of wall, ceiling and floor. This results in 73 object classes, with 8824 training images and 8844 testing images.

Figure 2.1 shows example objects taken from images of the NYU Depth v2 dataset. Each scene contains anywhere from 2 to 24 detected objects, with a median of 12 detected objects per scene.

Figure 2.1. Example objects extract from images from the NYU Depth V2 database. From left to right, the object classes are 'bag', 'microwave', 'monitor', 'television', and 'faucet'. These show some of the examples of the extreme range of pose and quality of object instances present. *Note*: Images do not print well because of low resolution; this also highlights the challenge of this dataset.

We evaluate performance using mean average precision (mAP) (Everingham et al., 2007). The SVM/deep-feature-based object recognition results in a mAP of 0.4018; adding static local pixel context increased mean mAP to 0.4118. Combining the SVM/deep features classifier, static local pixel context, dynamic scene and semantic context increases the mAP to 0.4581. The SVM/deep features classifier with dynamic semantic and scene context results in the highest performance with a mAP of 0.4779, an improvement of 0.0761 over the baseline system with no context. These results are summarized in Table 2.1.

Table 2.1. Recognition results for NYU V2 Scene database. The results are presented by the source of context and associated mean average precision (mAP).

Source of Context for NYU V2 Database	mAP
None	0.4018
Static Pixel	0.4118
Static Pixel, Dynamic Scene and Semantic	0.4581
Dynamic Semantic and Scene	0.4779

2.3.1.2.2 Results on Pascal VOC 2007

We also tested our approach on the well-known Pascal VOC 2007 dataset, which has 7844 training and 14976 testing images distributed across 20 different categories. The objects in this dataset are generally isolated, having been manually posed by a person with the intent of isolating a single (or a few) objects. For this reason, context is much more challenging, since there a median of 2 objects per image. In this experiment, we use the training and testing splits set up by the challenge. The results of this experiment are shown in Table 2.2. The results show a similar trend as the NYU Depth V2 experiment results. The baseline performance with SVM/deep features is 0.8481; when provided with static context the results are 0.8440. Dynamic context increases performance to 0.8682 with static context and 0.8931 without static context. This represents an overall improvement of 0.045 mean AP.

Table 2.2. Recognition Results for PASCAL VOC 2007. The results are presented by the source of context and associated mean average precision (mAP).

Source of Context for PASCAL VOC 2007	mAP
None	0.8481
Static Pixel	0.8440
Static Pixel, Dynamic Scene and Semantic	0.8682
Dynamic Semantic and Scene	0.8931

2.3.1.2.3 Discussion

Overall, the results indicate that cognitive priming is an effective way of providing context to improve object recognition. It learns associations between these objects in the external world guided by its own internal processes: what scenes and objects have been thought about together, as well as what objects have been thought about in close temporal proximity to one another. It provides a benefit across the two different datasets we consider here (Pascal VOC, 2007; NYU Depth V2) consistently in the range of 2.0–7.6% mAP. The improvement in Pascal VOC 2007 is also particularly impressive given the limited amount of semantic contextual information. If provided additional scene information, it is quite possible that this would also improve to a level that is more similar to the NYU Depth V2 dataset.

For comparison, Felzenszwalb et al. (2010) also augments the output of an object recognition classifier; in their case, they use a second classifier to revise the predictions of the first. While they have shown this to be effective in some situations, it has several potential issues. First, it is difficult to update contextual information dynamically, as we are able to do here. Second, while it does produce good results, it is not clear how this would respond to an object that might appear out of context, which our approach is able to handle given enough certainty in the original classification.

Interestingly, static local pixel context is much less useful than context from cognitive priming, despite having positive results on other datasets. Our explanation for this is that, in these more real-world datasets, objects appear from many different viewpoints and angles and, thus, objects appear in many different locations of the images. This makes it very difficult for static local pixel context to provide useful information. Consequently, object recognition with context performs better without the addition of static local pixel context.

2.3.2 Cognitive Priming for Anomaly Detection

We next turn to discuss how cognitive priming can help with the problem of anomaly detection. Anomaly detection is much like object recognition, in the sense that it is perceptually trying to make sense of the world; but, in this task, instead of recognizing objects *in* context, it focuses on trying to identify objects that are appearing *out of* context. At a high level, cognitive priming can assist with this task by providing information about what objects in the scene are *not* primed, instead of highlighting those that are.

To our knowledge, there are very few standardized databases that focus on anomalies stemming from objects that appear out of context. Therefore, we studied this problem by patrolling our own facilities over multiple days with a mobile robot, both to collect normative data and then, later on, to collect data of purposeful anomalies in the environment.

Because we do not want to limit anomalies to known objects, we reason here over deep features (as described above) instead of objects. As the robot patrols, it iteratively learns a dictionary of the deep features it encounters in the environment. In addition to the features, the robot also knows what scene it is in by classifying its current image. Using these features and scene labels, we can concurrently and iteratively construct a priming model to

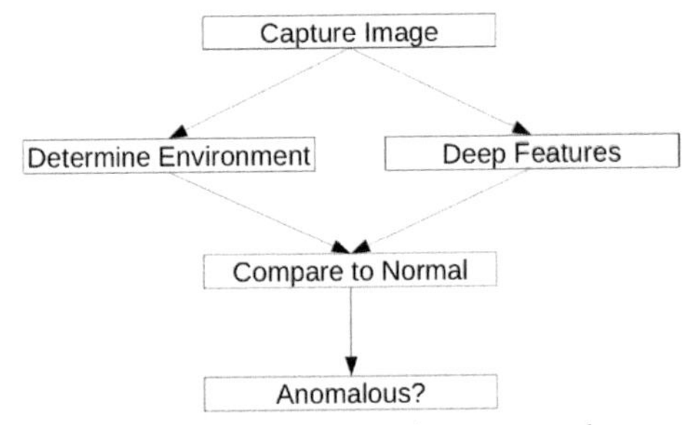

Figure 2.2. Overview of our approach.

provide scene context. As with the object recognition priming model, the features that appear in a scene (i.e., the features that are typically seen while the robot is in a hallway) become associated with that scene and are the basis for scene context.

We can then use the trained cognitive priming model to identify when we are looking at a part of a scene that is out of context. When an object is not expected, either because it matched no existing feature (i.e., a new feature) or because it did not receive much priming, it is marked as an anomaly. This overall framework is illustrated in Figure 2.2. Intuitively, this is an easy concept to understand. Consider a case where we expect to see a refrigerator in the kitchen. If it were moved out of the kitchen and into the hallway, this would be marked as an anomaly. In this example, we know about the refrigerator (it matches a feature), but we did not expect to see it in the hallway (from cognitive priming). Similarly, if someone were to park their motorcycle in the kitchen, it would also be marked as an anomaly. In this case, we would not actually know about the motorcycle (it does not match any known features), and so it is still (correctly) marked as an anomaly.

2.3.2.1 Building a Dictionary of Deep Features

The dictionary of deep features is iteratively built up as the robot patrols around the environment. To begin, each image is overlaid with a shifted grid, and image patches are extracted from each grid cell for further processing. The grid is of size $N \times N$, with a step size of $\frac{N}{2}$ to ensure a small overlap between grid cells. The appropriate size of N typically depends on the environment and the types of anomalies that will be detected. We found, however, that our approach worked well for a range of values of N since we typically see the objects and anomalies at various scales during normal interaction in the environment.

Each patch in the grid is then processed by a deep network, with the sole purpose of using the features to represent the appearance of the patch. For the deep network, we use the AlexNet architecture (Krizhevsky et al.,

2012), which was fully trained using ImageNet data. As was the case with the previous section, each patch is represented with deep features from the last fully connected layer (*fc7*).

Over time, as the robot learns about its environment, this approach can result in a dictionary that exceeds practical size: the robot typically captures at least 10,000 images during a patrol session, resulting in over 1.5 million patches from the shifted grid. Therefore, we organize the deep features into clusters. We use a streaming variant to *k*-means clustering (Duric et al., 2007), which builds clusters as it processes the data. When new features are seen, they are compared to existing clusters. If the features are sufficiently close to an existing cluster, it is labeled appropriately. If it exceeds a predefined distance (using Euclidean distance) to an existing cluster, a new cluster is formed. This is an appropriate approach to clustering for the iterative learning we use here because we can process information in a "stream" instead of having to store image patches in memory to reconsider later. Using this approach, the dictionary of clusters is built incrementally as the robot encounters them in its environment.

2.3.2.2 The Priming Model

Concurrently, the robot builds up a cognitive priming model that captures the scene context of its environment. As before, its internal state drives its learning about the external environment. Here, as the robot moves about in the world, it both iteratively sees feature clusters, and knows the scene label of the image as determined by the PlacesCNN (Zhou et al., 2014). As it thinks about clusters and scenes at the same time, it learns associations between these scene-cluster pairs. The strengths of the associations are then updated as appropriate, just as described in the earlier object recognition model. Ultimately, after training, feature clusters that are typical for a scene have very strong associations with that scene; feature clusters that are atypical for a scene have absent, or very weak, associations with it.

2.3.2.3 Detecting Anomalies

After the robot has undergone training, it can start to use what it has learned to detect anomalies. As it performs this detection, it considers both how well a patch's deep features match known dictionary clusters, as well as how strongly primed any potential matching clusters are given the current context. The equations for performing this detection are shown in Eqs. 2.5 and 2.6, where d_k is the distance from the new patch's features to dictionary cluster k, σ is a parameter that can be estimated from the expected distribution of the data, and $c_{k\alpha}$ is the priming from scene a to dictionary cluster k:

$$p_k = e^{-d_k/\sigma} \tag{2.5}$$

$$m = argmin_k(p_k \cdot c_{ka}) \tag{2.6}$$

Overall, these equations find the highest probability cluster, m, for the new patch's features, given both its perceptual match to existing clusters and its appropriateness in the current context. The robot detects an anomaly when $(p_m \cdot c_{ma})$ rises above a threshold.

2.3.2.4 Experiment and Results

As we mentioned above, to test this approach we first collected our own data from various areas, across various days, in our laboratory. To collect images, we drove our robot through three unique environments: hallway, cubicles, and outdoors. Figure 2.3 shows example images from each environment. The hallway environment is a typical office style hallway with two side corridors and multiple doors, both open and closed. The cubicle environment consists of multiple cubicles, chairs, and tables along with multiple reception areas,

Figure 2.3. Example images from each experimental environment. The top row is hallway, the middle row is cubicles, and the bottom row is outdoors.

a kitchenette, and a copy room. The outdoor environment contains mainly concrete and the outside of a building.

After collecting normative data, we then seeded the environment with purposeful anomalies. There are three sources of anomalies in this dataset. In the first, we add a new object to the environment that has not been seen before. The second class of anomalies come from modifications to existing objects. For example, a trashcan will be knocked over or a drawer will be left open. In the third class of anomalies, an object that belongs in the scene will be moved to a completely different scene. In each case, the anomalies were marked by the experimenter that introduced the anomalies in the scene.

The robot we used was a Pioneer 3AT mobile robot with a Carnegie Robotics S7 camera at 1024 × 1024 resolution at 15 FPS mounted on top. The S7 camera projects a circular image on a square field; we crop a centered middle square of 750 × 750 pixels. Data collection was performed by teleoperating the robot multiple times through each environment. Training data consisted of six runs through each environment spread over three days. Table 2.3 shows the number of images collected per environment. From these images, we extract features from 100 × 100 grid cells, using a step-size of 50 pixels; this results in a total of 169 evaluated patches per image. Interestingly, learning continues over all 3 days due to natural variations in the environment, as well as variation in the path taken by the robot, although the amount learned decreased over time. The PlacesCNN identified 47 different scenes which build up 11,203 associations between the scenes and the dictionary.

Table 2.3. Number of images collected during training for each environment.

Environment	Number of Images
Hallway	13206
Cubicles	8827
Outside	11669

During testing, each grid cell is evaluated according to how anomalous it is, using Eqs. 2.5 and 2.6 as described earlier. When a grid cell falls below a threshold, it is marked as an anomaly. We consider the anomaly to have been correctly detected when at least one grid cell overlaps the marked anomaly. Likewise, a false positive is whenever a grid cell activates on something that it is not on an anomaly. As we consider this from the perspective of alerting an operator, we consider a single detection on an anomaly sufficient for a true positive.

2.3.2.4.1 Results

Such an experiment can be both difficult and interesting to perform in an active office environment, since there are always small, subtle changes. In some cases, anomalies were detected, but since they were not one of the inserted anomalies, they are considered nuisances and are false positives. Some examples are office plants that were moved, chairs that were at a different orientation, trash bags that were placed differently, and paper towels that were left on a counter.

Figure 2.4 shows the ROC curve of the results, and Figure 2.5 shows some common detection and failure modes. Results at a selected threshold

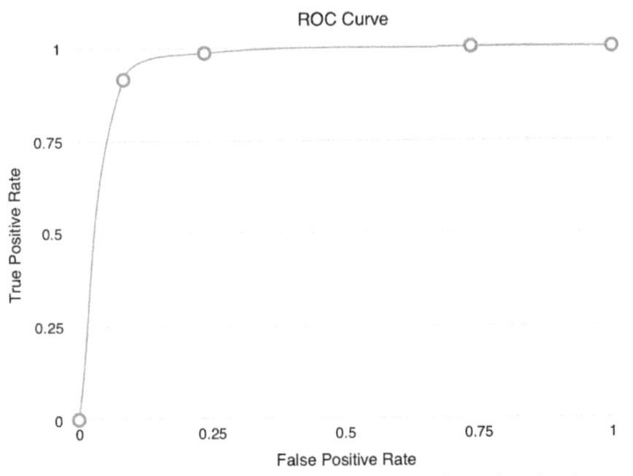

Figure 2.4. ROC curve showing the performance of the algorithm in the test environment.

Figure 2.5. Anomalies detected in the environments. In the figure, green rectangles show correctly detected anomalies; red rectangles show false positives.

Table 2.4. Accuracy of the approach at a selected threshold.

Environment	FPR	TPR
Office 1	11.82%	100%
Office 2	13.25%	87.5%
Hallway 1	4.27%	87.5%
Hallway 2	4.35%	88.9%
Outside 1	8.42%	93.33%
Outside 2	12.28%	88.24%

are shown in Table 2.4. The algorithm performs well on true positives, but, not surprisingly, struggles on non-rigid objects like sweatshirts. It is possible that evaluating this from different angles and perhaps different scales would increase the performance. Many of the false positives, particularly in the office environment, were at a distance. This was because of the increasing complexity when considering a lot of objects together; it is possible that we can eliminate this in the future by looking only at the immediately surrounding region. Finally, some of the biggest issues were related to lighting, due the auto-gain and auto-white balance of the camera. When near windows, or moving into and out of shadows, the picture changed dramatically. In all of the cases, the performance on the environment with constant light (hallway) was the highest.

2.3.2.4.2 Discussion

Despite the difficulties inherent in anomaly detection tasks, our experiments show that our technique learns quickly and is useful in a wide variety of environments, providing a low false positive rate while detecting most of the anomalies present. To confront the problem of having to recognize unknown objects, we learned about what objects are typically present by organizing what we have seen into feature-based clusters, and reasoning about whether new features fit into those clusters or are novel. Cognitive priming contributed to providing a normative model of which clusters are in context for different scenes, and which are out of place.

Interestingly, although the problems of object recognition and anomaly detection are, at the surface, opposite to one another, cognitive priming is able to assist with them in very similar ways. In addition, it is important to note that this approach to cognitive priming may also be applied to removed objects serving as the source of an anomaly. Our current experimental results were focused primarily on things added or modified in a scene; we did not explore the problem of things that are removed from a scene. However, if we were to look at removed objects, we could query the cognitive model to see which feature clusters we expected to see, and then provide an alert when a strongly primed feature cluster was not located.

2.4 Improving Sequential Decision-Making with Cognitive Context

We now turn to discuss our work using cognitive priming to help machines make more intelligent decisions. Like machine perception, imbuing an agent with the ability to decide what it will do next is a fundamental challenge of intelligent systems design; however, the knowledge and reasoning which the agent uses to make these decisions can differ greatly from that in machine perception. In some cases, an agent may leverage its experience or imitate an expert to guide its decisions. In others, an agent may apply general knowledge to better inform its exploration of choices. One of the challenges in these approaches is selecting the relevant experiences, or general knowledge, to base its decisions on. We argue that cognitive priming is an effective mechanism by which associations of past experience can inform decisions in the current context. For example, suppose an agent has experience indicating that building a bridge over lava makes it safe to move forward. Later, the appearance of lava would naturally prime the agent to consider building a bridge, without requiring the agent to explicitly search through all of its previous experiences of lava.

We investigate the benefits of priming for two sequential decision problems. In an *online* sequential decision problem of choosing a correct tool for a given task, we show that cognitive priming trained from a community wiki can improve the learning rate over a method that lacks priming. In an *offline* decision problem of choosing the next action in a maze navigation, we show that cognitive priming trained from expert traces can outperform incremental decision tree induction and performs comparably to machine learning approaches that were previously used for this task.

These two studies are situated in the game of Minecraft, for which Microsoft Research recently released an open-source platform called Malmo (Johnson et al., 2016). Minecraft is an open-world sandbox game where players choose their own objectives such as building structures, collecting resources, crafting, fighting enemies, or exploring. The world consists of 1 meter voxels (i.e., cube blocks) that can be broken to harvest resources, combined to build more sophisticated tools, or stacked to build structures. In the "vanilla" version of Minecraft, players rarely follow a scripted story line. Learning to play the game well involves trial-and-error as well as consulting other players or the Minecraft community wiki, which contains a plethora of tips and information on how to play the game successfully. Two central skills are navigating the world and using various tools to "break" blocks in the world in order to gather resources. Both of these skills require both knowledge and experience. For example, navigation requires knowledge about the immediate state around a player and experience about what is effective and what not to do (to avoid falling into lava, as an instance). Breaking blocks efficiently requires knowledge about the best tool for breaking a specific block and experience about when to apply that knowledge.

The open-ended nature of this world means that such details are learned over long periods of time, have a sparse reward signal, require learning by example, or require consulting other players or a wiki. While people excel at learning in these types of situations, these aspects can pose a significant challenge for off-the-shelf planning and learning algorithms, although recent efforts show promise at solving simplified tasks (e.g., Bonanno et al., 2016; Tessler et al., 2016; Abel et al., 2015; Usunier et al., 2016; Vinyals and Povey, 2012; Aluru et al., 2015). In particular, we leverage a system called ActorSim (Roberts et al., 2016a; Roberts et al., 2016b) that is primarily distinguished from these other systems in its implementation of goal reasoning, existing experiments using deep learning (Bonanno et al., 2016) and integration with additional simulators and robotics platforms.

Using Minecraft and ActorSim, we examine how cognitive priming can improve existing learning algorithms in this challenging domain. In particular, we demonstrate how cognitive priming can effectively and efficiently learn from *external* experience to produce an *internal* context model that enables an agent to make more informed decisions, both when choosing what tool to use to break a block, and when choosing what action to take to achieve a goal.

2.4.1 Cognitive Priming to Improve Tool Selection

We begin with our study of using priming to bias selection of the best tool to break a block in Minecraft. Blocks are the basic building material in Minecraft and consist of logs, sand, dirt, iron ore, cobblestone, etc. Some blocks can be *crafted* into items and tools. For example, logs can be crafted into wood planks, which can in turn be crafted into sticks, and sticks can be combined with cobblestone to craft a stone pickaxe, or combined with smelted iron to make an iron pickaxe, etc. Mining or breaking a block for its resource (e.g., breaking a tree to obtain a log) is often accomplished with tools; while a tree block can be broken by hand, it is much more efficient to break the tree with an axe. In Minecraft, selecting the incorrect tool for a block (e.g., using a sword on cobblestone) slows the breaking speed and damages the tool more, decreasing its durability. A player learns to select the correct tool through a combination of repeated interaction with the game, reading the wiki, or watching other players. Over time, the player learns to choose the best tool to minimize tool wear and speed up resource collection. We simulate this iterative process by formulating the tool choice as a multi-armed bandit problem: for each block to break (i.e., the reward), our simulated agent decides which tool (i.e., which arm) to try and learns from its experience. We train cognitive priming with data on the correct tools to use in different situations, and use these primed values to bias future tool selection (i.e., as a kind of *Bayesian prior*). Our research hypothesis is that machine learning and cognitive priming, together, will reduce regret and improve the learning rate over machine learning alone.

2.4.1.1 Minecraft Data for Tool Selection

We leverage data collected by Branavan et al. (2012a, 2012b), who created a planning system that could learn to plan action sequences in the game of Minecraft by parsing the community wiki. Although they did not specifically examine the tool selection problem, their data encodes information that can be used to train approaches for tool selection. The authors showed that, by parsing natural language from the community wiki, they could build and represent dependencies between different concepts in the wiki (where concepts can be materials, tools, etc.). They considered 74 of these concepts from Minecraft, and generated 694 potential dependencies over them.

Here, we leverage the dependencies they created that relate together tools and block types. We format the dependencies by stemming plural words and removing unique identifiers. Note that although Branavan et al. (2012a) excluded the concepts of dirt or gravel from their dependencies, we add them into our experiments later for completeness. The following sample sentences show some of the pair-wise dependencies listed below each sentence that we consider:

> *A chest is crafted from eight wooden plank, as shown below.*
> *(chest, plank) (plank, chest)*
>
> *Axes are tools used to ease the process of collecting wood, plank, chest and bookcases, but are not required to gather them.*
> *(axe, wood) (axe, plank) (axe, chest) (wood, axe) ... (chest, wood) (chest, plank)*
>
> *Sand can be mined easily by hand, although using a shovel is faster, and gives resources regardless of the tool used.*
> *(sand, shovel) (shovel, sand)*
>
> *Shovels are auxiliary tools used to ease the process of collecting dirt, sand, gravel, clay and snow.*
> *(shovel, sand) (shovel, clay) (sand, shovel) (sand, clay) (clay, shovel) (clay, sand)*
>
> *Shovels are not effective at destroying soul sand and mycelium.*
> *(shovel, sand) (sand, shovel)*

As these sentences illustrate, dependencies are the cross-product of the mentioned concepts without regard to a direction. Note that missing concepts can result in missing or incorrect associations. For example, dirt and gravel lack dependencies even though the sentences mention them and the last sentence produced an incorrect dependency for "sand" when the correct block was actually "soul sand". We do not correct these errors when we use the data to train the priming model.

2.4.1.2 The Priming Model

We use these parsed dependencies to build a cognitive priming model of the objects that can be mined with tools. When the priming model thinks about these dependencies during training, associations are created (or strengthened) from the first item in the pair to the second. Because dependency pairs exist in both directions, tools and materials symmetrically prime each other. Intuitively, cognitive priming learned stronger associations between items that are currently mentioned together in the wiki and weaker associations between items that are not mentioned together as often.

To determine how contextually relevant tools are to a given block, we inspect the model's representation of the block to be broken (e.g., sand), collect all the tool associations for that block, and consider the strengths of the associations. Here, we set the *maximum associative strength* parameter to 1 (see Eq. 2.1) and so the strengths were in the range of (0, 1) for each associated item. Table 2.5 (top) shows the association strengths produced from the model for each tool when primed with the block in the left column; Table 2.5 (bottom) will be discussed below. Some items have no training for the priming model even though they are included in this study; for example, the original concepts from Branavan et al. (2012a) did not contain dirt or gravel (i.e., the bottom rows of each table), so the bias value is uniformly zero.

2.4.1.3 Multi-Armed Bandit Problem Baselines

We formulated tool selection as an iterated multi-armed bandit problem, a simple reinforcement learning algorithm (Sutton and Barto, 1998; Kaebling et al., 1996). For the ith block, the tools represent a choice from K probability distributions $(D_{i,1}, \ldots, D_{j,K})$ with expected means $(\mu_{i,1}, \ldots, \mu_{j,K})$ and variances $(\sigma_{i,1}, \ldots, \sigma_{j,K})$. The correct tool is exactly known, so the mean of that correct tool is 1 while the mean of the incorrect tools is zero. Because there is no variation in the reward of choosing the correct tool, all variances for the tool distributions are also zero. At each time t and for the ith block, an algorithm selects a tool with index $i, j(t)$ and reward $r_{j(t)} = D_{i,j(t)}$, which is 1 for choosing correctly and 0 otherwise. A common way to evaluate an algorithm is with regret minimization. For I blocks the total regret of an algorithm is:

$R_T = \Sigma_{i=1}^{I}(T\,\mu_i^* - \Sigma_{i=1}^{T}\mu_{i,j(t)})$, where $\mu_i^* = \max_{j=1,\ldots,k} \mu_{i,j}$ and $T = 50$.

We consider here two baseline approaches to solving the multi-armed bandit problem, both of which can be biased by including priming. Our baseline approaches are EpsilonGreedy with an $\epsilon = 0.10$ and Softmax with

Table 2.5. Priming values used to bias tool selection (top) and final reward values for Softmax with priming (SM+Priming) after learning (bottom). Asterisks indicate the best tool for breaking a block, while bold indicates the highest weight for a block. The count in the far right of each column indicates how many times that tool was tried out of 50 attempts.

Values Provided by Priming

	hoe		axe		shears		pickaxe		shovel	
bars	0.36	0	0.40	0	0.00	0	**0.58** *	0	0.36	0
brick	0.14	0	0.17	0	0.00	0	0.25 *	0	**0.46**	0
chest	0.50	0	0.51 *	0	0.05	0	**0.53**	0	0.47	0
clay	0.15	0	0.17	0	0.00	0	0.33	0	**0.65** *	0
coal	0.00	0	0.14	0	0.00	0	**0.29** *	0	0.00	0
cobblestone	0.28	0	0.35	0	0.00	0	**0.42** *	0	0.41	0
door	0.50	0	0.57 *	0	0.06	0	**0.60**	0	0.52	0
fence	0.48	0	**0.54** *	0	0.05	0	0.51	0	0.49	0
furnace	0.19	0	0.29	0	0.00	0	**0.32** *	0	0.25	0
iron	0.19	0	0.24	0	0.10	0	**0.54** *	0	0.31	0
sand	0.25	0	0.20	0	0.08	0	0.26	0	**0.40** *	0
sandstone	0.00	0	0.00	0	0.00	0	0.00 *	0	**0.50**	0
stair	0.57	0	**0.64**	0	0.00	0	0.61 *	0	0.59	0
wood	0.40	0	**0.58** *	0	0.00	0	0.41	0	0.41	0
wool	0.15	0	0.20	0	**0.32** *	0	0.15	0	0.24	0
dirt	0.00	0	0.00	0	0.00	0	0.00	0	0.00 *	0
gravel	0.00	0	0.00	0	0.00	0	0.00	0	0.00 *	0

Learned Action Model After Learning for SM+Priming

	hoe		axe		shears		pickaxe		shovel	
bars	0.36	0	0.40	0	0.00	0	**1.00** *	50	0.36	0
brick	0.14	0	0.17	0	0.00	0	**1.00** *	49	0.00	1
chest	0.50	0	**1.00** *	49	0.05	0	0.53	0	0.00	1
clay	0.15	0	0.17	0	0.00	0	0.33	0	**1.00** *	50
coal	0.00	0	0.14	0	0.00	0	**1.00** *	50	0.00	0
cobblestone	0.28	0	0.00	1	0.00	0	**1.00** *	48	0.00	1
door	0.50	0	**1.00** *	49	0.06	0	0.00	1	0.52	0
fence	0.48	0	**1.00** *	49	0.05	0	0.00	1	0.49	0
furnace	0.19	0	0.29	0	0.00	0	**1.00** *	50	0.25	0
iron	0.19	0	0.24	0	0.10	0	**1.00** *	50	0.31	0
sand	0.00	1	0.20	0	0.08	0	0.26	0	**1.00** *	49
sandstone	0.00	0	0.00	0	0.00	0	**1.00** *	49	0.00	1
stair	0.57	0	0.00	1	0.00	0	**1.00** *	49	0.59	0
wood	0.40	0	**1.00** *	50	0.00	0	0.41	0	0.41	0
wool	0.15	0	0.20	0	**1.00** *	50	0.15	0	0.24	0
dirt	0.00	0	0.00	0	0.00	0	0.00	1	**1.00** *	49
gravel	0.00	4	0.00	3	0.00	3	0.00	2	**1.00** *	38

a temperature of 0.10; these are standard approaches to this problem. When more than one tool equals the maximum value, as can happen in the case of the priming values, EpsilonGreedy selects a random choice among the possible maximums while Softmax uses a weighted roulette wheel across the ranked choices.

2.4.1.4 Experiment and Results

We tested our approach to see whether it could learn, via sequential decision making, the best tool for breaking each of the blocks. To determine the ground truth for the best tool, we use the Minecraft community wiki, which contains a page (http://minecraft.gamepedia.com/Tools) with details on the block breaking mechanics. We parsed a table from this page to determine the best tools for the blocks we studied. We also excluded blocks from Branavan et al. (2012a) that can be mined with any tool.

After training the priming model, we inspected the priming model association strengths. The results are shown in Table 2.5 (top), where an asterisk identifies the best tool, while the highest activation is marked in bold. For 15 blocks, excluding dirt and gravel, priming alone correctly associates 10 of them; it incorrectly associates the best tools for brick, chest, door, sandstone, and stair.

We then ran our baseline algorithms with and without priming with a learning rate of 0.30. The evidence suggests that seeding the algorithm with priming reduces regret not only before learning begins but also over time. Table 2.5 (bottom) shows the final expected reward values for Softmax with Priming (SM+Priming). When compared with the top table, it is clear that relatively little exploration resulted in a good policy (note the number of times the incorrect tool was chosen). For example, consider the furnace, which is best broken with a pickaxe. The top table shows that priming identifies the correct tool, and the bottom table shows that Softmax plus priming correctly chose this tool 50 times. Even when priming is not exactly correct, however, the model can improve its exploration using priming. For example, the correct tool is not associated by the priming model for sandstone, but the knowledge of the model still leads to a single incorrect tool usage by Softmax. This contrasts with gravel, where no priming knowledge leads to the incorrect tool usage 12 times.

Figure 2.6 shows *cumulative* regret over time for learning and learning with priming, which are denoted as SM+Priming and EG+priming for Softmax and EpsilonGreedy with priming, respectively. Priming clearly improves the baseline learning. Both learning algorithms, when using priming, start with lower regret, and regret remains lower than their respective versions without priming. SM+Priming explores more at the start and shows higher regret at the start but seems to quickly learn a good policy compared to EG+Priming, which explores less at the beginning but continues to accumulate regret. Importantly, even for blocks for which priming had an incorrect association (e.g., chest or door), priming eliminates several poor choices (e.g., shears) and still results in lower regret for those items than a uniform selection.

For this tool selection problem, we showed that priming can substantially improve learning over common baselines by using data from a study by Branavan et al. (2012a) that collected conceptual dependencies from the Minecraft community wiki. The study shows a case where priming is particularly useful at learning from external experience and knowledge

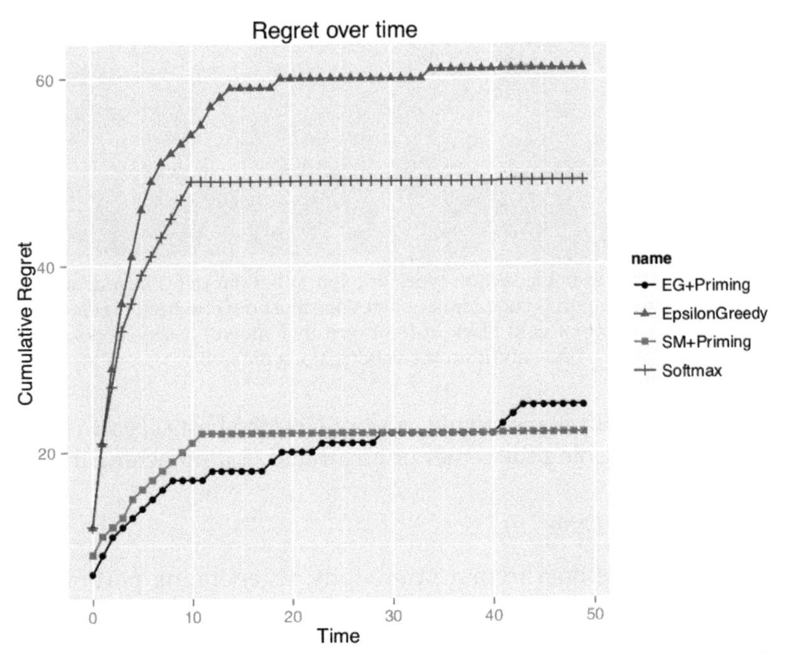

Figure 2.6. Cumulative reward (left) and cumulative regret for EpsilonGreedy and Softmax without priming and with priming (EG+Priming, SM+Priming).

(as *external* context) to make more informed choices and reduce regret over time via an internal context model.

2.4.2 Cognitive Priming for Action Selection

We now turn to examining the use of priming to choose the best subgoal (i.e., action) in the maze problem. In this problem, a character, Alex, is in a maze with various obstacles (e.g., Figure 2.7). The character needs to make it through to the other end of the maze without falling, going into lava, or drowning in water. We begin by describing the data we used, then give details of our priming model and baseline approach before discussing the results.

Figure 2.7 (left) shows four of the ten section types we use, including the arch, comb, pillar, and hill, while Figure 2.7 (right) shows a portion of a course with upcoming sections of a comb, swamp, lava and short wall. Not shown are tall walls (3 blocks high), deep ponds (water 3 blocks deep), and ponds (water 2 blocks deep). Each obstacle has an appropriate and ideal *subgoal* (i.e., action) choice. For lava or ponds, the best *subgoal* choice is to create a bridge if approaching the center or to go around if approaching the edge. For the short walls, the best *subgoal* is to create a single stair and

Figure 2.7. Left: four example section types (top left to bottom right) include arch, comb, pillar, and steps. Right: portion of a course where Alex must traverse from the emerald block behind it (not shown) to a gold block in front of it (not shown). Glass blocks on the top prevent Alex from walking along the wall.

step up. For the tall walls, combs, and pillars, the best *subgoal* is to mine through if approaching the center or go around if approaching the edge.

2.4.2.1 Minecraft Data

We again use here data from a prior study, regenerating portions of the data from Roberts et al. (2016b) using the ActorSim-Minecraft connector, which provides abstract methods for axis-aligned control for looking, moving, jumping, and destroying blocks. The connector can observe blocks directly around Alex and produce an alpha-numeric code to indicate the local position relative to Alex's feet "[lN | rN][fN | bN][uN | dN][Type]", where N is a non-negative integer, each letter designates **l**eft, **r**ight, **f**ront, **b**ack, **u**p, and **d**own, and Type is the block's type. For example, "f1d1" indicates the block directly in front and down one (i.e., the block that Alex would stand on after stepping forward one block), while "l1" indicates the block directly to the left at the same level as Alex's feet. Each position is also denoted with a single letter indicating the block's type: air (A); safe or solid (S); water (W); or lava (L). These codes serve as the features off of which Alex must make a decision, which consists of selecting one of five actions: step to (e.g., step forward), step around, build a bridge, build a stair one block high, and mine.

The data include traces of an Expert procedure guiding Alex through 30 courses, each of which is composed of 20 sections. The Expert procedure is a hand-coded decision loop that examines the state around the character to determine the appropriate action. The prior study had some corner cases where the Expert would get stuck, so we improved the Expert procedure so that it makes fewer errors. A run terminates when Alex reaches the goal; the revised Expert procedure never fails to complete a course. For each decision, several details were captured but we use only decisions where the chosen action succeeded. Together, these runs produced 5931 decisions (i.e., state-subgoal pairs).

2.4.2.2 Decision Tree Baselines

Our prior study using this dataset showed that this problem could be learned from expert traces using decision tree induction (Roberts et al., 2016b) or deep learning (Bonanno et al., 2016). To highlight the incremental nature of cognitive priming, we compare our work against incremental decision tree induction. For incremental decision tree induction, we use Hoeffding trees as implemented in WEKA (http://www.cs.waikato.ac.nz/~ml/weka/). For completeness, we also compare our approach to a batch decision tree classifier. For this, we use the J48 classifier, again as implemented in WEKA.

2.4.2.3 The Priming Model

The priming model learns which subgoals correspond to which features via an iterative learning process. During training, the model considers each decision Alex makes while also thinking about its state (i.e., the perceived blocks around Alex). Using this internal process as the basis for creating associations, the model then creates, or strengthens, associations from Alex's current state to the selected subgoal. Figure 2.8 (left) illustrates example model states during training. As more decisions are considered, the model learns differing strengths for the blocks around Alex and the selected subgoal (e.g., Figure 2.8, right).

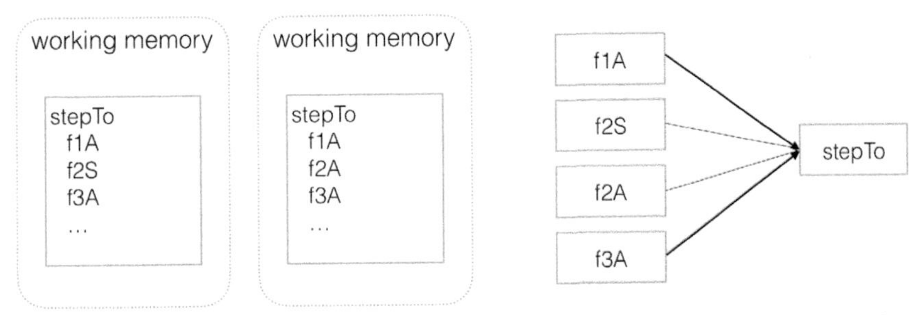

Figure 2.8. Associations after two snapshots of items in working memory. The association between the environment's features (f1A, etc.) and the items in working memory are directional. f1A and f3A more strongly prime StepTo (indicated by a darker line) because they have occurred together more often than f2S, f2A and StepTo.

Three important patterns emerge from the associative strengths between the coded features j and i subgoals (cf. Eqs. 2.1 and 2.2). First, coded features that are strongly indicative of certain subgoals become strongly associated with them because they occurred frequently with that subgoal but occurred infrequently with other subgoals, causing a high $f(N_iC_j)$ from Eq. 2.2 against a lower $f(C_j)$, and leading to a stronger overall associative strength. Similarly, features that are weakly indicative of certain subgoals have a higher $f(C_j)$ relative to $f(N_iC_j)$, resulting in a lower associative strength. Finally, features associated with many different subgoals spread association among many subgoals because they would have a high $f(N_iC_j)$ for each association, resulting in a lower overall associative strength for common features.

To test the priming model, we seed the model with the (location, block) values around Alex. These features then spread priming, according to their associative strength, to all connected subgoals. The subgoal with the highest association is selected.

2.4.2.4 Experiments and Results

The 5931 decisions from the Expert procedure are heavily skewed towards stepping forward (StepTo), the most common action. Therefore, we balance the data down to 1523 decisions which include 437 StepTo, 359 StepAround, 347 Stairs, 204 Bridge, and 175 Mine subgoals. We ran two different experiments for each of the decision tree baselines we use, incremental and batch.

For the incremental study, we focus on how well priming and the decision tree learn given different amounts of data. To approximate incremental learning, we create 10-folds to perform leave-one-out cross-validation. For each fold, we train on increments of 10%, 20%, ..., 100% of the data. At each increment, we observe the accuracy against the held-out test data. For incremental decision tree induction, we use a Hoeffding tree implementation from WEKA with default parameters. Figure 2.9 shows the average accuracy of decision trees learned by Hoeffding induction versus priming trained with the same data. While priming gracefully improves with more training data, it is clear that Hoeffding trees are more susceptible to errors given fewer training samples.

In our second study, comparing priming against the batch decision tree induction algorithm J48, we focused on understanding how both priming and decision tree induction were sensitive to errors in training or testing data, which equates to noisy observations in non-simulation environments. To do this, we construct "noisy" data and consider both how a decision tree learning and priming differentially perform when they are trained with pristine data but tested with increasingly mutated data, as well as how their performance compares with they are trained with increasingly mutated data and tested with pristine data. We use J48's default parameter settings in the study.

Here, we use sample sizes of 20 folds with 5% each, and we also mutate portions of the original data by randomly changing the values in the state. Figure 2.10 shows a performance comparison of the J48 decision trees and priming as the mutation rate increases. It is clear that priming and decision trees are functionally equivalent because there is little difference between

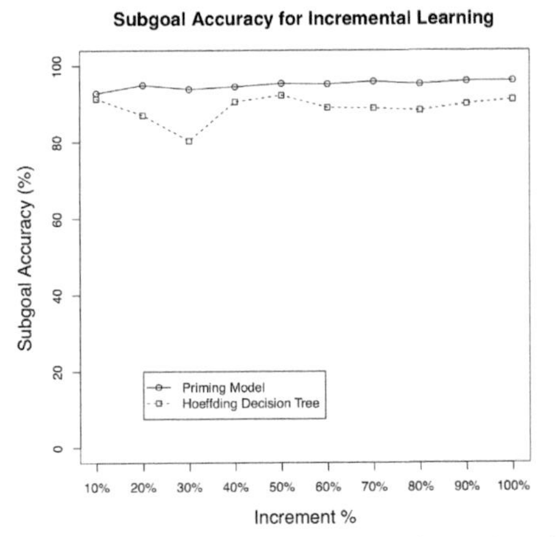

Figure 2.9. Accuracy for 10-fold cross validation training for 10%, 20%, ..., 100% of the data to simulate incremental learning.

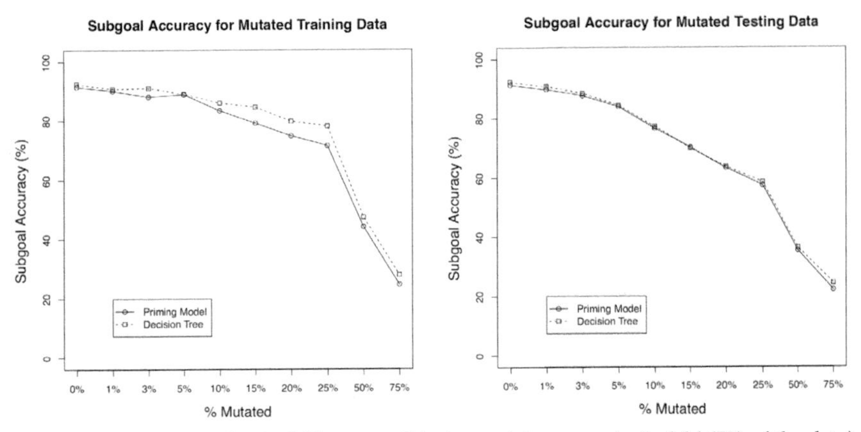

Figure 2.10. Accuracy for 20-fold cross validation training on a single fold (5% of the data) and testing on the other 19 folds (95%). The left plot shows mutated training data and the right shows mutated testing data.

the plots. In fact, the right plot shows nearly identical results. Still, this result is admittedly not as strong as we expected—we expected priming would be significantly more robust than decision tree induction because trained decision trees can be brittle to unseen examples that differ from the original problem distribution due to mutation, while priming has a more probabilistic interpretation of feature-class similarity.

2.4.2.4.1 Discussion

At a high level, the results show that priming dominates Hoeffding tree induction during incremental learning and performs less well against the batch learning of J48 tree induction under noisy observations. We offer some possible explanations that we will examine in future work. First, this problem is particularly amenable to batch tree induction because it has a high feature-to-class ratio, making this a difficult benchmark against which to compare priming. For example, there are 24 features (i.e., [location, block] pairs) that can be only 4 possible values and across only 5 classes. From that standpoint, it is impressive that priming was able to compete so strongly with J48.

Second, the noise we introduce may not mutate critical features. For example, building a bridge is relatively rare and only relevant when water or lava are present directly in front of Alex. Instead of randomly mutating across all 24 features, future work should mutate the handful of features critical to the decision tree or priming model. We believe priming will be more robust to such noise because it associates across many features, whereas the decision tree tends to focus on a small set of features. So, noise regarding the status of water or lava in front of Alex would hinder decision tree induction, whereas priming would consider details about features to the left and right as well.

Third, there is latent information in the kinds of errors each algorithm makes and what this means for improving them in an *online* setting—some homologous subgoals have negligible impact if chosen instead of the expert choice while other subgoals would result in great harm to Alex if executed. For example, in most situations StepTo and StepAround are applicable and have a nearly identical effect of moving the player closer to the target, but BuildBridgeTo when lava is in front of Alex is really the only appropriate decision. A quick glance at the confusion matrices as mutation increases reveals several trends: (1) both algorithms identify the StepTo/StepAround with similar accuracy, (2) priming identifies BuildStairsTo better than J48, and (3) J48 identifies BuildBridgeTo better than priming. Together, these results warrant continuing this line of work while moving toward more challenging *online* scenarios, better characterizing where each approach excels, and including a negative reward signal for damaging choices. Overall, in terms of context, both studies show that cognitive priming can effectively and efficiently learn from *external* experience to produce an *internal* context model that enables an agent to make more informed decisions.

2.5 General Discussion

We have examined the role of cognitive priming in improving classification in images, the detection of anomalies, sequential decision making, and action selection. In each of these cases, we have demonstrated that leveraging cognitive priming generally improves performance over systems that lack this form of context. Further, we have shown that context, in general, need not always be confined to and driven by the *external* world of the agent. Agents can, and probably should, use their *internal* state to drive their learning of context about the external world.

In addition to the results we have shown here, there are several important implications of this. First, as robots become more complex, and separate components of robots become more integrated (such as vision vs. task planning), we believe that this priming-based view of context will continue to be an important source of learning and reasoning for autonomous agents. Using priming's learning mechanisms results in easy, natural training of context since cognitive priming, like people, learns online as it goes about in the world, and can adapt quickly to new situations (Hiatt et al., 2016). The above approach also connects context with work on priming's influence on how people perceive and act in the world in other ways, such as explaining errors that people make on routine procedural tasks (Hiatt and Trafton, 2015a), and explaining induced cases of mind wandering (Hiatt and Trafton, 2015b). This can potentially increase the effectiveness of contextual models at not only understanding the world around them, but also relating that understanding to the behavior of human partners.

This approach to context also makes it possible to learn context over long periods of time. As we argue at the beginning of this chapter, guiding contextual learning with an agent's internal state limits the associations learned to those that have been useful in the past and so are likely to be useful in the future. We expect that this natural limiting will allow cognitive priming to provide useful context to agents operating and learning over long periods of time (Roberts et al., 2016b). Along those lines, we hope to extend our approach to agents operating independently over long periods of time, making decisions about what goals to pursue (Roberts et al., 2016a; Alford et al., 2016), and then understanding the world well enough to achieve them.

As part of this, we plan to extend priming to affect the machine's cognition at all levels concurrently: perception, task planning, and selecting goal strategies. When considering all of these together in an integrated approach to context via cognitive priming, the nature of cognitive priming to use its internal representations and processes to drive learning context about the external world becomes even more important. Tools relevant to a task performed in the past become associated with the task because they were thought about together—facilitating both the visual recognition and use of those tools when performing that task in the future. Goals that are associated with key features of the environment, even if those features are not explicitly recognized or interpreted, are primed when those features appear, helping the agent repeat good decisions in the past. In these ways, cognitive context has the potential to become a powerful tool that can combine with both low-level and high-level reasoning to greatly improve machine performance.

References

Abel, D., Hershkowitz, D.E., Barth-Maron, G., Brawner, S., O'Farrell, K., MacGlashan, J. and Tellex, S. 2015. Goal-based action priors. Proc. Int'l Conf. on Automated Planning and Scheduling.

Alford, R., Shivashankar, V., Roberts, M., Frank, J. and Aha, D.W. 2016. Hierarchical planning: eelating task and goal decomposition with task sharing. Proc. of the Int'l Joint Conf. on AI (IJCAI). AAAI Press.

Aluru, K., Tellex, S., Oberlin, J. and Macglashan, J. 2015. Minecraft as an experimental world for AI in robotics. AAAI Fall Symposium.

Anderson, J.R. 1983. A spreading activation theory of memory. Journal of Verbal Learning and Verbal Behavior 22(3): 261–295.

Anderson, J.R. 2007. How Can the Human Mind Occur in the Physical Universe? Oxford University Press, 2007.

Anderson, J.R., Bothell, D., Lebiere, C. and Matessa, M. 1998. An integrated theory of list memory. Journal of Memory and Language 38(4): 341–380.

Bonanno, D., Roberts, M., Smith, L. and Aha, D.W. 2016. Selecting Subgoals using Deep Learning in Minecraft: A Preliminary Report. IJCAI Workshop on Deep Learning for Artificial Intelligence.

Branavan, S., Kushman, N., Lei, T. and Barzilay, R. 2012a. Learning high-level planning from text. Proc. of the 50th Annual Meeting of the Association for Computational Linguistics (pp. 126–135). Jeju, Republic of Korea.

Branavan, S., Silver, D. and Barzilay, R. 2012b. Learning to win by reading manuals in a monte-carlo framework. J. of Art. Intell. Res. 43: 661–704.

Divvala, S., D. Hoiem, J. Hays, A. Efros and M. Hebert. 2009. An empirical study of context in object detection. Computer Vision and Pattern Recognition (CVPR).

Duric, Z., W.E. Lawson and D. Richards. 2007. Streaming clustering algorithms for foreground detection in color videos. International Conference on Computer Vision Theory and Applications (VISAPP).

Everingham, M., L. Van Gool, C.K.I. Williams, J. Winn and A. Zisserman. The PASCAL Visual Object Classes Challenge 2007 (VOC2007) Results. http://www.pascalnetwork.org/challenges/VOC/voc2007/workshop/index.html.

Felzenszwalb, P., Girshick, R., McAllester, D. and Ramanan, D. 2010. Object detection with discriminatively trained part based models. IEEE Transactions on Pattern Analysis and Machine Intelligence (TPAMI).

Galleguillos, C., Rabinovich, A. and Belongie, S. 2008. Object categorization using co-occurrence, location, and appearance. Computer Vision and Pattern Recognition (CVPR).

Harrison, A.M. and Trafton, J.G. 2010. Cognition for action: an architectural account for grounded interaction. Proceedings of the Annual Meeting of the Cognitive Science Society.

He, K., Zhang, X., Ren, S. and Sun, J. 2016. Deep residual learning for image recognition. In Proceedings of the IEEE conference on Computer Vision and Pattern Recognition.

Hiatt, L.M. 2017. A priming model of category-based feature inference. In Proceedings of the Annual Meeting of the Cognitive Science Society.

Hiatt, L.M.,Lawson, W.E., Harrison, A.M. and Trafton, J.G. 2016. Enhancing object recognition with dynamic cognitive context. In Workshop on Symbiotic Cognitive Systems (held in conjunction with AAAI).

Hiatt, L.M. and Trafton, J.G. 2015a. An activation-based model of routine sequence errors. In Proceedings of the International Conference on Cognitive Modeling.

Hiatt, L.M. and Trafton, J.G. 2015b. A computational model of mind wandering. In Proceedings of the Annual Meeting of the Cognitive Science Society.

Hiatt, L.M. and Trafton, J.G. 2017. Familiarity, priming and perception in similarity judgments. Cognitive Science 41(6): 1450–1484.

Johnson, M., Hofmann, K., Hutton, T. and Bignell, D. 2016. The Malmo Platform for Artificial Intelligence Experimentation. Proc. 25th International Joint Conference on Artificial Intelligence, Ed. Kambhampati S., p. 4246. AAAI Press, Palo Alto, California USA.

Kaebling, L., Littman, M. and Moore, A. 1996. Reinforcement learning: A survey. J. of Art. Intell. Res. 4: 237–285.

Krizhevsky, A., Sutskever, I. and Hinton, G.E. 2012. Imagenet classification with deep convolutional neural networks. In Advances in neural information processing systems, pp. 1097–1105.

Lawson, W., Hiatt, L. and Sullivan, K. 2016a. Detecting anomalous objects on mobile platforms. In CVPR Workshop on Workshop on Moving Cameras Meet Video Surveillance: from Body-borne Cameras to Drones.

Lawson, W., Hiatt, L. and Trafton, J.G. 2014. Leveraging cognitive context for object recognition. In Proceedings of the Vision Meets Cognition Workshop (held in conjunction with CVPR), pp. 381–386, Columbus, OH.

Lawson, W., Sullivan, K., Bekele, E., Hiatt, L.M., Goring, R. and Trafton, J.G. 2016b. Automated surveillance from a mobile robot. In Proceedings of the AAAI Fall Symposium on Artificial Intelligence for Human-Robot Interaction.

Li, C., Kowdle, A., Saxena, A. and Chen, T. 2010. Towards holistic scene understanding: Feedback enabled cascaded classification models. Advances in Neural Information Processing Systems (NIPS).

Medin, D.L., Goldstone, R.L. and Markman, A.B. 1995. Comparison and choice: Relations between similarity processes and decision processes. Psychonomic Bulletin and Review 2: 1–19.

Nosofsky, R.M. 1992. Exemplar-based approach to relating categorization, identification, and recognition. pp. 363–393. *In*: Ashby, F.G. (ed.). Multidimensional Models of Perception and Cognition. Lawrence Erlbaum Associates, Inc., Hillsdale, NJ, England.

Novick, L.R. 1990. Representational transfer in problem solving. Psychological Science, pp. 128–132.

Oquab, M., Bottou, L., Laptev, I. and Sivic, J. 2014. Learning and transferring mid-level image representations using convolutional neural networks. Computer Vision and Pattern Recognition (CVPR).

Oliva, A. and Torralba, A. 2007. The role of context in object recognition. TRENDS in Cognitive Sciences 11(12): 520–527.

Razavian, A.S., Azizpour, H., Sullivan, J. and Carlsson, S. 2014. CNN features off-the-shelf: an astounding baseline for recognition. Computer Vision and Pattern Recognition Workshops (CVPRW).

Redmon, J. and Farhadi, A. 2017. YOLO9000: better, faster, stronger. Computer Vision and Pattern Recognition (CVPR).

Roberts, M. and Hiatt, L. M. 2017. Improving sequential decision making with cognitive priming. In Advances in Cognitive Systems.

Roberts, M., Hiatt, L.M., Coman, A., Choi, D., Johnson, B. and Aha, D. 2016a. Actorsim, a toolkit for studying cross-disciplinary challenges in autonomy. Working Notes of the Symposium on Cross-Disciplinary Challenges in Autonomy, Fall Symposium Series.

Roberts, M., Shivashankar, V., Alford, R., Leece, M., Gupta, S. and Aha, D. 2016b. Goal reasoning, planning, and acting with actorsim, the actor simulator. Poster Proceedings of the Fourth Annual Conference on Advances in Cognitive Systems. Evanston, IL, USA.

Schneider, D.W. and Anderson, J.R. 2011. A memory-based model of hick's law. Cognitive Psychology 62(3): 193–222.

Silberman, P.K.N., Hoiem, D. and Fergus, R. 2012. Indoor segmentation and support inference from RGBD images. In Proceedings of the European Conference on Computer Vision.

Sutton, R. and Barto, A. 1998. Reinforcement Learning: An Introduction. MIT Press.

Tessler, C., Givony, S., Zahavy, T., Mankowitz, D.J. and Mannor, S. 2016. A Deep Hierarchical Approach to Lifelong Learning in Minecraft. Proc. of the 31st AAAI Conf. on AI, pp. 1553–1561.
Thomson, R., Harrison, A.M., Trafton, J.G. and Hiatt, L.M. 2017. An account of interference in associative memory: Learning the fan effect. Topics in Cognitive Science 9(1): 69–82.
Thomson, R., Pyke, A.A., Trafton, J.G. and Hiatt, L.M. 2015. An account of associative learning in memory recall. Proceedings of the Annual Meeting of the Cognitive Science Society.
Trafton, J.G., Hiatt, L.M., Harrison, A.M., Tamborello, F., II, Khemlani, S.S. and Schultz, A.C. 2013. ACT-R/E: An embodied cognitive architecture for human-robot interaction. Journal of Human-Robot Interaction 2(1): 30–55.
Usunier, N., Synnaeve, G., Lin, Z. and Chintala, S. 2016. Episodic Exploration for Deep Deterministic Policies: An Application to StarCraft Micromanagement Tasks. arXiv preprint arXiv: 1609.02993.
Vinyals, O. and Povey, D. 2012. Krylov Subspace Descent for Deep Learning. AISTATS, pp. 1261–1268.
Zhou, B., Lapedriza, A., Xiao, J., Torralba, A. and Oliva, A. 2014. Learning deep features for scene recognition using places database. In Advances in Neural Information Processing Systems (NIPS).

The Use of Contextual Knowledge in a Digital Society

Shu-Heng Chen[1,*] and *Ragupathy Venkatachalam*[2]

3.1 Introduction

How can knowledge that is scattered among millions of individuals in a society be mobilized and utilized in the best possible way? The importance of this question is rather self-evident. However, any reasonable answer to this question would depend on a variety of factors. To start with, the way in which a society is organized at the time of this consideration is an important factor. It may also depend on the state of relevant prevailing technological, institutional factors which govern how information is exchanged in a society. There may be difficulties that arise in unambiguously determining a unique best way to accomplish this monumental task in principle. Instead, one may have to be content with relatively better alternatives that are available in a society.

In the case of our current society, advancements in information and communication technology (ICT) have come to predominate the determination of the way in which we generate, store and communicate or exchange information. The development of the Internet, or more broadly Web 2.0, has enabled millions of individuals to interact, collaborate and

[1] Distinguished Professor, Department of Economics, Director of the AI-ECON Research Center, National Chengchi University, Taipei, Taiwan 11605.
[2] Lecturer in economics at the Institute of Management Studies, Goldsmiths, University of London.
* Corresponding author

share knowledge. These technologies have led to the production of digital data on a massive scale and variety that surpasses levels which existed at any point of time in history. This situation has often been referred to as the beginning of the big data era. In addition, developments in computational intelligence and machine learning have also placed a variety of tools at our disposal that enable us to process these massive volumes of data and synthesize into useful information.

The problem of knowledge or information aggregation is not unique to a digital society. In fact, this has been the subject of an important debate in the history of economic analysis that took place during the first half of the twentieth century, viz., the socialist calculation debate. This debate concerns the possibility of a planned economy performing (replicating) economic calculations (or computations) in relation to a market economy. Such calculations naturally involve a wide variety of information concerning relative scarcities, determining the optimal allocation of resources and prices in the economic system. We focus on an important contribution to this debate by Friedrich Hayek, which recast the fundamental nature of the economic problems facing a society as constituting a knowledge aggregation problem. He focused on the relative merits of markets and the efficacy of price as a coordinating device in aggregating knowledge that is dispersed among different actors. His broad conclusion was that with a centralized mechanism, it would be impossible to aggregate the relevant knowledge for it is often fragmented, incomplete (never available in its entirety to a single person), tacit and often contextual. For him, there was no rival mechanism that would have been effective like the decentralized market that coordinates the actions of individuals through price signals.

However, ICT-enabled technologies are increasingly seen as offering an effective way to mobilize knowledge that is dispersed among individuals. Big data and machine learning together offer individuals the ability to economize from the search of vast, relevant knowledge so that they can make a variety of decisions. Nevertheless, are there limits concerning the extent to which ICT-enabled digital technologies can aid in the utilisation of knowledge? Is this data-driven approach capable of handling all forms of knowledge that exist in a society? To what extent can they augment or replace other forms or modes through which decentralized knowledge has been aggregated in the past? These are questions that we aim to address in this chapter.

The chapter is organized as follows. In Section 3.2, we provide an overview of the socialist calculation debate and focus on Hayek's important contribution to this debate. In Section 3.3, we outline the developments that govern the way in which knowledge is used and exchanged in a digital society. We sketch how markets and many modes of ICT-enabled platforms can be viewed as information aggregation platforms. Section 3.4 explores the features that distinguish the ICT-enabled platform and explores some of its limitations, in particular, those that concern lack of disciplining constraints on the data generated by these platforms. Section 3.5 outlines the new possibilities that are offered by ICT-enabled platforms, such as matching, that constitute non-market coordination devices.

3.2 Hayek, Calculation Debate and Knowledge

Friedrich Hayek (1899–1992), considered to be one of the titans of economic theory in the twentieth century, made a seminal contribution to the socialist calculation debate that forms an important backdrop for the themes discussed in this chapter. Hayek advanced the Austrian tradition in economics in several original and important ways, but he is more than just an economist. He was an interdisciplinary scholar with works that inspired many later developments in a variety of areas such as complex systems, neural networks, philosophy, law, and methodological individualism. He was also awarded the Nobel memorial prize in economics in 1974. We will focus on his contribution to the idea of contextuality and knowledge, which are related to the theme of this volume. To understand this better, a brief description of the socialist calculation debate—an important theoretical debate in economic theory—may be apposite at this point.

3.2.1 Socialist Calculation Debate

The socialist calculation debate centered around the feasibility and relative efficiency of the socialist mode of economic organization. This mode relying on central planning and common ownership and its ability to perform (replicate) economic calculation on a par with a society organized in the capitalistic mode was questioned. The relative superiority of central planning vis-à-vis a decentralized market-based mechanism in the allocation of capital, labor and other resources was one of the topics that was intensely debated. It is considered to be one of the most important debates in economic theory, the debate broadly between Austrian economists, neoclassical economists and market socialists.[1]

The calculation debate started with Ludvig von Mises' essay (von Mises, 1920) entitled 'Economic Calculation in the Socialist Commonwealth', in which he challenged the feasibility of rational economic planning within socialism in the absence of private ownership. Among the responses, Oscar Lange in his essay 'On the Economic Theory of Socialism' (Lange, 1936, 1937) claimed that once the same set of information available to a theoretical market economy is presented to a central planning board, the latter will be able to devise mechanisms that replicate and achieve the

[1] For more on the Socialist Calculation Debate, see Levy and Peart, 2008; Boettke, 2000.

optimum allocation of resources. In doing so, Lange in fact utilized the Walrasian general equilibrium analysis, a benchmark theoretical model of the market-based economy, in making his case for such a possibility. Under the original Walrasian formulation, an economy can be seen as a set of equations. Thus, there should be no need for prices. Once the information about available resources and people's preferences are given, it should be possible to calculate the optimal solution for resource allocation. Lange even went on to propose a procedure in the form of an actual computing algorithm through which allocation could be accomplished by a centrally planned economy. In other words, his proposal was meant to be a non-market based alternative that would be able to achieve the optimal allocation of resources in an economy.

Among the many scholarly exchanges on this topic, Hayek later made an important contribution that is often regarded as the most 'influential single article in the debate' (Levy and Peart, 2008). A remarkable aspect of Hayek's contribution lies in placing the crucial role of knowledge, in particular, contextual knowledge, at the center of this debate. His paper is considered to be a classic even today and featured as one of the 20 most important articles in the American Economic Review in the 100 years since its inception (Arrow et al., 2011). This article (Hayek, 1945) has since been constantly re-examined in light of the subsequent advancements in information technology and computing (Cockshott and Cottrell, 1997; Lavoie, 1985). In the following subsection, we provide an overview of this article to set the stage for our later analysis.

3.2.2 The Use of Knowledge in a Society

In departing from the dominant theoretical outlook concerning the functioning of a decentralized market economy that underpinned the socialist calculation debate, Hayek ventured to completely reformulate the focus of the problem at hand. For him, there was a wedge and the economic problem that faced the society was quite different from the manner in which it was conceived within neoclassical economic theory. The problem was perceived by many who were participating in the debate as that of finding the best available use of the resources available in a society and the conditions under which this could be accomplished. Although the goal in itself may not have been impediment, the problem formulation was severely mistaken according to Hayek. It assumed that a problem solver was presented with all of the relevant information, the knowledge regarding all of the available means and starts with a given system of preferences. Conditional on this being available in its entirety to a single mind, the problem was to find the best possible use of the available means in a society.

Hayek argued that the actual economic problem was not the one above, but that it was instead a problem of the effective utilization of the knowledge in a society. If so, what might be the nature of this knowledge? Hayek pointed out that the data concerning the availability of resources, preferences, associated and relevant knowledge is never available to a single mind to perform the economic calculation outlined earlier. Hayek argued that, by its very nature, such knowledge is dispersed among a multitude of individuals in an economy, where everyone knows only bits and pieces, but never the knowledge in its entirety.

> The peculiar character of the problem of a rational economic order is determined precisely by the fact that the *knowledge of the circumstances of which we must make use* never exists in concentrated or integrated form, but solely as the dispersed bits of incomplete and frequently contradictory knowledge which all the separate individuals possess. The economic problem of society… is rather a problem of how to secure the best use of resources known to any of the members of society, for ends whose relative importance only these individuals know. (Hayek, 1945, italics added)

Hayek asked which of the available arrangements—market-based or a centrally planned economy—is better disposed to solving the knowledge problem. The answer to this question, of course, will depend on the kind of knowledge (scientific, contextual, tacit and so on) that we are talking about. In particular, when analyzing at the level of the whole society, the relative importance of different forms of knowledge becomes important. He argued that a substantial part of the knowledge that is relevant in this context is subjective, tacit (Polanyi, 1966) and often unorganized. Consequently, it would be infeasible for individuals to make all of this knowledge readily available to a planner to carry out computations. For him, 'the knowledge of the circumstances' is a crucial element to be confronted with when one tries to understand the economic problems facing the society. In our view, it is straightforward to interpret unorganized or dispersed knowledge—knowledge of the particular circumstances of time and place—as contextual knowledge. Hayek went on further to add that this form of knowledge cannot be easily summarized in an aggregate, statistical form since it is highly contextual and these important differences associated with context are necessarily obscured in trying to summarize them as statistical aggregates.

> ... I should briefly mention the fact that the sort of knowledge with which I have been concerned is knowledge of the kind which by its nature cannot enter into statistics and therefore cannot be conveyed to any central authority in statistical form. The statistics which such a central authority would have to use would have to be arrived at precisely by abstracting from minor differences between the things, by lumping together, as resources of one kind, items which differ as

> regards location, quality, and other particulars, in a way which may be very significant for the specific decision. **It follows from this that central planning based on statistical information by its nature cannot take direct account of these circumstances of time and place, and that the central planner will have to find some way or other in which the decisions depending on them can be left to the "man on the spot."** (Hayek, 1945, emphasis added)

If knowledge is in fact dispersed, contextual and often tacit as Hayek points out, then how does society tackle this problem of economic calculation, i.e., the efficient (at least reasonably) allocation of the available resources? To understand this, one can view the market as an institutional arrangement, which attempts to pool together or aggregate the 'knowledge of the particular circumstances of time and place' that is dispersed among many interdependent agents across space. This interdependence can be interpreted as a *distributed network* in modern parlance.

Even if we adopt the view that the market is an interdependent, institutional arrangement or a network of individuals, economizing knowledge (and resources) in a society involves a continuous, dynamic coordination of the actions of the various individuals who supply and demand these resources and products. Therefore, what the market eventually does with this pooled dispersed knowledge is important. Hayek, along the lines of von Mises, argues that markets generate prices (which are symbols, signals or indicators) that aid in economizing on the amount of knowledge required by any one agent. Prices, therefore, *emerge* as a result of the market activity that pools together the dispersed knowledge in an economy. They are *emergent* in the following sense: as Hayek states, these prices as numerical or quantitative indices "cannot be derived from any property possessed by that particular thing, but which reflects, or in which is condensed, its significance in view of the whole means-end structure." (Ibid). A market can be viewed as playing the role of aggregating information and generating "sufficient statistics" in the form of prices. These emergent prices, in turn, act as coordination devices that help actors to economize on the amount of knowledge that any particular individual needs to possess or know to make market decisions.

> Which of the events which happen beyond the horizon of his immediate knowledge are of relevance to his immediate decision, and *how much of them need he know*? (Ibid; Italics added)

> The most significant fact about this system is the economy of knowledge with which it operates, or *how little the individual participants need to know* to be able to take the right action. (Ibid; Italics added)

For instance, a decision maker or an entrepreneur in any one place does not have to concern himself with all of the potential causes that can alter the demand elsewhere and to seek the relevant knowledge on this. In Hayek's view, he will simply have to look at prices since they reflect the underlying scarcity in an economy. Although Hayek was writing at a time when ideas concerning network science were not readily available, he nevertheless incorporated some of these concepts implicitly. For example, the amount of search that an individual has to undertake can be seen as the network distance that exists between the 'man on the spot' and all of the knowledge that is relevant (and required) to make his decision. In a market economy, this involves only a single degree of separation. The dramatic shortening of network distance is due to the existence of emergent prices, indices or sufficient statistics (Figure 3.1). With prices, each decision maker "will have to consider only these *quantitative indices (or 'values') in which all the relevant information is concentrated*." (Ibid; Italics added) These indices

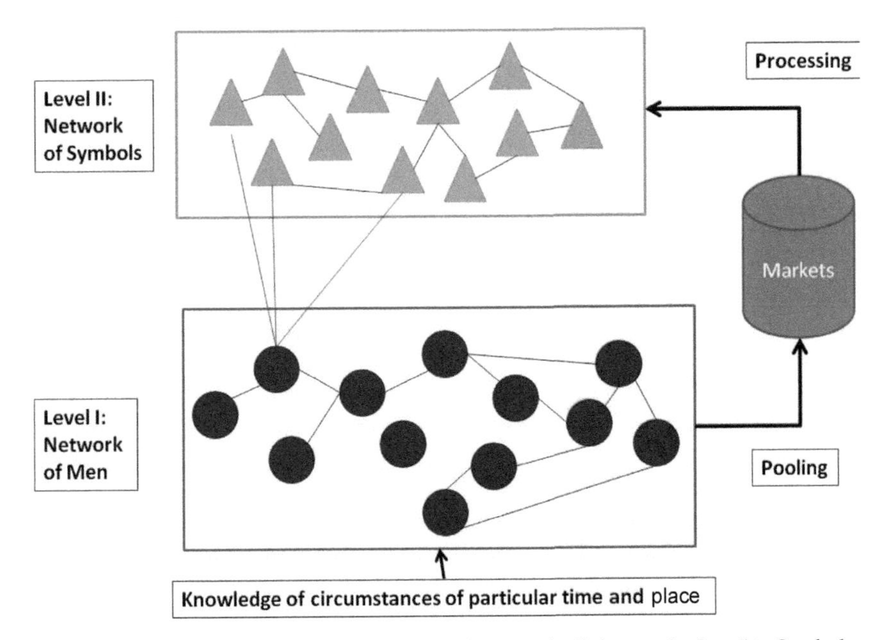

Figure 3.1. The Two-Level Knowledge Network: From the "Man on the Spot" to Symbols. In the figure above, the second-level knowledge network emerges from the bottom level. These include the emergence of the symbols (prices and indices) and their connections. At the first level, the "man on the spot" may be a long distance away from the other "man on the spot" who owns knowledge of specific circumstances at a particular time and place. However, given the availability of the indices at the second level, the man at the first level can get direct access to those symbols at the second level in one degree of separation. Hence, the appearance of the second-level network bridges the gap between individuals in the communication and use of knowledge.

incorporate all of the relevant knowledge, including contextual and tacit knowledge. Using this concept, Hayek advanced the argument that a free market system is well-equipped to solve the knowledge problem, while a centrally planned system is not.

We can interpret the entire system of prices as a second-level knowledge network that emerges from the underlying network (as shown in Figure 3.1). The first-level network can thus be seen as network of agents from which a second-level network of symbols emerge. These symbols compress relevant knowledge and help agents to deal with the dispersed, contextual knowledge and with it in a way to reckon with the complexity of the system in which they operate. However, the efficacy of this mechanism in this network formulation is contingent on whether there is a sufficient overlap among the agents in the first-level network. Leaving that aside, the idea that these symbols "can act to coordinate the separate actions of different people in the same way as subjective values help the individual to coordinate the parts of his plan." (Ibid). This referred to as the Hayek hypothesis, which has received considerable attention in the field of experimental economics (Smith, 1982).

In sum, Hayek's argument that markets pool contextual knowledge efficiently to generate prices and help to coordinate the actions of individuals is an important insight. For Hayek, there has not been an alternative or rival mechanism that effectively coordinates actions in the way that markets and prices do.

All that we can say is that nobody has yet succeeded in designing an alternative system in which certain features of the existing one can be preserved which are dear even to those who most violently assail it—such as particularly the extent to which the individual can choose his pursuits and consequently freely use his own knowledge and skill (Hayek, 1945).

However, are prices the only exclusive symbols that exist in a society? What has not been addressed in Hayek (1945) is the role and significance of non-price symbols, and how these non-price symbols interact with a 'man on the spot'. There are other coordination devices that exist in an economy which help align the actions of individuals (Manzini et al., 2009). On the other hand, several advancements in information technology have enabled us to re-conceptualize what we traditionally think of as data. In the next section, we will investigate the relevance of Hayek's view in the context of the socialist calculation debate to a society that is increasingly becoming digital and in which information access has advanced dramatically.

3.3 Knowledge and the Digital Society

A remarkable change in the society in which we currently live concerns the way in which information is stored, exchanged and utilized for making a variety of day-to-day decisions. Formidable developments in computational

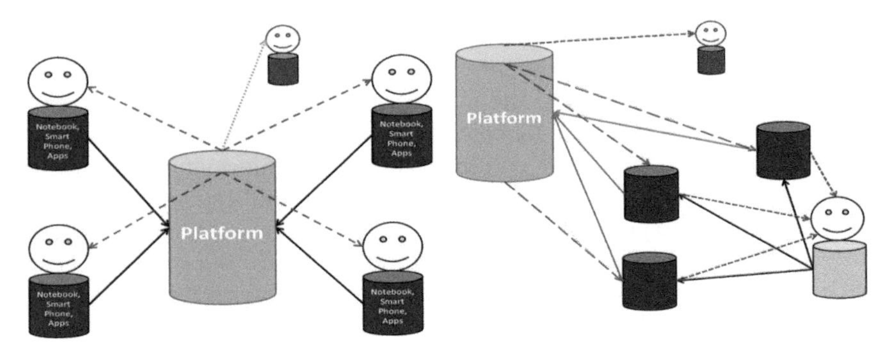

Figure 3.2. Ubiquitous Computing (Left Panel) and Internet of Things (Right Panel).

tools, information and communication technology have revolutionized the way in which we generate and exchange information, notably with the increasing penetration of the Internet throughout the world. Its increasing penetration has extended to various aspects of social life, ranging across the devices that we use, such as phones, watches, cars, washing machines, televisions, and personal assistants, among others. Computing information is no longer restricted to desktops alone; it has become ubiquitous across devices and space. This ubiquitous or pervasive computing has been partly facilitated by the rapid increase in the use of various hand-held devices and the growing popularity of social media. These developments have resulted in an information-rich environment where human beings now operate. Consequently, more and more aspects of our lives are being recorded and quantified in ways that were impossible to conceive of a few decades ago.

The manner in which information is exchanged and used within this larger framework has been succinctly synthesized in Figure 3.2. The left panel summarizes the way in which information or data is collected or pooled from a variety of electronic digital devices that individuals own, such as a computers, watches, tablets or smart phones. These devices are connected to a common platform via the Internet. However, information flow in this case is not just unidirectional since the data pooled from other agents can be aggregated or processed. This synthesized or processed information can be accessed by other individuals who might need it for their decisions. Some of the examples include the United Nations' project on Global Pulse, Google Flu Trends, Google Glass, and Street Bump (Table 3.1). In some cases, the information relayed back to the agent is merely informative as in the case of on-line restaurant reviews. In some other cases such as Environmental Teapot, BinCam, and smart mirror (Table 3.1), information signals are used to actively persuade or 'nudge' agents to behave or make decisions in a specific way.[2]

[2] This sort of technology is often referred to as persuasive technology (Fogg, 2002; Hamari et al., 2014).

In addition to ubiquitous computing, we find that the various physical digital devices constitute a network over which they routinely interact with each other and the individual (Figure 3.2, the right panel). Typically, stationary or mobile objects are endowed with a digital sensor that enables these objects to continuously collect information regarding their environment, ranging from temperature, humidity and suspended matter in the air to pedestrian intensity and even conversations. All of this information can be pooled into a platform that in turn is a source of very rich information.[3] This is popularly referred to as the Internet of Things (IoT) (Westerlund et al. 2014; Rajahonka, 2014). It facilitates real-time information sharing between agents and devices and this could be useful for increasing the quality of decision-making in a variety of ways. Some examples include BinCam, Environmental Teapot, smart mirror, smart carpet or smart belt, as well as intelligent personal assistants (Table 3.1).

Table 3.1. New data collection device.

Description	Related Research
Global Pulse	Kirkpatrick (2014)
Google Flu Trends	Ginsberg et al. (2009)
Google Glass	Ackerman (2013)
Street Bump	Schwartz (2012)
Traffic D4V	Picone et al. (2012)
BinCam	Comber and Thieme (2013)
Environmental Teapot	Marres (2012)
Smart Mirror	Pantano and Nacarato (2010)
Smart Carpet	Aud et al. (2010)
Smart Belt	Shieh et al. (2013)

3.3.1 Big Data

Another distinguishing feature related to increasing digitization in our society concerns the vastness of data in regard to the diverse domains of human activities that are available for analysis. This situation has to with advancements in data-collection and archiving processes. It can be referred to as the big-data phenomenon.[4] The amount of data available for academic and commercial research has seen an exponential growth in the past decade that has opened many new possibilities and avenues for rich

[3] It is important to acknowledge that there are obvious and serious ethical issues when such information is gathered without the active consent of the individual involved.

[4] The origins of the term Big Data can be traced back to the turn of the twenty-first century (Diebold, 2012).

data analysis. In addition to its commercial use, the popularity of this big data phenomenon has been significant across diverse academic disciplines such as computer science, epidemiology, economics, history, particle physics and sociology (Einav and Levin, 2014; Varian, 2014; Howe et al., 2008). The vastness of this data has increased the hope that a purely data-driven science is possible and that it is well within our reach.

A clear definition of what exactly is big data continues to be hotly de-bated. A widely accepted definition of big data focuses on the 3Vs: volume, velocity and variety. This definition supposedly characterizes and distinguishes big data from other forms of data.[5] The volume aspect in the 3V definition captures the continuously increasing magnitude of big data over time. Velocity points us to the high frequency at which the data is being generated and the remarkable increases in the speed at which we are able to process and analyze such data. Variety captures the immense diversity in the types of data that are available today. The idea of data is no longer limited to the structured, tabular or rectangular form as in the past. Data in the form of text, images, video, audio and other types of metadata are increasingly being considered on a par with traditional forms of data. Such non-traditional, semi-structured and unstructured data and parallel developments in computational techniques to analyze such data in terms of machine learning tools have facilitated unprecedented access to studying and understanding big data phenomena that was not possible before.

However, the 3V definition suffers from context specificity in that all of these dimensions stated are relative to a certain time and state of technology. Chen and Venkatachalam (2017b) alternatively provide a process-based definition of big data and argue that it is to be 'perceived as a continuous, unstructured and unprocessed dynamics of primitives, rather than as points (snapshots) or summaries (aggregates) of an underlying phenomenon'. In view of this definition, big data can be seen, in a given spatio-temporal domain, to be the archive of whatever people said, people did and even people thought. In other words, it is a microscopic, dynamic view of human activities. This view can also be found in Kirkpatrick (2014):

> Global Pulse is interested in trends that reveal something about human well-being, which can be revealed from data produced by people as they go about their daily lives (sometimes known as "data exhaust"). Broadly speaking, we have been exploring two types of data in the Pulse Labs. The first is data that reflects "*what people say*", which includes publicly available content from the open web, such as tweets, blog posts,

[5] Some scholars argue for including more Vs in broadening this characterization such as veracity, vincularity, and value, etc. See Kitchin and McArdle (2016) for a survey.

news stories, and so forth. The second is data that reflects *"what people do"*, which can include information routinely generated for business intelligence and to optimize sales in a private sector company. An example of "what people do" data is anonymised mobile phone traffic information, which can reveal everything from footfall in a shopping district during rush hour to how a population migrates after a natural catastrophe.

This view of big data helps us to see what is ontologically different about the nature of data available in this information era. On the one hand, data emanating from wearable devices, developments relating to Web 2.0, ubiquitous computing and social media provide unprecedented access to dynamic, organized and unorganized information on human beings at an individual level. On the other hand, advances in machine learning, ICT and artificial intelligence gives us ways to process the data gathered above. For instance, text mining and sentiment analysis are used in an attempt to discover patterns in non-conventional, user-generated forms of data and to aid in extracting useful information from them. According to Aggarwal and Zhai (2012), "[T]ext mining can be regarded as going beyond information access to further help users analyze and digest information and facilitate decision making." This is accomplished by a combination of tools and insights from natural language processing and computational linguistics with the aid of computational intelligence. In the case of sentiment analysis, user-generated data has been employed to understand shifts in the sentiments of the agents which could impact economic outcomes and indicators in the aggregate.

Together, these two, i.e., big data and computational intelligence tools, can be seen as accomplishing the function of *information aggregation* in a digital society. Take the example of Google Flu Trends. This pooled information uses a search engine that helps to collect unstructured posts, messages, searches, updates and tweets that are user-generated. This data is in turn used to predict influenza patterns ahead of the Centers for Disease Control and Prevention (CDC) which could potentially help in tackling of issues related to public health. Similarly, the Street Bump app (Table 3.1) relies on motion detectors that are available in our phones to trace the potholes in a city without having to rely on workers to patrol all of the streets. In other words, these applications can be seen as gathering important information possessed by the "man on the spot" (Hayek, 1945) which is then made available to a health official or a town planner.

In light of this, we can see a potential connection between the issues that Hayek raised for the role of markets in facilitating an effective use of knowledge in a society and the developments related to ICT in the big data era outlined above. They both perform the role of an information aggregation mechanism that is, in particular, related to the case in which information is distributed among many individuals in a society. Both of them feed information back to the society for the individuals who rely on it to make their decisions. This begs the question of the relation and relevance of Hayek's insight regarding the impossibility of socialist calculation to the digital society. This question will be discussed in the following section.

3.4 Calculation Debate in a Digital Society?

In the previous section, we argued that markets and aspects of digital society facilitated by ICT can both be viewed as platforms that aggregate information. Consequently, the digital society may be seen as introducing several non-market information aggregation mechanisms. It may be useful to sketch the process of information aggregation between traditional markets and ICT-enabled platforms. Although both deal with largely unstructured and unorganized, even contradictory, data, there are some essential differences. First, knowledge (or information) is not physically pooled in the case of traditional markets, even while it is placed in a concentrated form on the ICT platforms in one place that can be accessed. Second, in the case of markets, the available dispersed information is acted upon or processed by various 'men on the spot'. On the other hand, the pooled information can be processed by both human beings and intelligent algorithms in the case of ICT-enabled platforms. Third, the complex, decentralized act of processing dispersed information generates prices, which can be seen as an emergent phenomenon in the case of markets. ICT-enabled platforms, on the other hand, generate a variety of indexes and symbols (for instance, restaurant and movie ratings), but they are not necessarily price equivalents.

We therefore ask whether these ICT-enabled platforms together constitute a credible alternative to the conventional market mechanism? If this is answered in the affirmative, we further ask whether it can perform better than the market in the area of knowledge utilization and in aiding dynamic coordination among agents? Broadening the scope of the earlier calculation debate, we ask whether information aggregation can be effective enough to achieve the best possible resource allocation, bridging all gaps in supply and demand misalignments, and even avoiding accidents and traffic jams altogether? The relation between ICT and Hayek's insights on the socialist calculation debate can be investigated in the following ways:

- First, from the digital-society side, if ICT can enable the non-market platforms to aggregate information as markets do, then so would the unique ability of the market mechanism be challenged.
- Second, from the other side, if the socialist calculation debate is fundamentally ICT-irrelevant, i.e., the limitations pointed out by Hayek in this debate are independent of possible advancements in information and communication technology, then this debate can shed light on the

future prospects of the digital society. Consequently, it can guide us regarding the natural limitations concerning the role of ICT and the expectations that we can reasonably hold.

3.4.1 Data Discipline

In order to explore these connections, we need to consider the quality of information aggregation in these modes. Sunstein (2008) made an important observation concerning the quality of information aggregation in the context of blogs that is relevant here. In response to the claim by Posner (2004) that blogs have the 'potential to reveal dispersed bits of information', Sunstein argued that there are natural limitations to this mode of information aggregation. It is important to realize that the opinions shared in social media, blogs, and on Twitter are subjective and their contextual or tacit dimension of this knowledge cannot be truly overcome. Furthermore, Sunstein argued, unlike the case of Hayek, there are no price equivalents that perform the role of a coordination device in the case of blogs or other instances of text mining. The lack of effectiveness stems from the fact that price as a coordinating device aligns incentives and guides actions. On the other hand, ICT-enabled information aggregation does not necessarily guarantee that the incentives and actions of the players are always appropriately aligned in a dynamic context. This renders a natural weakness in the quality of knowledge aggregation and its efficacy is severely limited in terms of its role in disciplining adaptive agents.

> Participants in the blogosphere usually lack an economic incentive. They are not involved in any kind of trade, and most of the time they have little to gain or to lose. If they spread false-hoods, or simply offer their opinion, it may well be that they sacrifice little or nothing. Perhaps their reputation will suffer, but perhaps not; perhaps the most dramatic falsehoods or at least distortions will draw attention and hence readers. Most bloggers do not have the economic stake of those who trade on markets (Sunstein, 2008).

This phenomenon is not limited to blogs alone. It extends to other non-market information aggregation cases such as online reviews, where a lack of robust incentives and possibilities for adverse reinforcement and persistence of sub-optimal options exist (Vriend, 2002; Chen and Venkatachalam, 2017a,b). One way to evaluate the efficiency of information aggregation arising from markets and non-market platforms would be to compare and contrast the success of the predictions, where possible. Let us turn to prediction markets, which provide a relevant example.[6]

[6] For more on prediction markets, see Arrow et al. (2008); Wolfers and Zitzewitz (2004; 2006).

Prediction markets for electoral outcomes is a case in which information is often private. Agents in such prediction markets trade contracts for which the associated pay-offs typically depend on the future outcomes that are currently unknown (say, election results). The price of trading options is, in a perfect scenario, supposed to act as an aggregate indicator of all of the private information, knowledge or preferences dispersed among different individuals (Berg et al., 2008; Rothschild, 2015). Note that this is a market for information aggregation with a built-in coordination device.

In addition to prediction markets, we have traditional methods of information aggregation such as opinion polls. The success of their respective predictions of different outcomes give us one way to evaluate the efficiency of information aggregation of these different modes. Prediction markets are seen to increase the accuracy of forecasts concerning diverse social phenomena vis-a-vis traditional forecasting methods. However, these prediction markets are not infallible and there have been several instances in the past where they have proved to be inaccurate. There are perhaps natural limits to both these cases; therefore, it could be argued that their *relative* efficiency in information aggregation should be judged and not their absolute efficiency in relation to actual outcomes. Since our focus here is on ICT-enabled non-market platforms, in particular, the possibility of overcoming the shortcomings of traditional opinion polls (platforms) becomes important. In this case, the expanding, near-universal penetration of social media can provide a way to aggregate opinion in real-time. However, the lack of a credible coordination device pointed out by Sunstein (2008) equally plagues ICT-enabled platforms as well. Huberty (2015) casted doubt on the ability of ICT-enabled platforms such as social media to meaningfully aggregate information to forecast accurately given their nature.

In addition to providing incentives, the prices and other statistics emanating from the markets are subject to several implicit disciplining mechanisms. Some instances of such discipline exerted by markets on the data generated take the form of budget constraints and stock-flow consistencies, which cannot be violated. For instance, let us consider an economy underpinned by agents who have a specific distribution of beliefs and wealth, possess certain learning behaviors and are embedded in a social network of a specific kind, all of which are assumed to be approximated by actual observable data. Let these agents interact in an actual financial market and we can then observe them and generate several statistics concerning prices, their volatility, spread, volume of transactions and so on. A theory of information aggregation starting with the same underlying assumptions about agents and attempting to relate to actual aggregate data and statistics must necessarily be constrained by these disciplining mechanisms of the market on the generated data. Therefore, it is possible to identify the checks and balances while building such a theory. Although ICT-enabled platforms generate various sentiment indices, statistics and non-price symbols, whether they impose discipline on the data that they generate is unclear.[7] Even if such mechanisms were to exist in principle, they do not seem to be readily discernible.

3.5 Information Aggregation and Non-market Alternatives

We have seen that the power of markets in knowledge aggregation is intimately tied to the presence of an in-built coordination device, namely a price signal, which aligns incentives. It also successfully matches participants with different needs, who possess different knowledge or expertise in ways that may be mutually beneficial. A question that then arises is whether it is possible to 'calculate' or compute without prices, instead of through other symbols. In the big-data era, we do see that some symbols are successfully used as the 'equivalent' of price. Similarly, they also help in matching different individuals across space who are unified in their need to connect, with a potential for exchange. We will examine these possibilities in this section.

Let us take the example of the environmental teapot (Marres, 2012). This presents a case of augmented technology that connects an everyday object like a teapot to the electricity grid. The pooled information takes into account the real-time surge in electricity consumption and gives signals as to when it is a good time to brew one's tea. Here, the green and red lights act as non-price signals that help to coordinate the actions of the individuals who are sensitive to the overall, collective demand that prevails at any given point in time. A similar example can be found in the case of the BinCam (Comber and Thieme, 2013). This is another persuasive technology which enables the possibility to record and share images of waste that is thrown away onto a network consisting of other individuals. This signal, which can be seen as a disciplining device, leads to self-reflection and a re-evaluation of the actions by those who are concerned about the environment.

The idea behind these two experiments is to combine the use of social norms and persuasive technology. Instead of solely relying on prices to reflect the composition of preferences or the supply and demand of various resources, the norms and technology allow preferences to be modified or changed to be consistent with prices through the use of persuasive technology. There are surely some ethical concerns about the use of these or variants of these technologies in manipulating behavior. However, they highlight the potential for the use of digitally-enabled persuasive technology

[7] See Chen and Venkatachalam (2017a) for a discussion on the limitations of sentiment analysis and text mining tools in the context of knowledge aggregation.

to help facilitate the economic calculations traditionally left to prices and the market. In doing so, it is not simply an economic calculation, given preferences, as in the case of the economic calculation as seen in the socialist calculation debate that Hayek was criticizing. Instead, it resorts to the active formation or evolution of preferences mediated by social norms and social influences that is even paternalistic in a sense. Our intention here is not to justify such paternalism or to reject it outright, but instead, to highlight the role of non-price equivalents in shaping behavior that influences resource utilization.

3.5.1 Matching

If one takes a slightly broader view of markets, one quickly realizes that some of the markets that we encounter today do not in fact observe prices as co-ordination devices. This brings us to the idea of matching. At least since Gale and Shapley (1962), an impressive literature has developed on matching and market design. There are applications of this idea to marriage, dating platforms, school admission, college admissions, organ exchange and donation (Roth and Sotomayor, 1992; Abdulkadiroglu and Sönmez, 2013). In these markets, there are no explicit prices that are available to coordinate and solve the 'economic calculation' and they may not even be necessary. Instead, these platforms work by relying on other signals to achieve successful pairing between those who seek and those who are willing to supply an item in question. Even though many ICT-enabled information aggregation platforms cannot generate prices, they may still be very helpful in enhancing and widening these 'markets without a price' by providing a platform to achieve productive matching.

In addition, there are also examples of matching which bring together people with common goals. This knowledge concerning individuals with such common goals or interests is highly contextual and dispersed, and is typically not mediated through price signals. These arrangements often help achieve common goals that are not possible with individual effort alone. Examples of such arrangements include peer production, crowdsourcing and crowd funding. In peer production, individuals voluntarily get together, collaborate and produce goods, services and knowledge. Crowdsourcing involves pro-social behavior and is also a collaborative model of production that brings together services and knowledge from dispersed people across space. Crowd funding mobilizes resources and investors in novel ways to search, communicate and reach out to people across space. These modes of interaction, facilitated by ICT and the World Wide Web, are increasingly popular.[8]

[8] See Chen et al. (2017) for a detailed discussion.

Overall, the digital society allows a wide variety of matching processes to proceed at a finer level through web data crawling and mining. This fine granularity enhances the matches among crowds that may extend the boundary of traditional markets and thereby increase our reliance on non-market activities. In these activities, the fine granularity that implies a lower effort or threshold for engagements that makes it easier for any agent with a minimal or reasonable level of intrinsic motivation to act or engage. These platforms do not simply pool information and compute, but they do something more. They bring together an intrinsic motivation for contributing to public goods and efficient channels to respond to extrinsic motivating factors.

3.6 Conclusion

In the years since Hayek's famous work of 1945 was published, society has changed in many remarkable ways, especially in the way that agents connect, communicate and share information. In particular, developments in digital technology and computing have altered the mediums and ways in which knowledge sharing occurs among people. Given these developments, this chapter has examined the relevance of Hayek's classic contribution to the digital society. Insights concerning the impossibility of the socialist economic calculations and the superiority of market-based solutions have been used as a framework to understand the possibilities and limitations of ICT-enabled technologies and computational tools in knowledge aggregation.

In addition to the technological aspects, the discipline of economics has also changed in many ways since Hayek's paper was published. Ideas such as pro-social behavior, fairness, reciprocity and tools like social network analysis and complex systems have garnered a lot of attention in economic research. Furthermore, advances in behavioral economics have focused on the importance of context in decision-making. Despite these developments, Hayek's creative reformulation of the economic problem facing our society still has immense relevance.

We have systematically contrasted the differences between markets, ICT-enabled information platforms, prices and non-price coordination devices. Based on this contrast, we have argued that the developments in computational intelligence digital technology and ICT have the potential to help or enhance knowledge or information aggregation in novel ways. They have facilitated the formation of a variety of information platforms as alternative decentralized mechanisms (markets without prices). However, ICT alone will not be able to surpass the limitations pointed out by Hayek. The problem of contextual, subjective knowledge and tacit information still cannot be completely solved. Furthermore, increased effectiveness in knowledge utilization depends on the combination of technology and the ability to tap ICT-enabled intrinsic motivation through effective matching. These technologies and recent developments in economics together present new and interesting windows to understand and appreciate Hayek's important insights.

Acknowledgements

An earlier version of the paper was presented as a plenary speech at the Conference on Complex Systems, Sep. 28–Oct. 2, 2015, and as an invited talk at AAAI Spring Symposium, March 27–29, 2017. The authors are grateful for the suggestions and comments given by the conference participants; in particular, we thank William Griffin and William Lawless for their generous invitations and kind arrangement. This paper was then further expanded with the financial support granted by the Taiwan Ministry of Science and Technology (MOST) in the form of Grant No. MOST 106-2410-H-004 -006-MY2, which the authors greatly acknowledge. We are also indebted to William Lawless for his painstaking review of this chapter; however, all remaining errors are solely authors' responsibilities.

References

Abdulkadiroglu, A. and Sönmez, T. 2013. Matching markets: Theory and practice. In Advances in Economics and Econometrics 1: 3–47.

Ackerman, E. 2013. Google gets in your face. IEEE Spectrum 50(1): 26–29.

Aggarwal, C. and Zhai, C. 2012. An introduction to text mining. *In*: Aggarwal, C. and Zhai, C. (eds.). Mining Text Data, Chapter 1. New York: Springer.

Arrow, K. et al. 2008. The promise of prediction markets, Science 320(5878): 877–878.

Arrow, K. et al. 2011. 100 years of the American Economic Review: The top 20 articles. American Economic Review 101(1): 1–8.

Aud, M. et al. 2010. Developing a sensor system to detect falls and summon assistance. Journal of Gerontological Nursing 36(7): 8–12.

Berg, J., Forsythe, R., Nelson, F. and Rietz, T. 2008. Results from a dozen years of election futures markets research. Handbook of Experimental Economics Results 1: 742–751.

Boettke, P. 2000. Socialism and the Market: The Socialist Calculation Debate Revisited. edited. 9 volumes. New York: Routledge.

Chen, S.H., Chie, B.T. and Tai, C.C. 2017. Smart societies. *In*: Frantz, R., Chen, S.-H., Dopfer, K., Heukelom, F. and Mousavi, S. (eds.). Routledge Handbook of Behavioral Economics. New York, NY: Routledge 250–265.

Chen, S.H. and Venkatachalam, R. 2017a. Information aggregation and computational intelligence. Evolutionary and Institutional Economics Review 14: 231–252.

Chen, S.H. and Venkatachalam, R. 2017b. Agent-based modelling as a foundation for big data. Journal of Economic Methodology 24(4): 362–383.

Cockshott, W. and Cottrell, A. 1997. Information and economics: A critique of Hayek. Research in Political Economy 18(1): 177–202.

Comber, R. and Thieme, A. 2013. Designing beyond habit: Opening space for improved recycling and food waste behaviors through processes of persuasion, social inuence and aversive affect. Personal and Ubiquitous Computing 17(6): 1197–1210.

Diebold, F.X. 2012. On the Origin (s) and Development of the Term 'Big Data'. Penn Institute of Economic Research Working Paper 12–037.
Einav, L. and Levin, J. 2014. Economics in the age of big data. Science. 346(6210): 1243089.
Fogg, B. 2002. Persuasive Technology: Using Computers to Change What We Think and Do. Morgan Kaufmann.
Ginsberg, J. et al. 2009. Detecting inuenza epidemics using search engine query data. Nature 457(7232): 1012–1014.
Hamari, J., Koivisto, J. and Pakkanen, T. 2014. Do persuasive technologies persuade?—A review of empirical studies. pp. 118–136. *In*: Spagnolli, A. et al. (eds.). Persuasive Technology. New York: Springer.
Hayek, F.A. 1945. The use of knowledge in society. The American Economic Review 35(4): 519–530.
Howe, D. et al. 2008. Big data: The future of biocuration. Nature 455(7209): 47–50.
Huberty, M. 2015. Can we vote with our tweet? On the perennial difficulty of election forecasting with social media. International Journal of Forecasting 31(3): 992–1007.
Kirkpatrick, R. 2014. A conversation with Robert Kirkpatrick, Director of United Nations Global Pulse. SAIS Review of International Affairs 34(1): 3–8.
Kitchin, R. and McArdle, G. 2016. What makes Big Data, Big Data? Exploring the ontological characteristics of 26 datasets. Big Data & Society 3(1): 1–10.
Lange, O. 1936. On the economic theory of socialism: Part one. Review of Economic Studies 4(1): 53–71.
Lange, O. 1937. On the economic theory of socialism: Part two. Review of Economic Studies 4(2): 123–142.
Lavoie, D. 1985. Rivalry and Central Planning: The Socialist Calculation Debate Reconsidered. Cambridge: Cambridge University Press.
Levy, D.M. and Peart, S.J. 2008. Socialist calculation debate. *In*: Steven N. Durlauf and Lawrence E. Blume (eds.). The New Palgrave Dictionary of Economics, second edition. London: Palgrave Macmillan.
Manzini, P., Sadrieh, A. and Vriend. N.J. 2009. On smiles, winks and hand-shakes as coordination devices. Economic Journal 119(537): 826–854.
Marres, N. 2012. The environmental teapot and other loaded household objects: Reconnecting the politics of technology, issues and things. *In*: Harvey, P., Casella, E., Evans, G., Knox, H., McLean, C., Silva, E., Thoburn, N. and Woodward, K. (eds.). Objects and Materials: A Routledge Companion. London: Routledge.
Pantano, E. and Nacarato, G. 2010. Entertainment in retailing: The influences of advanced technologies. Journal of Retailing and Consumer Services 17(3): 200–204.
Picone, M., Amoretti, M. and Zanichelli, F. 2012. A decentralized smartphone-based traffic information system. In IEEE Intelligent Vehicles Symposium (IV), pp. 523–528.
Polanyi, M. 1966. The Tacit Dimension. Chicago: University of Chicago Press.
Posner, R. 2004. Introduction to the Becker-Posner blog. http://www.becker-posner-blog.com/archives/2004/12/introduction to 1.html, December.
Roth, A.E. and Sotomayor, M. 1992. Two-sided matching. In Handbook of Game Theory with Economic Applications 1: 485–541.
Rothschild, D. 2015. Combining forecasts for elections: Accurate, relevant, and timely. International Journal of Forecasting 31(3): 952–964.
Schwartz, A. 2012. Street Bump: An app that automatically tells the city when you drive over potholes. Fast Company.
Shieh, W.-Y., Guu, T.-T. and Liu, A.-P. 2013. A portable smart belt design for home-based gait parameter collection. In: Proceedings of IEEE Conference on Computational Problem-Solving (ICCP), pp. 16–19. IEEE Press.
Smith, V. 1982. Markets as economizers of information: Experimental examination of the Hayek Hypothesis. Economic Inquiry 20(2): 165–179.
Sunstein, C.R. 2008. Neither Hayek nor Habermas. Public Choice 134(12): 87–95.

Varian, H.R. 2014. Big data: New tricks for econometrics. Journal of Economic Perspectives 28(2): 3–27.

von Mises, L.E. 1920. Economic calculation in the socialist commonwealth. Translated. S. Adler. *In*: Hayek, F.A. (ed.). Collectivist Economic Planning. London: Routledge, 1935.

Vriend, N. 2002. Was Hayek an ACE? Southern Economic Journal 68(4): 811–840.

Westerlund, M., Leminen, S. and Rajahonka, M. 2014. Designing business models for the internet of things. Technology Innovation Management Review 4(7): 5–14.

Wolfers, J. and Zitzewitz, E. 2004. Prediction markets. Journal of Economic Perspectives 18(2): 107–126.

Wolfers, J. and Zitzewitz, E. 2006. Interpreting prediction market prices as probabilities. National Bureau of Economic Research, No. w12200.

4

Challenges with Addressing the Issue of Context within AI and Human-Robot Teaming

Kristin E. Schaefer,[1,*] *Derya Aksaray,*[2] *Julia Wright*[3] and *Nicholas Roy*[4]

4.1 Introduction

As the future of human-robot interaction moves towards interdependent teaming initiatives, developing efficient complex decision-making processes is an essential part of the design and development of autonomous robots. The underlying principles for such decisions have moved beyond that of simple decision trees used by preprogrammed systems. New models for inference and planning algorithms require access to real-world data to account for dynamic and uncertain tasks and environments. But it is not enough to only focus on the development of a world model or the underlying algorithms that allow the robot to make highly-complex decisions under

[1] Engineer, Human Research & Engineering Directorate, US Army Research Laboratory, Aberdeen Proving Ground, MD.
[2] Assistant Professor, Department of Aerospace Engineering and Mechanics, University of Minnesota Twin Cities, Minneapolis, MN.
[3] Research Psychologist, Human Research & Engineering Directorate, US Army Research Laboratory, Orlando, FL.
[4] Professor, Department of Aeronautics and Astronautics and Computer Science and Artificial Intelligence Laboratory (CSAIL), Massachusetts Institute of Technology (MIT), Cambridge, MA.
* Corresponding author: kristin.e.schaefer-lay.civ@mail.mil

various circumstances. Within the structure of interdependent teams, it is also important to recognize that a human team member often needs to interpret the actions and behaviors of its robot partner. Apart from the actual system design capabilities, the human team member's situation awareness is often what drives expectations for the interaction. If the expectations do not match the robot's actions, there can be a degradation of trust that can directly impact the effectiveness and performance of the team (Schaefer et al., 2017). Therefore, teaming will benefit from both advanced inference and planning algorithms that support decision processes in dynamic environments, and by incorporating agent-based transparency to develop situation awareness during joint operations. Context and the interpretation of context play a key role in these technological developments required for advanced human-robot teaming.

4.1.1 Importance of Context to Human-Robot Teaming

The eventual goal in human-robot teaming is for the robot to become a team member, rather than just a tool (Phillips et al., 2011). In order to do so, the robot must be able to function in cooperation with their human counterparts in real-world uncertain environments. Cooperative functionality requires both the ability to perform independent tasking, as well as the development of common ground with team members (Cooke et al., 2013) while being able to critique errors and suggest alternative actions (Sukthankar et al., 2012). As such, the robot's ability to participate in joint activities with the team is dependent upon its interpredictability, its ability to maintain common ground with its teammates, and its redirectability (Klein et al., 2005). In addition, it is important for the robot to be able to share knowledge with its team members. Teaming research has shown that team members who share knowledge tend to make similar decisions and actions, and have been shown to be more effective (Cannon-Bowers and Salas, 2001; Cooke et al., 2000; Cooke et al., 2013; Sukthankar et al., 2012). Thus, the robot must have a clear understanding of the team and task needs including goals, membership, capabilities, and tasking.

It is reasonable to assume that for the near-future, humans and robots faced with the same circumstances will not make the same decisions, even under the same set of apparent constraints, nor will they necessarily have the same consequences resulting from those decisions. For example, when comparing human and algorithmic path planning, research has shown that there is more than one 'human way' of solving a planning problem which may or may not match an algorithm (Perelman et al., 2017). Since each team member has a unique perspective, no one team member has a complete understanding of the context, and this disparity in understanding increases as the size of the team increases. Each team member contributes their understanding of context—spatially, historically, and functionally—to the intra-team dialogue wherefrom team cognition evolves. By integrating all team members' perspectives, a global team understanding of context evolves. When team members cannot contribute to this intercourse, team cognition and performance suffer, and team effectiveness is reduced (Cooke et al., 2013).

Team cognition supports shared situation awareness, which in turn enables the team's ability to effectively coordinate actions toward the accomplishment of overall and individual goals (Bolstad and Endsley, 2003; Endsley, 1995; Endsley and Jones, 1997). The development of shared situation awareness is crucial to teaming in that it can assist in the development of team cognition, as well as the calibration of trust between the team members (Schaefer et al., 2017). As such, incorporating context could result in faster and more accurate inference, which could lead to team performance gains. However, the mechanism by which context is integrated into the development of team cognition is an open research question.

4.1.2 Developing Team Cognition

One of the driving reasons behind human-agent teaming is to develop the ability to expand the human's scope of influence, particularly in dangerous settings. As such, robots are being developed specifically for tasks where they may not be co-located with their human team members. Teleoperated and low-level autonomous agents do not have the capability to recognize, much less disambiguate changes in context, and their teammates' understanding of its action may be limited by the agent's capability for transparent communication (Chen et al., 2018). Consequently, the robot's actions may seem mysterious to its remote teammates (Linegang et al., 2006). Humans interacting with highly autonomous systems encounter multiple challenges including understanding the current system state, comprehending reasons for its current behavior, and projecting what its next behavior will be (Sarter and Woods, 1995). This lack of transparency is exacerbated by a lack of observable or communicated contextual information to the human team member, limiting the capability for the human team member to disambiguate problems (Stubbs et al., 2007). It also leads to increased difficulties in understanding the robot's underlying rationale for its decision-making (Linegang et al., 2006). Additionally, if the agent encounters difficulties in executing its tasks, its team will be unable to offer effective assistance. Remote team members must be capable of conveying context to their teammates to support team cognition. In order to support team cognition, robots should not simply push information to their teammates without then assessing the team's understanding of that information.

Incorporating contextual cues and contextual understanding into the development of an autonomous robot can advance the robot's decision-making capabilities, but may also help improve the human's understanding of those decisions to improve team effectiveness. Context enables effective high-level information exchange between humans and robots that can directly enhance the human team member's capability to better comprehend the robot's intent, performance, future plans, and reasoning process (Chen et al., 2014). Therefore, context-driven artificial intelligence (AI) could better support the three major facets of human-robot interaction: mutual predictability of team members, shared understanding, and the ability to redirect and adapt to one another (Lyons and Havig, 2014; Sycara and Sukthankar, 2006).

4.1.3 Context-driven AI

Developing the ability for AI to incorporate contextual cues and understanding, complete with reasoning how such cues can be used to predict changes in the task and environment, is a key challenge to advancing teaming potential. In order for any AI to operate effectively in dynamic and uncertain environments or scenarios, generalization is crucial, i.e., the AI must be able to understand different states, situations, and information in the surrounding world in order to make appropriate decisions and behave accordingly. Context-driven AI could help to achieve this goal. For example, incorporating task-based context is likely to result in faster and more accurate inferences, leading to performance gains. Further, context-driven AI can advance an agent's understanding of the situation in the environment in order to better communicate with human team members. Thus, integrating aspects of context into inference and planning may help improve not only the efficiency of the underlying decision-making processes, but also assist in communicating intent-reasoning so that actions better match human expectations (i.e., increased transparency and shared reasoning; Chen et al., 2018). As with human team members, it is crucial that a robot has the ability to infer, represent and reason about the world in order to develop shared situation awareness to effectively communicate intent. Even for a single robot carrying out a higher-level task, spatial and temporal contextual cues have been identified as significantly improving the performance of many components of a robotic system, including perception, inference and planning (e.g., Heitz and Koller, 2008; Posner et al., 2009). By incorporating these contextual elements, the robot's reasoning process is more easily understood by and communicated to the human team members.

4.1.4 Chapter Outline

To date, intelligent agents have the ability to recognize contextual knowledge only through some pre-programmed rules or user-defined models. However, they have limited capability to accurately recognize context and detect changes in context for arbitrary settings. As a result, human subject studies exploring the effects of teaming with context-aware agents on human performance and attributions of the agent are quite limited. This chapter provides a review of prior research regarding the integration of context into robot autonomy in order to identify challenges of developing context-driven AI and the potential benefits for future human-robot teaming. To address this topic, Section 4.2 outlines the context terminology used to identify context-driven AI in terms of spatial, temporal and task-specific properties. Section 4.3 reviews current AI models and representations in order to explain how context is used today and in developing future AI. Section 4.4 concludes the chapter with the open research challenges. Overall, context-driven AI can support the capability for generalizable operations of robotic systems, advance the transition of a robot from tool to team member through the advancement of independent and interdependent decision-making processes, and improve the speed (efficiency) and accuracy (effectiveness) of inference, which makes it essential for time-critical operations.

4.2 Context Terminology

One of the most cited definitions of context was stated as "any information that can be used to characterize the situation of an entity. An entity is a user, a place, or a physical or computational object that is considered relevant to the interaction between a user and an application, including the user and application themselves" (Dey et al., 2001). In this definition, context can fundamentally be stated as an abstraction of physical properties of space, time, the mission and history of the robot behaviors. As a result, context can have different scales or meanings depending on the activity, mission or environment, and it may also be a result of other abstract notions such as a human teammate's intentional or emotional state (Scheutz et al., 2005). Therefore, it is not sufficient to infer a single context variable, but a truly robust robot should instead be able to reason across different spatial, temporal and mission scales, as well as the capability to reason about context as a component of a larger world model.

4.2.1 Spatial Context

Spatial context is hereby defined as the relationship among the locations of different objects in a scene that adds meaning to the environment. Similarly, environmental context is the larger knowledge of the scene beyond the location of objects, to their functional relationships, the types of activities that occur in the scene, etc. As such, this type of context can provide inferences on what objects could occur in the scene or potential relationships or behaviors that need to occur in order to meet task or mission objectives. One of the advantages of incorporating spatial and environmental context into AI is enhancing the capability for the robotic system to automatically learn features (i.e., derive knowledge about the context). This is especially important in machine learning paradigms where the use of deep networks for automatic representations learning has become increasingly effective. In addition, spatial and environmental context could assist in the management of ambiguity by allowing the robotic system the capability to interpolate the contextual information and construct the current context (e.g., a room next to the kitchen would probably be an indoor environment). Further, this type of context can help filter irrelevant data allowing the algorithm to have more accurate and faster inference capabilities.

4.2.1.1 Spatial Context and Teaming

Effective human-robot teaming must take into account the spatial and environmental context in which operations take place. Features and cues from the environment can be used to provide essential information required to convey knowledge and make appropriate operational decisions. The main teaming challenge here is the different types of inferences that human and robotic team members make in the same environment. This type of context can directly (either positively or negatively) impact the development of shared situation awareness between human and robot team members.

At present, human team members often make more informed decisions in dynamic and uncertain environments than their robotic team members. People are good at sorting through large domain spaces and making inferences about that space in regards to task or mission needs (i.e., they can extract meaning from the world to make decisions). This can be in part due to the fact that people can perceive space through multiple modalities including visual, tactile, acoustic, smell and even temperature sensations to make informed decisions based on the physics of space, spatial relations, and universal domain properties that are not reliant on specific situations (Freska, 1991).

To enhance teaming efforts, a number of research efforts into robotic design have incorporated elements of spatial context. Many of these algorithms are inspired by human cognition. On a larger theoretical level, Kuipers (1983) suggested that there are five types of spatial information that can be used to inform the design of a conceptual formulation of a cognitive map. These include topologies, distances, routes, fixed features, and observations. These elements can be used to help a robot conceptualize and reason about a space (Vasudevan et al., 2007), as well as classify representations such as space and objects (Yeap and Jefferies, 2001). Previous studies have looked at specific issues for teaming, such as navigation (e.g., through door openings; Anguelov et al., 2004; Kortenkamp et al., 1992; Stachniss et al., 2005), and communication (e.g., gesture-based communication whereby the spatial orientation to the axis of the hand can be used to disambiguate the gestures; Strobel et al., 2002). However, such algorithms are not robust to disturbances such as any change in environment, robotic platform, or task. Thus some studies in the literature focus on how to infer spatial context and incorporate it into the development of AI, which is reviewed further in the following section.

4.2.1.2 Research on Spatial Context and AI Development

There are two major areas of research that support the integration of spatial or environmental context into AI development. These include semantic mapping, and context-based classification including scene recognition. These areas of research are important to the development of future AI to execute missions efficiently in arbitrary world configurations. Research in semantic mapping is the process of labeling objects in the environment, such as assigning high-level information to robotic path mapping (e.g., Kostevelis and Gasteratos, 2015; Wolf and Sukhatme, 2008). A major benefit of semantic mapping is to provide an abstraction of the environment and a means of human-robot communication. Current research uses computer vision (e.g., Heitz and Koller, 2008; Meger et al., 2008) or language (e.g., Walter et al., 2009) to accomplish this type of mapping typically using probabilistic models.

While current research efforts have demonstrated advancements to algorithms that can rely on less training data and speed up the inference process, there are still research challenges, such as the dependence on domain-specific knowledge (e.g., feature selection), perception systems (e.g., resolution of the sensed image), and the lack of robustness to environmental changes (e.g., the algorithms for indoor missions may not be used for outdoor scenarios). Contextual information can create or activate schemata for object arrangements or locations in particular scenes. These schemata contain important cues for humans to detect objects, so much so that humans can identify violations within milliseconds. However, these activated schemata may also bias the recognition of entities improbable to the scene. For example, an object that is out of place within a scene may lack the environmental cues necessary for the agent to appropriately detect the object (Biederman et al., 1982). In a similar sense, with context-based classification and scene recognition, a specific set of features can be used to improve detection and classification of regions and objects in an environment (e.g., Kumar and Hebert, 2005). The benefits of this type of classification are efficiency and robustness of image labeling and object detection under various settings. However, the major challenge is the incorporation and access of feature sets by the robotic system. For example, it is not feasible to incorporate every possible feature set needed for uncertain or novel environments. Moreover, the inability to generalize from training data to new scenes is another issue.

4.2.2 Temporal Context

For the purpose of this chapter, temporal context is defined as the information that is available across consecutive scenes. This includes things such as recognizing and tracking consecutive scenes and constructing hypotheses regarding the current situation at a given time. For example, this could include the apparent distinction between an object and its surroundings at a specific moment of time.

4.2.2.1 Research on Temporal Context and Teaming

Teams often conduct time-critical operations for when decisions are made based upon rapid assessments of the environment whereby team members use incoming information from team members, as well as prior knowledge, to assist in making these decisions. Therefore, temporal context can directly impact the perception and associated actions over time by providing additional meaning to a scene, task or mission (e.g., map displays that depict information over time; Ellis and Dix, 2007), including an ordering of events (Humphrey and Adams, 2010). Time estimates can provide additional information to a static scene thus providing contextual information for appropriate decision-making. The process for making time-critical decisions are influenced by a number of factors. For example, previous experiences can directly influence how people make decisions, as well as the time required for informed decisions that impact future decisions. Therefore, there are trade-offs that occur between making informed decisions taking longer perception times versus making time-critical decisions by perceiving enough information.

While people are relatively good at using temporal context to inform decision-making, minimal work has looked at the flow of time within human-machine teaming. Some specific types of teaming initiatives that could benefit from temporal context include turn-taking (Iizuka and Ikegami, 2004), and natural language communication (e.g., converting natural language sentences into temporal logic to reason about future commands; Dzifcak et al., 2009; Kamp and Reyle, 2013; Moldovan et al., 2005). Overall, human-like time perception and processing capabilities could bring robotic cognition closer to human cognition and thereby help address challenges with timescale (i.e., understanding past, present and future timeline).

4.2.2.2 Research on Temporal Context and AI Development

Integrating temporal context is important when developing robots because it enables the capability of interpreting the state or recognizing the current situation by observing the partial or entire state history at any time. Such inference from history is essential for a robot to develop appropriate situation awareness. While research on this topic is still rather limited, there has been work aimed at developing computational cognitive models that integrate time. Maniadakis and Trahanias (2014) provided a review of this early work related to self-organized time perception for decision-making (Maniadakis et al., 2009), grounded temporal lexicon linking space and time (Schulz et al., 2011), links between timing and motor activity (Addyman et al., 2011), representations of duration (Maniadakis and Trahanias, 2013), time perception including interval timing, attention, perception and learning (Taatgen et al., 2007), internal state dynamics and memory of past, present and future (Choe et al., 2012), recall and memory (Hasselmo, 2009), temporal properties of learning (Howard, 2014), reshaping memory, such as forgetting (Oberauer and Lewandowsky, 2011), and links between events, temporal context, and memory (Sederberg et al., 2011). While this work has shown the potential benefits of time in consciousness, memory, and human-robot interaction, it also highlights how new this field of research currently is and the challenges to implementing temporal context.

4.2.3 Tasks and Context

Robotic systems are often developed for a specific task or set of tasks. Task specifics in and of themselves can provide contextual understanding to the larger operation in that tasks include goals, sub-goals, planning, actions, and feedback. However, in future robotic systems, spatial, environmental and temporal context can be used to make inferences that can be used to update task goals and sub-goals so that robotic systems will be able to better support planning and actions within collaborative teams. Without the capability for contextual understanding, these robotic systems will be limited in scope within uncertain and dynamic environments. Further, as task complexity and independent functioning requirements increase, robotic systems will need more advanced language and communication capabilities, a larger range of goals and adaptive behavior, as well as the ability to be able to change plans and actions.

Context can also change as a result of human behavior, as well as the presence or absence of humans (Sukthankar et al., 2012). For example, an agent that can disambiguate between similar scenes with differing human activity would be much more useful in reconnaissance tasks than one that simply acts as a remote camera. In addition to external sources of context, there is within-team context to consider. Team composition could change, either intentionally (e.g., reducing staffing on the overnight shift, or increasing team size ahead of a challenging task) or by accident (e.g., a team member is injured or separated from the squad). Understanding differences in team composition and how those differences affect the team's goals, member roles and task assignments is essential to agents ultimately becoming autonomous teammates.

4.3 Context: Modeling and Reasoning in AI

Most of the existing work in incorporating context into robotic systems has assumed either a fixed or known ontology of semantic context. With a few exceptions, the research has focused on how to incorporate context into the perception, inference or planning processes. The literature indicates that there is a general sense that context is a state variable that can be used to select relevant parts of the state space, or select relevant perceptual feature functions for object recognition. This implies that context represents a discrete selector variable, possibly with a distribution attached to it. For example, the context of an outdoor environment may lead the robot to reason about a different state space and object classes than the context of an indoor environment. In this sense, the context is an ontological model (Suh et al., 2007). On the other hand, the context may be *a priori* knowledge over concepts in the environment, rather than a selector variable; in this sense, context requires a probabilistic model (Zender et al., 2008).

4.3.1 Graphical Models

One way to incorporate context into the robot decision-making process is via graphical models, where the structure of the contextual knowledge is represented by the topology of the graph. In this respect, the nodes of the graph refer to the entities such as agents (e.g., humans or other robots), places, physical objects, or physical properties, whereas the edges represent the relationship among the entities. An important aspect in constructing a graphical model is the decision of what nodes will refer to and which edges will exist among the nodes. To this end, the graph can be generated by using expert knowledge or it can be constructed automatically. For example, the nodes can be derived from the input image of a perception system (e.g., the regions in an input image are mapped to a building, road or car) and the edges exist based on the object relationships (e.g., a road frequently appears in front of a building; Myeong et al., 2012). Graphical model representation also allows for context reasoning. One prevalent way to achieve this goal is to train probabilistic graphical models with a set of data and formulate an inference problem. For example, Markov random fields, relational Markov networks and Bayesian networks can be used to infer contextual knowledge such as object relationships or object occurrences from input images (Heitz and Koller, 2008; Kollar and Roy, 2009; Limketkai et al., 2005; Posner et al., 2009; Ranganathan and Dellaert, 2003). Similarly, factor graphs can be used to understand language cues to interpret directions (Kollar et al., 2010) and instructions (Duvallet et al., 2014) based on context. In probabilistic graphical models, contextual information is provided mostly through local interactions between predictions. While such models can be used to store a large amount of data, there may be scalability issues when incrementally adding new contextual knowledge to the graphical model.

4.3.2 Logic Based Models

In a logic based context model, the context can be defined as rules and propositions. A proposition is a statement that has a binary truth value (e.g., being in a classroom) and a rule is a condition on which a fact may be derived from another fact (e.g., if existence of chairs in multiple rows is true, then being in a classroom is true). One of the first logic based context modeling in AI was proposed by McCarthy (1993) where mathematical abstractions were used to formalize common sense phenomena. Later, more formal expressions were used to model and reason context. For example, the syntax and semantics of a general propositional logic of context was proposed in Buvač and Mason (1993). First-order predicate logic was used in Gray and Salber (2001) to formulate sensed context information. Overall, logic based models can be used to generate high-level context from low-level context. They rely on formulating a context in terms of rigid rules and relations between entities, which are difficult to list in practice. They also depend on the application so they are not generalizable or transferrable to other domains.

4.3.3 Ontology Based Models

An ontology can be stated as the shared understanding of some domains, which typically includes a set of vocabulary and their meanings. The degree of formality by which the vocabulary is created and the definitions are specified typically depends on the application. For example, it can vary from natural language to formal semantics or theorems. Ontology based models can be used to model domain knowledge and structure context based on the relationships defined by the ontology. One drawback of such models is that the structure of the models and information retrieval from them may get complex as the amount of context data increases.

4.3.4 AI Models for Human-Robot Teaming

In general, there does not exist a widely agreed upon representation of what constitutes context, or the types of models that best support human-robot interaction. There is in general a tension between those that naturally support inference from data, and those that are more easily interpreted by human team members. The most progress in robot autonomy in recent years has been driven first by the graphical models described above, and more recently by deep networks. In either case, these models are very specific to the sensor data and control task, and therefore very amenable to incorporating the spatial and temporal context described by the sensor data. On the other hand, such models are frequently completely non-transparent to human partners. The logic and ontology based models are frequently much more intuitive and communicable to human team members as they can incorporate human context delivered through natural language, gesture or other human-centered modalities. But, they are frequently divorced from the sensor data and as a result they typically become incomplete descriptors of the world. Developing the appropriate models is further complicated by determining which contextual elements are important to represent, and which elements have a higher weight in the decision-making process. This tension represents an open challenge in developing autonomous systems that can work effectively within collaborative teams.

4.4 Conclusions

Overall, it is not sufficient to infer a single context variable, but a truly robust robot should instead be able to reason across different spatial, temporal and mission scales, and reason about context as a component of a bigger world model. This work will help to begin to reconcile these very different views of context leading towards a future development of a coherent representation that can be used across dynamic, uncertain teaming environments. However, there are a number of open research challenges to the development and technical advancement of context-driven AI.

4.4.1 Challenge 1: Infer Context from Sensor Data

In the current literature, two frequently used sensor data to infer context are visual images and languages. However, methods to derive context from such sensor data are not applicable to various scenarios (e.g., a context model used for indoor cannot be used for outdoor or vice versa). Even if the missions are fixed to only indoor or outdoor, the models used for context inference are not robust to disturbances, such as changes in the image resolution (the models used with an image taken from a standard camera may not be used with images taken from a satellite) or variability in language commands (generally it is challenging to understand arbitrary natural language sentences by AI).

One future goal is to extend the application areas of existing methods and models for generalization. Another goal is to use sensor data other than vision or language to infer context (e.g., temperature, humidity, smell, audio, etc.). To reach this end, the ideas from sensor fusion might be used. Nonetheless, the challenge is the identification of which combination of sensory information gives an efficient way of understanding a context. The main difficulty is that the type of sensor data in terms of mission specifications for efficient context inference is not known (e.g., for reconnaissance missions, sensing smell may say something about whether there exists dangerous chemicals in a room).

4.4.2 Challenge 2: Development of Learning Classification Schemes

Incorporating context into the classification problem can be very useful to improve classification performance because it can filter out the irrelevant data from the state-space and aid the selection of features. However, one challenge is the proper selection of the contextual knowledge for this purpose because including excessive amount of contextual knowledge into the problem can increase the complexity of the model, data or representation. Another challenge is how to obtain the useful contextual knowledge automatically (see Section 4.4.1) and how to incorporate it into the classification problem. Since context models are diverse, it is not clear how to transform them as useful data into the classification problem. For example, will they refer to binary features or parametric equations?

The major challenge is data complexity. The inability to use context to allow generalization across domains or tasks forces the systems to relearn from domain to domain and from scratch. AI systems that can use context to select relevant components of previous concepts will automatically learn more efficiently and more transparently.

4.4.3 Challenge 3: Addressing New and Unknown Context Variables

Given a finite set of possible contexts (e.g., spatial areas, indoor versus outdoor, etc.), it is possible to learn the current context. However, pre-specified context variables are unlikely to be sufficient for long-duration robots operating in populated environments. People naturally develop their understanding of context over time, and a robot must be able to do the same. Even for a single spatial context variable that behaves as a selector for state or perception features, new contexts may be encountered, reasoned about, and added to the representation. As new missions are encountered or as human teammates demonstrate new activities, the representation of context must adapt or expand in concert.

Extracting new contextual knowledge online and incorporating it into context models are two crucial issues to improve the performance of human-robot collaboration. Especially, when new information is added, creating the contextual model from scratch is not practical as a mission is progressing. The goal is to recycle the current context model and update it in a scalable way by the incremental addition of new data. However, it is challenging to obtain a common solution to this problem since the context models are diverse and application dependent.

4.4.4 Challenge 4: Classification of Appropriate Representations for Contextual Elements

The difference in semantics for context has implications for how it is both represented and used by the relevant perception, inference and planning algorithms. For example, the notion of context in a human-robot dialogue is more about the state of the world and its history, rather than a selection of parts of the model. In this sense, context is not a model, but an instantiation or possibly a history (or summary or statistic) of the world. Reconciling these very different views of context and developing a coherent representation that can be used in different ways is an open research question.

4.4.5 Challenge 5: Teaming

The final challenge is in developing contextual models that allow human-robot teams to operate more effectively. Humans invariably incorporate any number of components, whether they are logic concepts, or prior sensor data, or time histories of state estimates, into their inference and reasoning processes. For effective, long-term collaboration, robots need to be able to incorporate similar components into their own inference reasoning processes. More crucially, any contextual variables that drive the inference process must be transparent to human team members, to explain future decisions based on context derived from sensor data experienced long ago.

Acknowledgments

Research was sponsored by the Army Research Laboratory and was accomplished under Cooperative Agreement Numbers W911NF-16-2-0216 and W911NF-10-2-0016. The views and conclusions contained in this document are those of the authors and should not be interpreted as representing the official policies, either expressed or implied, of the Army Research Laboratory or the U.S. Government. The U.S. Government is authorized to reproduce and distribute reprints for Government purposes notwithstanding any copyright notation herein.

References

Addyman, C., French, R., Mareschal, D. and Thomas, E. 2011. Learning to perceive time: A connectionist, memory-decay model of the development of interval timing in infants. In Proc of the 33rd Annual Conference of the Cognitive Science Society (COGSCI) 33: 354–359.

Anguelov, D., Koller, D., Parker, E. and Thrun, S. 2004. Detecting and modeling doors with mobile robots. In Proc of the International Conference on Robotics and Automation (ICRA). Doi: 10.1109/ROBOT.2004.1308857.

Biederman, I., Mezzanotte, R.J. and Rabinowitz, J.C. 1982. Scene perception: Detecting and judging objects undergoing relational violations. Cognitive Psychology 14(2): 143–177.

Bolstad, C. and Endsley, M. 2003. Measuring shared and team situation awareness in the Army's future objective force. In Proc of the 4th Annual Meeting of the Human Factors and Ergonomics Society (HFES) 369–373.

Buvač, S. and Mason, I.A. 1993. Positional logic of context. In Proc of the AAAI Conference 412–419.

Cannon-Bowers, J.A. and Salas, E. 2001. Reflections on shared cognition. Journal of Organizational Behavior 22(2): 195–202.

Chen, J.Y.C., Procci, K., Boyce, M., Wright, J., Andre, G. and Barnes, M. 2014. Situation Awareness-Based Agent Transparency (ARL-TR-6905). US Army Research Laboratory: Aberdeen Proving Ground, MD.

Chen, J.Y.C., Lakhmani, S.G., Stowers, K., Selkowitz, A.R., Wright, J.L. and Barnes, M.J. 2018. Situation awareness-based agent transparency and human-autonomy teaming effectiveness. Theoretical Issues in Ergonomics Science, in press.

Choe, Y., Kwon, J. and Chung, J.R. 2012. Time, consciousness, and mind uploading. International Journal of Machine Consciousness 4(1): 257–274.

Cooke, N.J., Salas, E., Cannon-Bowers, J.A. and Stout, R.J. 2000. Measuring team knowledge. Human Factors 42(1): 151–173.

Cooke, N.J., Gorman, J.C., Myers, C.W. and Duran, J.L. 2013. Interactive team cognition. Cognitive Science 37(2): 255–285.

Dey, A.K., Abowd, G.D. and Salber, D. 2001. A conceptual framework and a toolkit for supporting the rapid prototyping of context-aware applications. Human-Computer Interaction 16: 97–166.

Duvallet, F., Walter, M.R., Howard, T., Hemachandra, S., Oh, J., Teller, S., Roy, N. and Stentz, A. 2014. Inferring maps and behaviors from natural language instructions. pp. 373–388. *In*: Hsieh, M., Khatib, O. and Kumar, V. (eds.). Experimental Robotics. Springer.

Dzifcak, J., Scheutz, M., Baral, C. and Schermerhorn, P. 2009. What to do and how to do it: Translating natural language directives into temporal and dynamic logic representation for goal management and action execution. In Proc IEEE International Conference on Robotics and Automation (ICRA), 4163–4168.

Ellis, G. and Dix, A. 2007. A taxonomy of clutter reduction for information visualisation. IEEE Transactions on Visualization and Computer Graphics 13(6): 1216–1223.

Endsley, M.R. 1995. Toward a theory of situation awareness in dynamic systems. Human Factor 37(1): 32–64.

Endsley, M. and Jones, W.M. 1997. Situation awareness information dominance & information warfare (AL/CF-TR-1997-0156). US Air Force Armstrong Laboratory: Dayton OH.

Freska, C. 1991. Qualitative Spatial Reasoning. pp. 361–372. *In*: Mark, D.M. and Frank,A.U. (eds.). Cognitive and Linguistic Aspects of Geographic Space. Kluwer Academic Publishers: Dordrecht.

Gray, P.D. and Salber, D. 2001. Modelling and using sensed context information in the design of interactive applications. pp. 317–335. *In*: Little, M.R. and Nigay, L. (eds.). Engineering for Human-Computer Interaction. Springer.

Hasselmo, M.E. 2009. A model of episodic memory: Mental time travel along encoded trajectories using grid cells. Neurobiology of Learning and Memory 92(4): 559–573.

Heitz, G. and Koller, D. 2008. Learning spatial context: Using stuff to find things. pp. 30–43. *In*: Forsyth, D., Torr, P. and Zisserman, A. (eds.). Lecture Notes in Computer Science. Springer, Berlin, Heidelberg.

Howard, M.W. 2014. Mathematical learning theory through time. Journal of Mathematical Psychology 59: 18–29.

Humphrey, C.M. and Adams, J.A. 2010. General visualization abstraction algorithm for directable interfaces: Component performance and learning effects. IEEE Transactions on Systems, Man and Cybernetics Part A: Systems and Humans 40(6): 1156–1167.

Iizuka, H. and Ikegami, T. 2004. Simulating autonomous coupling in discrimination of light frequencies. Connection Science 16(4): 283–299.

Kamp, H. and Reyle, U. 2013. From discourse to logic: Introduction to modeltheoretics semantics of natural language, formal logic and discourse representation theory. Springer.
Klein, G., Feltovich, P.J. and Woods, D.D. 2005. Common ground and coordination in joint activity. pp. 139–178. *In*: Rouse, W.R. and Boff, K.B. (eds.). Organizational Simulation. Wiley & Sons, NJ.
Kollar, T. and Roy, N. 2009. Utilizing object-object and object-scene context when planning to find things. In Proc of the IEEE International Conference on Robotics and Automation (ICRA). Doi: 10.1109/ROBOT.2009.5152831.
Kollar, T., Tellex, S., Roy, D. and Roy, N. 2010. Toward understanding natural language directions. In Proc of the 5th ACM/IEEE International Conference on Human-Robot Interaction, 259–266.
Kortenkamp, D., Baker, L.D. and Weymouth, T. 1992. Using gateways to build a route map. In Proc of the IEEE/RSJ International Conference on Intelligent Robots and Systems (IROS).
Kostevelis, I. and Gasteratos, A. 2015. Semantic mapping for mobile robotics tasks: A survey. Robotics and Autonomous Systems. Doi: 10.1016/j.robot.2014.12.006.
Kuipers, B.J. 1983. The cognitive map: Could it have been any other way? In Spatial Orientation: Theory, Research, and Application, 345–359. Plenum Press, NY.
Kumar, S. and Hebert, M. 2005. A hierarchical field framework for unified context-based classification. In Proc of the 10th International Conference on Computer Vision (ICCV).
Limketkai, B., Liao, L. and Fox, D. 2005. Relational object maps for mobile robots. In Proc of the International Joint Conference on Artificial Intelligence (IJCAI).
Linegang, M.P., Stoner, H.A., Paterson, M.J., Seppelt, B.D., Hoffman, J.D., Crittendon, Z.B. and Lee, J.D. 2006. Human-automation collaboration in dynamic mission planning: A challenge requiring an ecological approach. In Proc of the Human Factors and Ergonomics Society Annual Meeting 50(23): 2482–2486.
Lyons, J.B. and Havig, P.R. 2014. Transparency in human-machine context: Approaches for fostering shared awareness/intent. In Proc of the International Conference on Virtual, Augmented and Mixed Reality, 181–190.
Maniadakis, M., Trahanias, P. and Tani, J. 2009. Explorations on artificial time perception. Neural Networks 22: 509–517.
Maniadakis, M. and Trahanias, P. 2013. Self-organized neural representation of time. In Lecture Notes in Computer Science, 74–81. Springer-Verlag, Berlin Heidelberg.
Maniadakis, M. and Trahanias, P. 2014. Time models and cognitive processes: A review. Frontiers in Neurorobotics 8(7): 1–6. doi: 10.3389/fnbot.2014.00007.
McCarthy, J. 1993. Notes on formalizing context. In Proc of the 13th International Joint Conference on Artificial Intelligence 555–560.
Meger, D., Forssén, P.E., Lai, K., Helmer, S., McCann, S., Southey, T., Baumann, M., Little, J.J. and Lowe, D.G. 2008. Curious George: An attentive semantic robot. Robot and Autonomous Systems 56(6): 503–511.
Moldovan, D., Clark, C. and Harabagiu, S. 2005. Temporal context representation and reasoning. In Proc of the 19th International Joint Conference on Artificial Intelligence.
Myeong, H., Chang, J.Y. and Lee, K.M. 2012. Learning object relationships via graph-based context model. In Proc IEEE Conference on Computer Vision and Pattern Recognition (CVPR).
Oberauer, K. and Lewandowsky, S. 2011. Modeling working memory: a computational implementation of the time-based resource-sharing theory. Psychonomic Bulletin & Review 18(1): 10–45.
Perelman, B.S., Evans, A.W. and Schaefer, K.E. 2017. Mental model consensus and shifts during navigation system-assisted route planning. In Proc of the Human Factors and Ergonomics Society 61(1), doi:10.1177/1541931213601779.
Phillips, E., Ososky, S., Grove, J. and Jentsch, F. 2011. From tools to teammates: Toward the development of appropriate mental models for intelligent robots. In Proceedings of the Human Factors and Ergonomics Society 55: 1491–1495. doi: 10.1177/1071181311551310.

Posner, I., Cummins, M. and Newman, P. 2009. A generative framework for fast urban labeling using spatial and temporal context. Autonomous Robotics 26: 153–170.
Ranganathan, A. and Dellaert, F. 2003. Semantic modeling of places using objects. Computational Perception & Robotics, Georgia Tech Library.
Sarter, N.B. and Woods, D.D. 1995. How in the world did we ever get into that mode? Mode error and awareness in supervisory control. Human Factors 37(1): 5–19.
Schaefer, K.E., Straub, E.R., Chen, J.Y.C., Putney, J. and Evans, A.W. 2017. Communicating intent to develop shared situation awareness and engender trust in human-agent teams. Cognitive Systems Research 46: 26–39.
Scheutz, M., Schermerhorn, P., Middendorff, C., Kramer, J., Anderson, D. and Dingler, A. 2005. Toward affective cognitive robots for human-robot interaction. In Proc for the American Association for Artificial Intelligence (AAAI) 1–66.
Schulz, R., Wyeth, G. and Wiles, J. 2011. Are we there yet? Grounding temporal concepts in shared journeys. IEEE Transactions on Autonomous Mental Development 3(2): 163–175.
Sederberg, P., Gershman, S.J., Polyn, S.M. and Norman, K.A. 2011. Human memory reconsolidation can be explained using the temporal context model. Psychonomic Bulletin Review 18(3): 455–468.
Stachniss, C., Martínez-Mozos, O., Rottman, A. and Burgard, W. 2005. Semantic labeling of places. In Proc of the International Symposium of Robotics Research (ISRR).
Strobel, M., Illmann, J., Kluge, B. and Marrone, F. 2002. Using spatial context knowledge in gesture recognition for commanding a domestic service robot. In Proc of the 11th IEEE Workshop on Robot and Human Interactive Communication, 468–473.
Stubbs, K., Hinds, P.J. and Wettergreen, D. 2007. Autonomy and common ground in human-robot interaction: A field study. IEEE Intelligent Systems 22(2): 42–50.
Suh, I.H., Lim, G.H., Hwang, W., Suh, H., Choi, J.-H. and Park, Y.-T. 2007. Ontology-based multi-layered robot knowledge framework (OMRKF) for robot intelligence. In Proc of the IEEE International Conference on Intelligent Robots and Systems (IROS), 429–436.
Sukthankar, G., Shumaker, R. and Lewis, M. 2012. Intelligent agents as teammates. pp. 313–343. *In*: Salas, E., Fiore, S.M. and Letsky, M.P. (eds.). Theories of Team Cognition: Cross-Disciplinary Perspectives. Routledge.
Sycara, K. and Sukthankar, G. 2006. Literature review of teamwork models (CMU-RI-TR-06-50). Robotics Institute Carnegie Mellon University: Pittsburgh, PA.
Taatgen, N., Van Rijn, H. and Anderson, J. 2007. An integrated theory of prospective time interval estimation: The role of cognition, attention and learning. Psychological Review 114(3): 577–598.
Vasudevan, S., Gächter, S., Nguyen, V. and Siegwart, R. 2007. Cognitive maps for mobile robots—An object based approach. Robotics and Autonomous Systems 55: 359–371.
Walter, M.R., Hemachandra, S., Homberg, B., Tellex, S. and Teller, S. 2009. Learning semantic maps from natural language descriptions. In Proc of Robotics: Science and Systems.
Wolf, D.F. and Sukhatme, G.S. 2008. Semantic mapping using mobile robots. IEEE Transactions on Robotics 24(2): 245–258.
Yeap, W.-K. and Jefferies, M.E. 2001. On early cognitive mapping. Spatial Cognition and Computation 2(2): 85–116.
Zender, H., Martínez-Mozos, O., Jensfelt, P., Kruijiff, G.-J.M. and Burgard, W. 2008. Conceptual spatial representations for indoor mobile robots. Autonomous Robots 56: 293–502.

5

Machine Learning Approach for Task Generation in Uncertain Contexts

Luke Marsh,[1] *Iryna Dzieciuch*[2,*] and *Douglas S. Lange*[2]

5.1 Introduction

We are using the Intelligent Multi-UxV Planner with Adaptive Collaborative Control Technologies (IMPACT) system. It is a collection of technologies used for the purpose of aiding users in the command and control of multiple Unmanned Vehicles to achieve different tasks (Rowe et al., 2015). The IMPACT system models Unmanned Vehicles, their sensors and the environment within which they operate. The unmanned vehicles can be tasked via "plays" to perform certain actions such as scanning and tracking other objects in the simulation. Given the potentially large number of unmanned vehicles in a simulation, human operators can become overwhelmed attempting to construct the context by maintaining situational awareness and control over the vehicles. To address this problem, our team has developed a Task Manager module for IMPACT with the aim of

[1] Computer Scientist, Defence Science and Technology Group, Edinburgh, Australia 5111; luke.marsh@dsto.defence.gov.au
[2] Space and Naval Warfare Systems Center-Pacific (SPAWAR), San Diego, CA.
* Corresponding Author: Iryna.Dzieciuch@navy.mil

assisting human operators (Lange et al., 2014; Gutzwiller et al., 2015). Still, however, even with the assistance of the task manager, human operators can be overloaded at times in managing and creating tasks in the system. Additionally, the IMPACT task manager can only assist the operator if it has a representation relating the operational context for the task space being maintained.

Task space is a discrete space that we know how to manage. Complications arise when trying to relate the continuous multivariable space of UxV operations and the continuous space of operator attention with the discreteness inherent to task space. In this chapter, we outline the task manager module of IMPACT; for now, we describe approaches to discretize the operational context relative to the selection of task types to instantiate by the task manager. We also outline initial efforts to apply machine learning techniques to automatically generate tasks for the task manager, and we discuss a model to optimize the scheduling and queuing of tasks under different complexity levels in IMPACT to reduce the cognitive load on human operators. The final section of this chapter concludes with an outline of the future efforts planned for this work by our team.

5.2 Task Manager

The computational system offers a task manager with assistance that ranges from attention management services to fully automated control. The core of the task manager contains a task model that allows it to act as the human operator's intelligent assistant's manage and even to perform tasks on behalf of the user. These tasks may include assigning missions to Unmanned Vehicles (UxVs), evaluating the performance of a mission against established criteria, making decisions at checkpoints during a mission, or changing the resources (assets) available to a team (Lange et al., 2014). When the system detects that the human is or is becoming overloaded and the computer has the tools available to perform that task, the computer can assume responsibility for that task to assist the human operator. Figure 5.1 represents an example task model to illustrate that there can be multiple methods of achieving a task when it is selected (Lange et al., 2014). Tasks represent elements that must be performed to execute a chosen method performed by either a human operator or the computer depending on the capabilities of both and also the operator's availability. Methods offer alternative approaches to completing a task. While tasks taken by the computer to assist human operators in managing the human's workload during a simulation, the humans are required to create many of the tasks. This instantiation by the human can also become onerous, especially when a human operator is almost already overloaded.

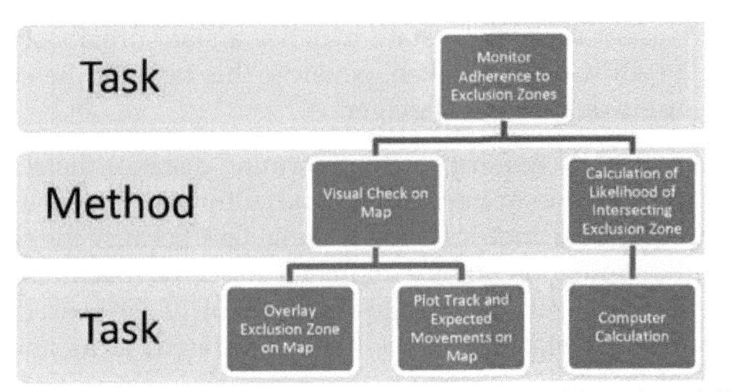

Figure 5.1. An example of the Task Model. The recursive nature of the task model's structure creates alternating levels of tasks and methods.

5.3 Machine Learning Task Generation

In this section we discuss our initial ideas and efforts to automatically generate tasks for users of the Task Manager system based on machine learning techniques.

All of the data in the IMPACT simulation is stored in different states. The most relevant states in IMPACT include air, ground and sea vehicle states which store live data for all of the vehicles involved in a simulation. Other data comes from the active sensors of the UxVs. All of this data is stored in camera, video stream and radio-state variables. Each vehicle's state is comprised of many variables such as: current location, velocity, acceleration, current heading, available energy, energy usage rate, payload lists and current tasks. This data is available to be used by the machine learning techniques we have employed to determine whether a task should be generated.

In using task manager for IMPACT, users create their own tasks for a job at any point in time. The task could be of any type, but the user has the ability to create new types of tasks previously known or unknown to the system. The trigger for the creation of the tasks is, therefore, based on the information available in the simulation. The operators have no prior knowledge about the data to be used in the simulation to make a task work properly. The task chosen is likely based on any or all of the information in the system. Therefore, the initial machine learning approach taken is the K-Nearest Neighbor algorithm (kNN) based on simplicity and applicability to many of the tasks the simulation problem entails (Fix and Hodes, 1951).

This high-level approach is comprised of these three steps:

1. Record the states of the IMPACT simulation during the period when a task is created.
2. Continuously monitor all of the IMPACT simulation states.

3. If the current simulation state matches a state previously recorded when a task was undertaken, generate this task for the current and future users of the Task Manager.

The core of kNN requires us to determine distance metrics between states of the simulation to match them and, ultimately, to generate a task. But determining these metrics is not a trivial task because the states of the vehicles in the simulation rarely ever match perfectly. To address this issue, the feature standardization technique by Peterson (2009) is used to remove the bias caused by state variables having different units and different measurement scales.

Initial implementations of the approach in the method have highlighted for us a number of important design questions: Should variables in each state in the distance measurement be weighted evenly? Are vehicles to be treated as homogenous with states comparable to the states of other vehicles? Should we combine the distance measurements for all individual states or should we contrast subsets of states? Finally, what is the error threshold for the similarity between states before a task can be automatically generated?

Additionally, we can record the simulation state for the task creation data for prior simulations and use this data for training and classification. However, what if different numbers and types of vehicles are used in the simulations? If we take an action, does the data remain relevant?

In the future, our research will entail experimenting with different options in an attempt to resolve the questions noted above. Given the difficulties in determining the distance metrics between the states for this problem, other more abstract machine learning techniques such as artificial neural networks (ANN) that may be better suited for this problem should be explored. But, an ANN likely requires significantly more training data from cases to train them effectively given the number of variables in each state. To address this issue more fully in the future, we will also explore utilizing a time-windowed history of IMPACT simulation states for task generation.

5.4 Task Optimization

In this section, we discuss the complexities of the IMPACT system that causes a human operator to experience an information overload. Once identified, we outline a model to optimize the scheduling and queuing of tasks in IMPACT with the aim of reducing these cognitive loads for operators.

5.4.1 Data Collection

All of the raw data collection from the IMPACT system yields two basic types of data: user generated chat messages along with sensor data.

Table 5.1. Example of the key information obtained from chat messages.

Time	Point (ROI)	Duration (time)	Asset (Sensor, UxV)
45:00	NAI - 4	4 min	Listening Post/ Observation Post (LPOP)
0:00:00	Ammo Dump	40 min	Imagery
00:17:00	Chow Hall	infinity	360 degrees (imagery)
20:00	Gate 2	10 min	Force (vehicle)
30:00	Gate 2	5 min	Force (vehicle)
message arrival	Point Quebec	15 min	Force (vehicle)
53:00	Barracks	infinity	360 degrees (imagery)
message arrival	Current collation	3 min	Flight line
message arrival	Current location	infinity	UGV, 53% fuel left

Messages from users come in the form of natural language; aspects of these messages must be parsed into four data elements: time of message, location, duration and asset. These representations of data are summarized below in Table 5.1.

Time in column 1 in Table 5.1 is the time stamp that a message was issued or an action is scheduled to take place. The Point column is the location for the region of interest (ROI) where an asset is planned to appear. Duration is the time of job completion when the simulation's trigger is stopped. Asset is the name of the sensor or vehicle capability used to complete an assigned task.

The idea behind the rapid retrieval of structured chat messages from the database is to quickly distinguish the key semantics extracted from the natural language into a computer readable form. But for this chapter, this stage requires much more development and will not be covered at this time.

5.4.2 Complexity in IMPACT

For the user whose job it is to simultaneously control multiple autonomous agents, higher rates of multitasking increase complexity. For our purposes, complexity reduces when many simpler tasking states are grouped together into a sequence that can be automated.

Timing is important in these simulations, as events often occur randomly. For example, tasking can be routine, such as ground monitoring, but the user's attention is diminished after we introduce additional information or events. Complexity is different for each user. What complexity means for each individual is a topic of further investigation; however, for our approach, tasks can be qualitatively categorized linguistically as low, medium or high. Because an accumulation of simple tasks at a high rate is overwhelming for human operators, timing and the rate of task occurrence in the user task queue is a key factor.

5.5 Environmental Events

Environmental events routinely take place outside of a user's control. Events cause a user to react, and that reaction requires new actions in the IMPACT system to respond to the environmental events. Example environmental events can include these observations: Gate runner, Mortar fire and a chat message. These events can evoke quick reaction responses and need to be monitored by users of the IMPACT system.

5.5.1 Sensor Data

Variables in the IMPACT system also include data supplied by a sensor from the unmanned vehicles (UxV). These UxV's operate in time and space domains, carrying sensor performance characteristics; for example: Airspeed, Energy Rate, Altitude and Latitude/Longitude coordinates. Users are constantly updated with sensor information as they perform a data retrieval after a chat message query is issued. Data representation is summarized below in Table 5.2; notice that it is also composed of time, ROI information and duration for task completion; accompanying this data is the status for each of these: UxV, sensor and vehicle.

Table 5.2. Example of UxV sensor data in the IMPACT system.

Time	Point (lat/long/alt)	Duration (time)	Sensor Characteristics
00:00	30.471585; 87.181458; 650	0	Airspeed: 300 Energy rate: 100 Pitch Angle: 0.5 Sensor Type: IR camera
50:00	30.446; 87.150706; 0	50/100 complete	Airspeed: 23 Energy rate: 98 Pitch Angle: 0.5 Sensor Types: IR camera

5.5.2 Play

A play is a representation of a multi-vehicle mission within the IMPACT system. It commonly consists of a pre-programmed number of simpler task actions that are allowed to run their course while still being monitored by the user. Operators have tasks to call and monitor plays. These tasks are stimulated by chat messages, requirements of operating plays, and by the operator's own awareness within the tactical situation.

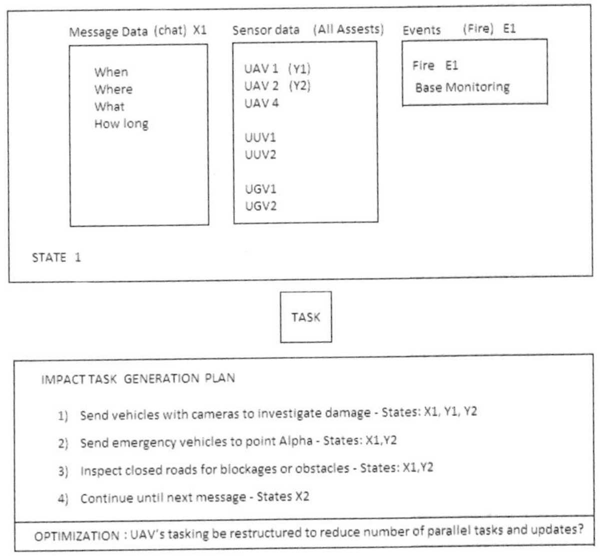

Figure 5.2. Proposed model for decision making process.

The common attributes in the data presented and summarized above regard time and space. All events and tasks occur at a specific ROI and point in time. Suppose, for example, that the basic problem to be considered is in the state space *S* at a time and location of these variables: sensor, chat (user) and environmental events. Let the control space be composed of the sequence of decisions for tasking in the domain *C*. The optimization problem for a data-task in the IMPACT system can be stated: Of the sequences available, assuming the given states, can the least complexity in the sequence of tasks be found that needs to occur successfully and within a specified duration?

We can represent state space for chat messages by a variable $X = \{x1, x2, x3, x4\}$, where $x1$ is time, $x2$ is the location point, $x3$ is the time duration, and $x4$ is the tasked asset. Similarly, $Y = \{y1, y2, y3, y4\}$ is sensor data for the corresponding arguments. Given that, currently, one user can control a number of UxVs, each representation can be written as $X1$ and $Y1$ for UxV1. The events occurring in a scenario can be represented as $E = \{e1, e2, e3, e4\}$; each of these data describe the time, location, duration and event type. The events trigger sequences of tasks C suggested so that a user can reach a favorable mission result under varying levels of complexity. For the purpose of illustration in this chapter, for a chosen variable complexity tasking, we plan presently three main settings of complexity: low, medium and high; e.g., low complexity occurs when there are less than three events that require user attention; medium complexity occurs when there are three events that require user attention; and high complexity occurs when there are five or more events in the tasking queue that require user attention.

For example, a sequence of tasking decisions C chosen by an operator after a single event *E* could be the following:

1. Send two vehicles with cameras to investigate the damage to an object or the environment—States: X1, Y1, Y2.
2. Send emergency vehicles to the point Alpha—States: X1, Y2.
3. Inspect roads closed by blockages or obstacles that might impair passage—States: X1, Y2.
4. Proceed until receiving a new message—States X2.

With the occurrence of single events, the rule set is straightforward but even when simultaneous simple events occur, a situation becomes complex. Imagine an overloaded user's reaction to multiple, simple emergencies. That makes it essential to prioritize a sequence of decisions or tasks for different levels of complexity. We can then formulate the control problem as follows: The information available in time and space is optimized to find a sequence of decisions C that produces the maximum result (i.e., a successfully completed task) under varying complexity levels (*CL*) from low to medium and high. The main problem to properly control and evaluate states for different complexities then becomes a minimization-maximization problem.

As we have proposed for the IMPACT planning problem, it can be generalized for an optimal selection of task *max C*, given a message state space *X*, sensor data *Y* and event *E* for a complexity level *CL*. The optimal scheduling and queuing of tasks is meant to minimize (reduce) complexity in the time and space domain. This reduction is described mathematically in Eq. 5.1 as:

$$min\ CL\,(t, s)\ [\max C\,(t, s)\,(X1, \ldots, Xn) \cup (Y1, \ldots, Yn) \cup (E1, .., En)] \quad (5.1)$$

Table 5.3. Example decision table. X and Y represent chat and sensors, respectively, to provide data at certain time intervals in a simulation.

Time	UxV1		UxV2		Ground Patrol 1		Ground Patrol 2	
	X1	*Y1*	*X2*	*Y2*	*X*	*Y*	*X*	*Y*
00:10	X						X	X
00:14		X	X					
00:21					X	X		

Decision trees have been computed for the IMPACT system by creating subcategories. The repetitive decision for variable complexity can be described and structured in a decision table (e.g., Table 5.3 below). For example, this question might be posed: are there chat messages and/or sensor data in the 10 minute window period overlapping in the time and space domains? Does this information influence the tasking state? And, can the tasking for a UxV or UxVs be restructured to minimize the number of parallel tasks and updates? This question means that we must construct a decision table for each case (Table 5.3).

After building a table of optimized tasks, we can observe the behavior of separate UxVs and their task load in the time and space domain. In Table 5.3, upon examination, we can observe the two chat messages that were sent at time 00:10, task UxV1 and Ground Patrol 2. At time 00:14 (i.e., 4 seconds later), the sensor data from UxV1 and message at UxV2 is sent for Task 1. We can group chat message data and sensor data happening at about the same time or space. We repeat the procedure for consecutive Task 2. Clustering sensor and chat message status is the first step that reduces operational complexity for users; this reduction can be used to investigate whether the data can model the control system under different levels of complexity. This approach, we postulate, helps to optimize tasking decisions and reduce complexity for human operators. For this step, we use existing data feeds to group tasks in time and space domains.

Some of the clustering techniques, such as the k-nearest neighbor classifier, help us to understand task creation dynamics. When complexity in a context increases, as when four vehicles perform a task for two types of events, our approach decreases the number of tasks in the queue.

5.6 Conclusion

The context of the command and control of unmanned vehicles in the IMPACT system is cognitively intensive. Human operators of this system are required to multitask in uncertain environments, to process situational data and yet to efficiently use autonomous agents under varying levels of complexity across multiple regions of interest (ROIs), a challenging context. These requirements often produce an information overload that can adversely affect the success of missions.

In this chapter, we have described our initial efforts to apply machine learning to generate tasks automatically for operators of IMPACT; we have also outlined a model to optimize the scheduling and queuing of tasks in IMPACT that reduces cognitive loads for human operators. In the future, we plan to restructure data feeds for the time and space domain for optimization with machine learning; to apply machine learning to generate tasks in the system automatically; and to explore optimizing the scheduling of tasks to further reduce the cognitive loads for human operators.

References

Fix, E. and Hodes, L.J. 1951. Discriminatory analysis, nonparametric discrimination: Consistency properties. Technical Report 4, USAF School of Aviation Medicine, Randolph Field, Texas.

Gutzwiller, R.S., Lange, D.S., Reeder, J., Morris, R.L. and Rodas, O. 2015. Human-computer collaboration in adaptive supervisory control and function allocation of autonomous system teams. In Proceedings of the 7th International Conference on Virtual, Augmented and Mixed Reality, 447–456. Los Angeles, Calif.: Springer International Publishing.

Lange, D.S., Verbancsics, P., Gutzwiller, R. and Reeder, J. 2012. Command and Control of Teams of Autonomous Units. In Proceedings of the 17th International Command & Control Research & Technology Symposium, Fairfax, VA.: International Command & Control Research & Technology Symposium.

Lange, D.S., Gutzwiller, R.S., Verbancsics, P. and Sin, T. 2014. Task models for human-computer collaboration in supervisory control of teams of autonomous systems. In Proceedings of the 2014 IEEE International Inter-Disciplinary Conference on Cognitive Methods in Situation Awareness and Decision Support, 97–102. San Antonio, USA: IEEE.

Peterson, L.E. 2009. K-nearest neighbor. Scholarpedia 4(2): 1883.

Rowe, A., Spriggs, S. and Hooper, D. 2015. Fusion: a framework for human interaction with flexible-adaptive automation across multiple unmanned systems. In Proceedings of the 18th Symposium on Aviation Psychology, 464–469. Dayton, Ohio.: International Symposium on Aviation Psychology.

Rasmussen, S.J., Shima, T., Mitchell, J.W., Sparks, A.G. and Chandler, P. 2004. State-space search for improved autonomous UAVs assignment algorithm, 2004 43rd IEEE Conference on Decision and Control (CDC).

Creating and Maintaining a World Model for Automated Decision Making

Hope Allen[1] and *Donald Steiner*[2]

6.1 Introduction

This chapter presents a computational approach for representing knowledge in a world model, establishing the context for decision-making by autonomous agents and cognitive decision aids. We begin with the context of our work and some basic terminology for the scope of this chapter with the realization that other chapters may have different meanings for the same terms.

We establish the context of our work as enabling mission-level cognitive autonomy across all five domains of warfare: air, land, maritime, space, and cyberspace [1]. Thus, we address both 'autonomy at rest' and 'autonomy in motion' as described by the Defense Science Board [2]. We focus on the 'mission-level' aspect rather than the 'vehicular control' aspect; that is, we address how autonomous systems work together in human/machine teams to achieve commander's intent. By 'autonomy,' we mean the ability to achieve mission objectives without external control by humans or

[1] Nvidia Corporation, Product Manager, Autonomous Vehicles, Santa Clara, CA.
[2] Product Area Architect for Autonomy and Cognition, Technical Fellow, Northrop Grumman Corporation, McLean, VA.

machines while sensing and adapting to changes in the environment.[1] While definitions of 'cognitive' abound, in this context, we mean the ability to *learn* from own experience and from others to detect *unforeseen* changes and to reason about *unplanned* events to achieve mission goals. Thus, the system should be able to deal with achieving mission objectives in *uncertain* environments. We have designed and implemented a reusable software framework automating the OODA loop[2] for each individual autonomous system, enabling a seamless transition path from *advanced automation*[3] to more cognitive autonomous systems. Figure 6.1 depicts the Cognitive Autonomy Engine (CAE™)[4] framework in more detail. Communication and collaboration among autonomous systems and human teammates are also directly incorporated in the CAE framework.

We refer to autonomous systems as agents. 'Intelligent' agents are those that exhibit some degree of cognitive ability. We recognize that, as with biological systems, there are varying dimensions and degrees of intelligence

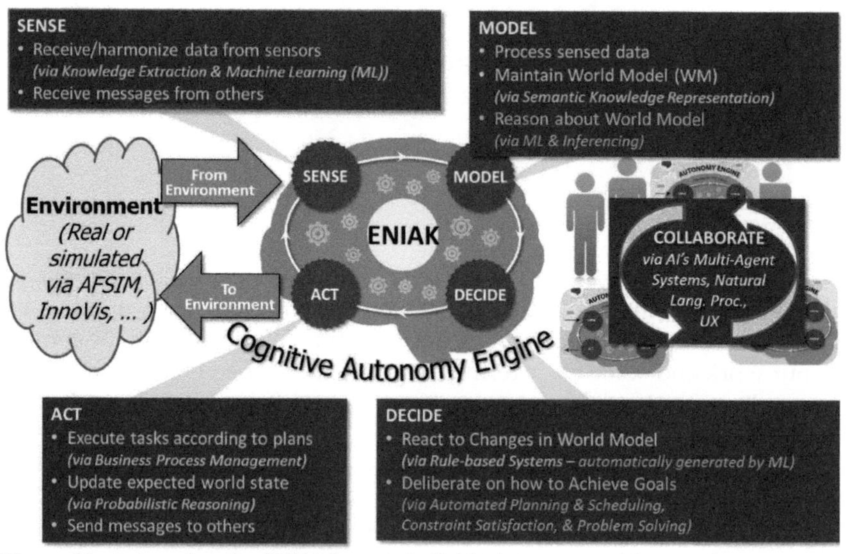

Figure 6.1. Automating the OODA Loop in individual systems enables seamless migration from Advanced Automation to Cognitive Autonomy.

[1] Note: For the purposes of this chapter, our definition of autonomy includes automation. We use this definition because even current, highly complex autonomous systems, such as self-driving cars, can still be considered to be automated, being driven by fixed, albeit complex, rules at any given time.

[2] Observe-Orient-Decide-Act loop – This moves the human from *in* the loop to *on* the loop.

[3] Recently, the term 'Robotic Process Automation' [5] has become popular to denote the automated execution of pre-defined processes by software.

[4] Pronounced 'kay'.

or cognition. Our purpose is to continually improve the ability of machines to complement and support human decision making and activities and not necessarily to achieve "Artificial General Intelligence." [3, 4].

Section 6.2 introduces the components of the world model, which is based on semantic knowledge representation of and reasoning over ontologies of entities and their relationships. Every assertion (belief) in the world model is provided with a time stamp,[5] a (variable) confidence factor[6] and associated reasons for belief.[7] The world model comprises not only knowledge about the environment in which the agent operates, but also knowledge about the agent itself (sensors, actuators and associated capabilities), external actors (agent/human teammates and adversaries) and the mission(s) (objectives, plans, tasks, etc.) of which the agent is aware.

Section 6.3 outlines our approach to building and managing the world model based on sensor input and reasoning about information in the model. Sensor drivers map the output of sensors into information provided to the world model. Sensor drivers may be more or less complex, depending on the sensor. Information from 'smart' sensors that already derive actionable, semantically rich intelligence from raw data is inserted directly into the world model. Otherwise, the sensor drivers must apply analytics to raw data from 'dumb' sensors to update the world model. Deep learning (on either simulations or real world experiences) can be used to automatically generate and enhance the ontology, minimizing the brittleness that is often associated with manually generated ontologies. Finally, computational techniques such as information fusion, anomaly detection, threat classification and predictive analysis are applied to the world model to derive further information supporting automated decision-making that, in turn, is also asserted into the world model.

Section 6.4 outlines how the world model provides the context that enables computational decision-making, leading to actions performed by actuators. Actuator drivers control the actuators to execute the decisions. Reactive behavior is enabled by executing pre-defined mission 'plans', triggered by changes in the world model; plans are sequences of tasks achieving mission goals represented through the Business Process Management Notation (BPMN 2.0) standard [6]. Deliberative (goal-directed) behavior is accomplished by using automated planning and scheduling techniques to generate mission plans and allocate resources based on pre-conditions and expected effects of actions as represented in the world model.

---

[5] Tagging all beliefs with the time they are considered to be valid (or invalid) enables reasoning about not only the current world state, but also past and future world states.

[6] Information may be unreliable due to faulty sensors, misleading informants, and inaccurate predictions.

[7] These include local sensors, other agents/humans, and local inferences and analytics based upon existing beliefs.

Section 6.5 explores how this approach, together with the agent communication and collaboration standards specified by the Foundation for Intelligent Physical agents (FIPA) [7], directly enables collaboration among autonomous systems and humans in performing joint missions.

The Cognitive Autonomy Engine

An autonomous system (agent) achieves its mission objectives without external control while sensing and adapting to changes to its environment. Though autonomous agents operate independently, their power is enhanced when they interact with others through social exchanges such as negotiation, cooperation, and coordination in support of mission objectives. An agent-based system is highly flexible, resilient, and efficient, allowing for dynamic introduction and discovery of new capabilities from different sources while using collaboration protocols, such as market-based bidding and negotiation, to effectively allocate resources and execute services. By learning from experience and others, cognitive agents can better represent their capabilities to the system.

Northrop Grumman's Cognitive Autonomy Engine (CAE™) is a reusable software framework that provides each agent with:

(1) an automated OODA loop so that it can make decisions in changing or uncertain environments

(2) the ability to interact with other agents and humans in order to achieve objectives.

The Cognitive Autonomy Engine employs knowledge representation and reasoning techniques that enable the agent to semantically model its external environment, its mission, and itself so that the agent can reason over its current situation to solve problems and make decisions. Additionally, the Cognitive Autonomy Engine complies with the Foundation for Intelligent Physical Agents (FIPA) standards for agent interoperability, thus providing a common framework for agents to discover and collaborate with each other. The Cognitive Autonomy Engine provides the core agent functionality on which mission-specific capabilities can then be layered.

The CAE software architecture comprises four major components: Sense, Model, Decide, and Act. Information flow across these components and processing is managed by the Embedded Networked Intelligent agent Kernel (ENIAK™) software implementation. The ENIAK core may have different implementations (for example, using Java™ software, C++, or Robotic Operating System (ROS)) to support deployment of CAE-based agents on enterprise and embedded platforms. The CAE world model is designed to be reusable across ENIAK implementations.

Use Cases

The concepts and methods discussed in each section of this chapter are motivated by an analysis of two disparate use cases that leverage the reusable CAE framework in order to model and act in an uncertain and ever-changing operational context. The first use case addresses 'autonomy in motion' (i.e., physical or kinetic autonomy) in the undersea mission domain where actionable intelligence collected by an Unmanned Undersea Vehicle (UUV) supports decision aiding to commanders. The second use case addresses 'autonomy at rest' in an enterprise operations environment to automatically detect and report system component failure, reducing operational costs while increasing system performance. This demonstrates the applicability of the underlying framework to a wide variety of domains. A high-level overview of each use case follows.

Use Case 1: Autonomous Mine Hunting

Blue Forces want to navigate their fleet through a body of water presumed to contain naval mines, but the number, type and location of these mines are unknown. A UUV is deployed to survey the dangerous waters, identify threats and report the information back to the Command Center, where a plan of action will be generated that accounts for the mines and provides safe passage through the water (Figure 6.2). The UUV is a stock system that has been equipped with a sophisticated synthetic aperture

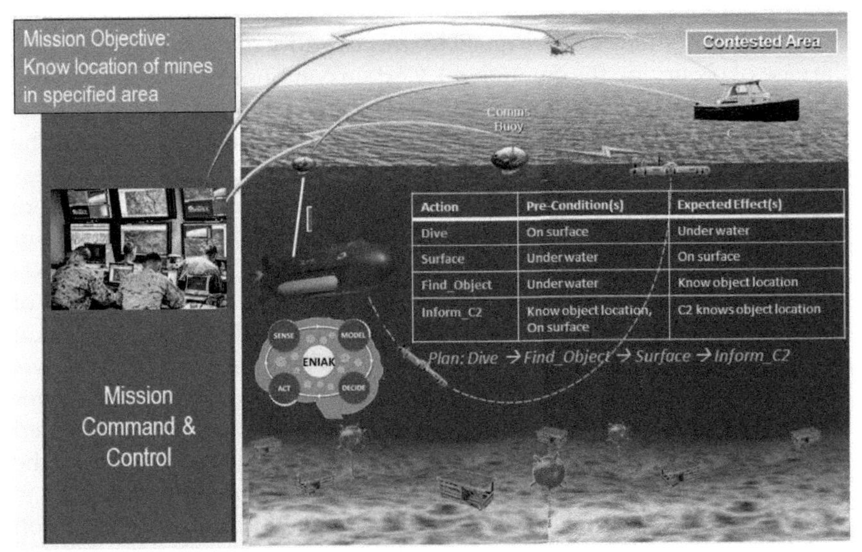

Figure 6.2. In the Autonomous Mine Hunting use case, an agent deployed onboard the UUV autonomously controls the UUV to survey an area and report threatening mine-like objects to the Command Center, supporting decision-making and planning activities.

sonar [8], a processing package to detect, classify and localize mines and a communication mechanism that allows it to transmit information when it is at the surface of the water.

Outside of these capabilities, the UUV has basic vehicle and payload controllers but does not have a built-in mission controller that manages the responsibilities of finding and reporting mines or even basic keep-alive capabilities, such as mine avoidance. To support this mission, an autonomous agent has been developed according to the CAE framework and is deployed onboard the UUV. This UUV agent processes information from the UUV's sensors to build a model of the current world state, identifies the best course of action to take to achieve mission goals and interfaces with the UUV platform to execute the selected plan.

Use Case 2: Automated Fault Detection and Isolation

Even though mission-critical information systems are designed to meet high availability and reliability requirements, unexpected component failure is inevitable. These systems are generally engineered to fail in a way that minimizes operational risk, employing techniques such as redundancy and fault monitoring. While most systems currently use some form of automated fault detection and alerting, these methods often result in simple visual or audible alerts that still require a human in the loop to monitor, record and troubleshoot. This situation results in a delay in the detection and diagnosis of issues.

Additionally, components in the information system are monitored by staff on schedules that correlate to the criticality of the component and the ease of observing the component. Certain components may be checked only once per shift, resulting in a lag of up to several hours before an issue is reported. Some equipment may be checked less frequently due to the remoteness in location and the security risks involved with providing access to maintenance staff. Even if other components may be monitored more frequently, another potential lag is introduced when a fault is detected but a technician is unavailable to troubleshoot, resulting in disruptive wake-up calls and waiting for a technician to become available. An analysis of the maintenance manuals for several subsystems indicates that the fault detection and fault isolation performed by maintenance personnel follows predefined procedures that are straightforward to execute. These procedures are defined by a set of finite and bounded steps with observable expected effects. At a high level, procedures such as these are well suited for the advanced automation capabilities of the Cognitive Autonomy Engine.

To demonstrate this use case, we present a simplified fault detection and isolation scenario. A power supply assembly provides power to a

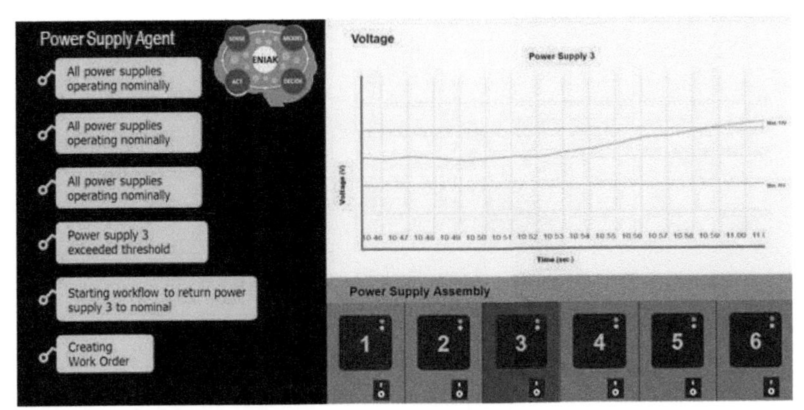

Figure 6.3. In the Automated Fault Detection and Isolation use case, an agent continuously monitors several power supply modules. In the event of abnormal behavior, the agent automatically executes a process to mitigate immediate impact to the system, isolate the fault and create a detailed work order for maintenance staff to promptly address the issue. L and H indicate low and high thresholds, respectively.

mission-critical system component. This power supply assembly comprises several power supply modules; while there are redundancy and safeguards in the system to mitigate emergencies, it is generally unfavorable for the voltages of any of the power supply modules to be outside of upper or lower operating limits. A 'Power Supply' agent has been developed using the CAE framework; it monitors voltage measurements from each of the power supply modules, and, in the case of abnormal measurements, it automatically isolates the fault and creates a detailed work order in an enterprise work order management system so that the issue is promptly resolved (Figure 6.3).

6.2 A World Model for Automated Decision Making

In this section, we introduce the components of the world model, which is based on semantic knowledge representation of and reasoning over ontologies of entities and their relationships. Typically, a semantic knowledge base is governed by one or more ontologies [9] that represent a taxonomy of entries in the knowledge base as well as relationships among them. The ontologies also define sets of attributes that elements in the taxonomy may have. We provide a 'core' ontology (a specific set of entities, their relationships and attributes) common to all agents in our system, independent of the domain and mission. An agent may then also be configured or loaded with ontologies representing a particular domain as well as the missions in the domain (Figure 6.4).

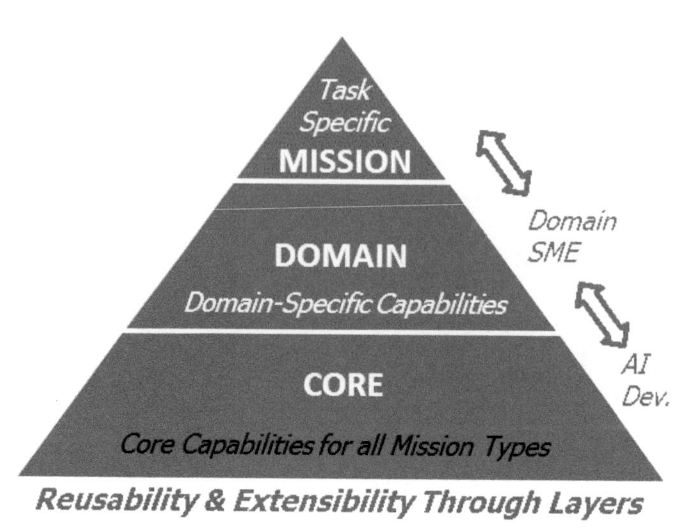

Figure 6.4. The world model of every autonomous agent implemented via the CAE framework is based on the same approach. All of the agent's knowledge is represented in the world model. The core model is common to all agents and domains, while the domain and mission models are configured pre-mission.

Core Agent Ontology

Core Attributes

Semantic knowledge bases in agent systems are generally used to represent facts (or beliefs) about the world model representing the environment of the agent. The Cognitive Autonomy Engine's world model is a semantic knowledge base modeled in the Web Ontology Language (OWL) [10], where beliefs are represented in the form of subject-predicate-object triples. Typical semantic knowledge bases often simply overwrite these beliefs when updated. As this does not lend itself to learning from experience or reasoning about future events, we also associate every assertion (belief) in the world model with the following additional attributes:

Time stamp: The timestamp indicates the time, as believed by the agent, that the fact holds. This enables reasoning about not only the current world state, but also past and future world states. Note that this is not necessarily the time at which the agent first learned about the belief. For example, suppose an incident I occurs at time T_0 but the agent learns about it at time $T_1 > T_0$. The agent will add a belief in I (B_I) with timestamp T_0. The agent may also add a belief B_{LI} that it learned about I occurring at T_0. B_{LI} will have timestamp T_1. Agents can also represent future world states by associating a future timestamp with a belief.

***Confidence factor*:** Clearly, future world states are not 100% certain. Furthermore, information may be unreliable due to faulty sensors or misleading informants. Thus, we associate every belief in the knowledge base with a degree of belief or confidence. This is, obviously, particularly important when dealing with future world states and possibly inaccurate predictions. Currently, the confidence factor is represented as a single variable. Future implementations may represent the confidence factor as a probability distribution function over time.

***Reason(s) for belief*:** Finally, in order to maintain the confidence factors across associated data, the agent needs to represent the justification for the belief. The justification may come from a variety of sources, in particular, sensors, other agents/humans[8] and local inferences and analytics based upon existing beliefs; therefore, an agent may have several reasons for its belief. For example, the information may have been reported by a local sensor or confirmed by another agent. In this case, the confidence in the belief may be strengthened. Alternatively, an agent may have conflicting evidence, in which case the confidence factor is reduced.

Note that domain-specific ontologies may incorporate additional attributes, such as position, etc. The above attributes are those that are necessary to all beliefs across all agents.

Core Objects

In order to effectively reason about the mission, the world model comprises not only knowledge about the environment in which the agent operates, but also about the agent itself, other actors and the mission (Figure 6.5).

The following entities are represented by the core ontology:

***Actor*:** Actors are entities that may effect changes in the environment. The agent itself is an actor; it is denoted by the special object Self. Other actors include human, autonomous teammates, adversaries and organizations or teams of agents.

***Goal*:** A goal is a specific desired world state that the agent strives to achieve. Goals have priorities about which the agent reasons to determine which goals to pursue.

***Capability*:** The ability of an actor or team of actors to effect changes on the environment resulting in a given end state as represented in the world model. There are composite capabilities and unit capabilities (actions). An agent can include the capabilities of other actors and teams of actors in its world model. All capabilities have preconditions and expected effects.

[8] Note: Information from other agents and humans also comes through sensors, but we highlight the distinction between local sensors and other agents here.

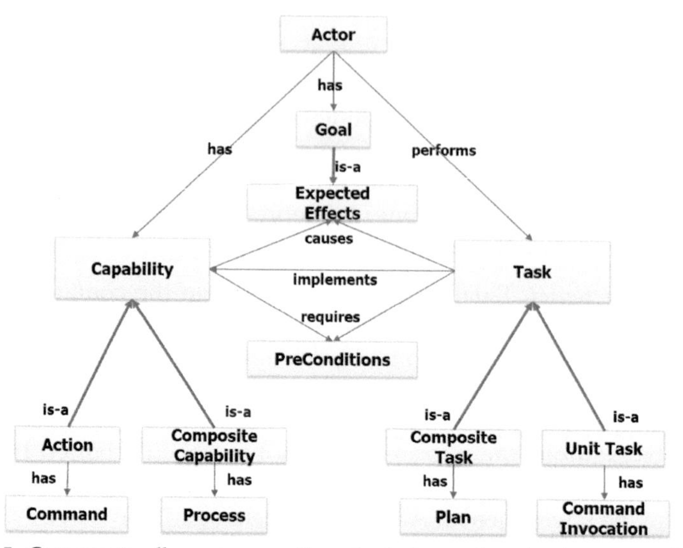

Figure 6.5. Common to all agents regardless of mission or domain, the core world model describes the fundamental relationships between concepts such as goals, plans and capabilities.

Preconditions: The conditions in the world state (both external and internal to the agent) that must hold before the actor can perform the capability. Given that the world state may be uncertain, the preconditions may be associated with a minimum confidence level that they hold prior to performing the capability.

Expected Effects: The expected world state resulting from performing the capability. There is no guarantee this end state will actually occur upon performance of the capability, so we use the term 'expected'. This may be associated with a probability that the state is achieved. Independent of mission or domain, capabilities always have expected effects representing the status of the execution of the capability over time (e.g., initiated, ongoing, done, failed). As with all beliefs, each of these expected effects have timestamps and confidence levels.

***Composite Capability*:** A type of capability that is derived from combining other capabilities (that possibly involve other agents).

Process: The process defines how an agent achieves the expected effects corresponding to this capability by representing a series of further capabilities using BPMN 2.0 notation. Each of the capabilities in the process is either a composite capability or an action. These capabilities may end up being performed by other actors.

***Action*:** Also called a Unit Capability, an action is the most fundamental non-decomposable capability that an actor can achieve on its own.

Command: Command represents the action that is executed by the actuator driver in order to trigger the actuator (see Section 6.4). This is a method, function or service description in a standard programming language with corresponding variable arguments.

Task: A task is a specific reification of a capability to achieve a goal. Like capabilities, tasks have preconditions and expected effects and may be either composite tasks or unit tasks.

Initiating Actor: The actor that is responsible for initiating the task.

Schedule: The specific time the task is to be initiated, along with the expected duration.

Composite Task: A composite task is a specific reification of a composite capability that is managed by a business process engine (e.g., jBPM) according to the process of the corresponding composite capability.

Plan: A plan is an invocation of the process underlying a composite capability to implement a goal and comprises a set of tasks that may be either composite or unit tasks. While the initiating actor invokes the plan, other actors may be allocated to perform subtasks within the plan.

Unit Task: A unit task is a specific reification of an action.

Command Invocation: A specific reification of a command that is executed by the actuator driver of the performing agent as discussed in Section 6.4. This would be a specific invocation of a method, function or service call. By the time the command is invoked, all arguments must be instantiated, though they might not have been fully instantiated when originally added to the schedule.

Capabilities and actions are independent of specific world state or environment of an agent. Tasks and unit tasks are concrete and specific to a particular mission and goal.

Use Case 1: Autonomous Mine Hunting

Prior to deployment, undersea and mission subject matter experts configure the pre-defined domain- and mission-specific sections of the UUV agent's world model. Automatic ontology generation and existing domain models reduce the overhead caused by manually developing the ontologies.

Ontologies pertaining to the undersea domain are overlaid onto the CAE core world model. This includes concepts such as mine-like objects (MLO) and non-mine-like objects (NOMBO) as well as items such as bathymetric maps, communications buoys and other objects and resources the agent may encounter. The sensors and actuators of the UUV platform

Table 6.1. Example capabilities of the UUV agent as semantically represented in its world model.

Capability	**Precondition(s)**	**Expected Effect(s)**	**Process/ Command**
Dive (z)	I am located on surface (z=0)	I am located below surface at depth z	Command UUV to go to depth = z
Surface	I am located below surface	I am located on surface	Command UUV to go to depth = 0
FindObject(object)	I am located below surface; I do not know where object is located	I believe object is located at (x, y, z)	Command UUV to perform ladder search
Inform(agentX, fact)	I believe fact; I can communicate; agentX unaware of fact	agentX believes fact	Send 'Inform(fact)' message to agentX
Inform_ ObjectLocation (object, agentX)	I do not know where object is located	agentX believes where object is located	Find_Mine
Loiter	I am located at (x, y, 0) at time = t	I am located at (x, y, 0) at time = t+1	Null

are also represented in the ontology with references to their associated sensor and actuator drivers.

The agent's capabilities are also represented semantically and integrated into the world model, including the preconditions, expected effects and actions that the agent will affect on the environment via the actuators. Refer to Table 6.1 for a representative list of the UUV agent's capabilities. Capabilities can include parameters; for example, the Dive capability takes a depth *z* as input and commands the UUV to dive to the specified depth z.

Section 6.3 shows how rules are used to identify and trigger mission plans to achieve goals based on the processes in this use case. The details of the Find_Mine process are described in Section 6.4.

Use Case 2: Automated Fault Detection and Isolation

As in the Mine Hunting use case, subject matter experts, supported by automatic ontology generation techniques, configure the Power Supply agent's world model ontologies prior to the mission.

The domain-specific ontologies represent information relating to the system in which the agent will be operating and include concepts such as

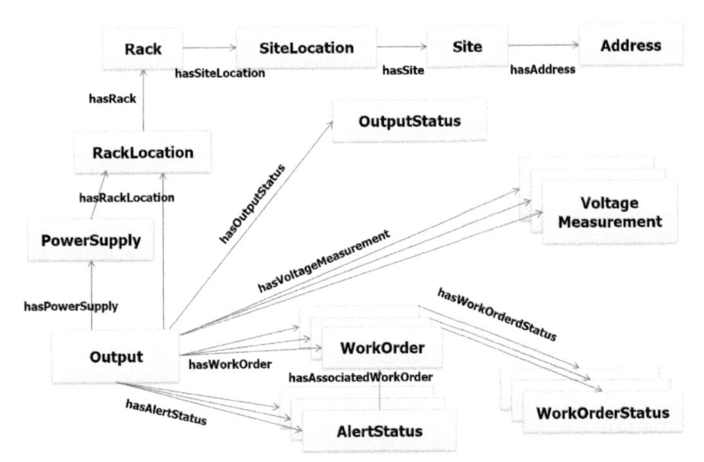

Figure 6.6. The Power Supply agent's world model comprises a domain-specific ontology.

power supplies, asset numbers, voltage thresholds and the other systems and agents that comprise the enterprise. Figure 6.6 shows the relationships among various components in the domain.

The agent's world model represents sensors, such as voltmeters, that provide information to the Power Supply agent as well as actuators, such as interfaces to the power supply modules and the work order management system, that affect the environment per the agent's commands, along with their associated drivers.

The agent's capabilities are also represented semantically in the world model. An example is shown in Table 6.2.

6.3 Creation and Management of the World Model

In this section, we outline our approach to building and managing the world model based on sensor input and reasoning about information in the model. Knowledge in the world model must come from either outside the agent via a sensor or from within the agent through reasoning.

Sensors

All knowledge about the external environment of an agent must come into the world model via a sensor. Sensors can take on many forms; they sense attributes of the physical or virtual (i.e., cyberspace) environment, including incoming communication[9] (receipt of messages) from other actors. In all cases, however, they serve the purpose of providing information about the environment to the agent.

[9] Inbound communication is a special form of sensing that will be addressed more thoroughly in Section 6.5.

Table 6.2. Example capabilities of the Power Supply agent as semantically represented in its world model.

Capability	Precondition(s)	Expected Effect(s)	Process/ Command
Open_Alert(ps_id)	Voltages of Power Supply ps_id are not nominal; No Active Alerts for Power Supply ps_id	Active Alert for Power Supply ps_id	Update Alert status for Power Supply ps_id in WM to Active
ShutdownPowerSupply(ps_id)	None	Power Supply ps_id is Off	Turn off Power Supply ps_id
CreateReplacePowerSupplyWorkOrder(ps_id)	Power Supply ps_id has active alert; Work Order does not exist	Work Order exists	Create/ Replace Work Order for Power Supply ps_id
WaitForNominalOutput(ps_id)	Power Supply ps_id has active alert; Work Order has been created	Voltages of Power Supply ps_id are nominal	Null – pausing BP execution until human can replace power supply
Close_WorkOrder(wo_id)	Voltages of Power Supply ps_id are nominal; Power Supply ps_id has active alert; Work Order wo_id is open	Work Order wo_id is closed	Close Work Order wo_id
Close_Alert(alert_id)	Voltages of Power Supply ps_id are nominal; Active Alert for Power Supply ps_id; Work Order is closed	Alert alert_id is closed	Update status of Alert alert_id in WM to closed
Replace_PowerSupply(ps_id)	Voltages of Power Supply ps_id are not nominal; No Active Alerts for Power Supply ps_id	Voltages of Power Supply ps_id are nominal	Replace_PowerSupply_Process

The details of the Replace_PowerSupply_Process will be defined in Section 6.4.

An agent needs to be aware of its sensors, thus, the information about the sensors available to an agent is itself represented in the world model. It would be desirable to have an ontology of sensors to allow for the ready incorporation of sensors across agents. Sensor attributes to be maintained in the world model include:

- ***Environment*:** The type of environmental data the sensor monitors (e.g., temperature, current, electro-magnetic spectrum).
- ***Data format*:** The format of the data output from the sensor.
- ***Access mechanism*:** Port/socket information (or URI/URL) to access the data output of the sensor.
- ***Sensor accuracy*:** Clearly, sensors can go bad and occasionally deliver inaccurate information. The belief of the accuracy of the sensor must also be represented so that reasoning mechanisms can regularly assess the information provided by the sensor to evaluate its accuracy.

Sensor Drivers

Each sensor has a corresponding 'sensor driver' that is a piece of software code responsible for receiving the data from the sensor, converting it to a form suitable for the world model and adding it to the world model as a new belief along with its corresponding required attributes (timestamp, confidence and source). The level of detail provided by the sensor driver depends on the level of modeling the agent programmer desires. Some sensors may output raw data that is then transformed by analytics in the sensor driver to the beliefs to be added to the world model. On the other hand, some sensors may be 'cognitive' in their own right and may already derive actionable intelligence from the raw data that they observe. In this case, the sensor driver may only need to translate the sensor output into the ontology of the world model and add it (Figure 6.7).

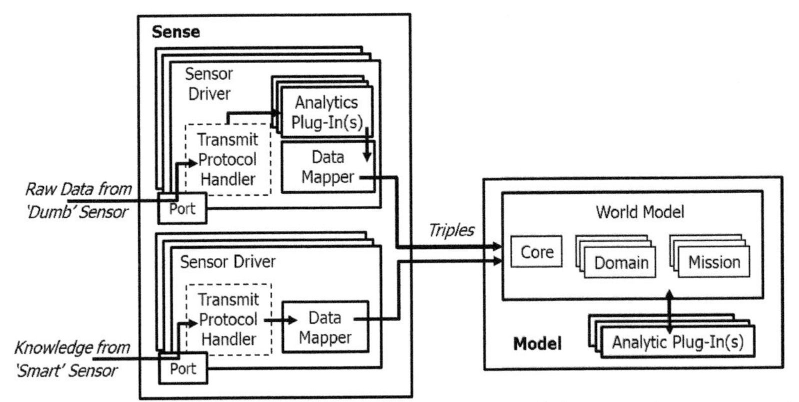

Figure 6.7. The CAE sensor drivers manage the interfaces with the system's sensors, receive data from the sensors and map it into the world model.

Inference

Instead of reasoning about an object to deduce data type or class, state-of-the-art ML systems rely on enormous datasets comprising thousands of examples to learn how to categorize new instances of similar data. However, large datasets containing examples of specialized interest are often lacking or are not available, leading to poor performance and an inability to generate actionable intelligence from novel data. Semantic data representation, combined with hierarchal reasoning methods, provide an opportunity to overcome challenges presented by novel data. Deep learning on either simulations or real world experience can also be used to automatically generate and enhance the ontology, minimizing the brittleness that is often associated with manually generated ontologies.

Computational techniques such as information fusion, anomaly detection, threat classification, predictive analysis and lifelong learning can be applied to the world model to derive further information required for automated decision making that is also asserted in the world model.

Common semantic mapping techniques enable increased situational awareness by automatically combining data from multiple sources into a single knowledge base that provides a full probabilistic picture of the evolving situation. Represented as a relational, probabilistic semantic graph, the knowledge base facilitates integrating multi-modal data by providing a common flexible data model capable of expressing the objects, relationships, activities, events, and uncertainties discovered from examination of any data modality. This is achieved through probabilistic relational models (PRMs), an extension to Bayesian networks composed of hierarchical classes where each class defines attributes of objects, possible relationships, and probabilistic dependencies.

In order for agents to autonomously adapt to changes in their environment, they must have the ability to detect these changes in real-time first so that they can then reason about the optimal course of action. The reusability requirement of the Cognitive Autonomy Engine as well as the dynamic nature of real-world environments mandates a flexible solution where models of normalcy are automatically derived and anomalous behavior is automatically detected. Northrop Grumman's Real-time Anomaly Detection in Heterogeneous Data Streams (RADISH™) system provides this required functionality by 'simultaneously analyzing incoming data streams to learn patterns of normal behavior and, in the context of this learned behavior, searching for anomalous activity that portends abnormal system behavior' [11].

It is generally recommended to localize processing as close to the edge as possible. For example, while processing sensor data could be done in the world model, it is more efficient to do the processing of individual sensor data in the sensor driver and even more efficient in a 'smart' sensor.

'Cognitive' aspects of autonomy can reside in each of the sensor, sensor driver and world model components.

Use Case 1: Autonomous Mine Hunting

The UUV's synthetic aperture sonar and associated processing package serve as a 'smart' sensor to output processed information related to signal detections and classifications. Specifically, this sonar sensor detects, classifies and localizes threatening mine-like objects (MLO) and benign non-mine-like objects (NOMBO) and makes this information available to the payload controller.

In order for the UUV agent to interface with the payload controller and receive the sonar output, a CAE sensor driver is implemented and is registered with the world model. This sensor driver leverages a reusable streaming sensor driver design pattern and requires minimal configuration to interface with the payload controller. The sensor driver's mapping component is configured to translate the sensed MLOs and NOMBOs into triples and insert them into the environment model of the agent's world model.

Other sensors on the UUV may provide vehicle status information such as current location, velocity and energy levels. This knowledge is mapped to the world model along with corresponding timestamps, confidence levels (hopefully high!) and reasons for belief (sensor input).

Use Case 2: Automated Fault Detection and Isolation

Unlike the previous use case where the UUV agent obtains information from a 'smart' sensor, the Power Supply agent collects raw voltage measurements from each of the monitored power supply modules in the assembly. This collection is enabled by a sensor driver that has been developed using the streaming sensor design pattern and configured to map voltage measurements from a voltmeter into the Environment Model of the agent's world model.

There are many power supply assemblies with different voltage outputs located throughout this system, many with different nominal voltage thresholds. It is desirable to develop a single Power Supply agent that can automatically configure itself for the specific assembly it is monitoring. The power supply sensor driver is further extended with the capability to model the voltage output, at agent startup, of the monitored power supply and to calculate upper and lower bounds for nominal behavior. These derived thresholds are then inserted into the world model, preventing the need for hard-coding, unnecessary versioning and the potential for human error.

6.4 Computational Decision-Making and Execution Based on the World Model

Key to any autonomous system is the ability to make decisions independently and then act on those decisions. This section outlines how the world model provides the context enabling computational decision-making and acting. Decision-making is all about deciding what to do, so first we need to address how an agent acts on the environment.

Actuators

All effects on the external environment of an agent must occur through an actuator. Like sensors, actuators can take on many forms. They may affect the physical environment or the virtual environment (cyberspace), including outgoing communication[10] (sending of messages) to other actors. In all cases, however, they serve the purpose of making changes to the environment by the agent. An agent needs to be aware of its actuators, so information about them, including availability and reliability, is stored in the agent's world model.

Actuator Drivers

Similar to sensors, every agent has an actuator driver that is software code mapping the process of an action (i.e., the singleton command as represented in the world model) to the application programming interface of the actuator. In particular, when a unit_task is created, any arguments to the command are instantiated and the command invocation (method call) is scheduled to execute at the corresponding time stamp. Before the command invocation actually happens, the preconditions of the corresponding unit_task are checked. As soon as the command is invoked, the expected effects of the corresponding unit_tasks are updated in the world model with appropriate timestamps (and reason for belief—i.e., unit_task invoked).

Hybrid Actuator-Sensor Drivers

There are some cases where a sensor and actuator may be tightly coupled. For example, a command-line interface in an operating system comprises both an actuator (the command) and a sensor (the return result). It is important that this functionality be clearly distinguished and supported with both the sensor driver and the actuator driver (where the expected effect of the action is the sensor response). Another example where

[10] Outbound communication is a special form of acting that will be treated more thoroughly in Section 6.5.

hybrids occur is with sensors that must be queried in order to receive data (information pull). The corresponding actuator driver represents the query and the response is managed by the corresponding sensor driver. The link between query and response is handled between the two drivers.

Decision Making

As defined in Section 6.1, the fundamental driver of autonomy is the achievement of mission objectives. All actions are performed in pursuit of a goal and the CAE framework supports goal-driven agents. As such, the world model forms the basis for the CAE Model component to represent and create goals and then for the CAE Decide component to decide how to achieve the goals. This happens in a two-step process:

1. Incoming events from sensor drives representing changes in the world model may cause a goal to be added to the world model.
2. A new goal triggers the agent to decide which actions it (or other agents) should perform to achieve the goal.

This process is managed by using a production rule system with a very small number of domain-independent rules using the semantics of the world model to match the rules. For example (simplified), for Step 1, a rule might be:

IF X is broken THEN
Add Goal('X is normal')

What it means exactly to be broken can be defined in the ontology; for example, it could be a system monitored by a sensor, whose value exceeds a certain threshold. As new events come into the world model, inferencing will determine if they match the rule to create a goal.

Similarly, for Step 2, there is a rule that states:

IF G is a goal AND
G does not conflict with other goals THEN
Find task T to achieve G AND
Execute T

Reactive Behavior

Reactive behavior is enabled by executing pre-defined mission plans corresponding to goals that are triggered by changes in the world model. Thus, a capability must already be defined whose preconditions match the current world state and whose expected effects match the goal (desired world state). Then, it is simply a matter of creating a task that implements the capability by mapping the process corresponding to that capability into a plan and executing the task. This can be done very efficiently.

Deliberative Behavior

Deliberative (goal-directed) behavior is enabled by using automated planning and scheduling techniques to generate mission plans and allocate resources based on pre-conditions and the expected effects of actions as represented in the world model.

Often, an autonomous system will not have a capability directly available from which a task can be generated to achieve a desired goal. In that case, it must compose a task to achieve a goal from its existing capabilities (and possibly capabilities of teammates). This task can be created via several means:

- **Deep learning applied to a simulation managed by the world model:** Use a modeling and simulation (ModSim) environment coupled with predictive analysis to generate and evaluate 'what-if' scenarios. The ModSim may be based on an underlying physics model or a history of experiences. In fact, a separate agent may provide the ModSim as a service. In any case, rather than arbitrarily generating what-if scenarios, the system may apply deep reinforcement learning that builds a model of actions. These models may then be 'engrained' in the agent, moving them to a reactive behavior. This situation is similar to when we move to a new location, it takes us a while to figure out the best path to the grocery store, but after a while, it becomes routine.
- **Deep planning applied to explicitly represented semantics of actions as stored in the world model:** Commonly referred to Goal-Driven Autonomy [12,13], the system uses forms such as logic programming to derive the sequence(s) of tasks to achieve the goal world state from the current world state. This method is clearly more complicated, since the preconditions and expected effects must not only be well represented, but the beliefs must also be taken into consideration.

Plan Execution (Acting)

Once a task is identified (either through reactive or deliberative means) to achieve a goal, it is passed to the CAE Act component to be executed. The Act component first verifies that the preconditions of the task still hold in the current world state and then executes the plan corresponding to the task as follows:

If the task is a unit_task, then its corresponding plan is a command invocation that is passed to the actuator driver that triggers the actuator. This will affect the physical or virtual environment of the agent or may be communication to other agents that ultimately affect their world model as described in Section 6.5. In any case, whenever the unit_task plan is executed, the world model is immediately updated with the expected effects of the task, including

- 'task status is initiated' with current timestamp, high confidence, and reason being actuator driver invoked
- 'task status is complete' with timestamp in the future, some degree of confidence, and reason being actuator driver invoked.

If the task is not a unit_task then its corresponding plan is a sequence of further tasks (represented in BPMN) and the Act component uses a BPMN-compliant engine to follow the plan, recursively executing the tasks in the plan until they all turn into unit_tasks.

Use Case 1: Autonomous Mine Hunting

The commands that are relayed from ENIAK to the vehicle controller relate to the capabilities in the agent's world model that can be achieved by the UUV platform, such as Dive, Surface and FindObject.

The use case starts with the UUV loitering at the surface of the water in range of a communications buoy as the UUV agent waits for its goals. The agent receives a message from the command and control (C2) center that asks the UUV agent to notify C2 of a mine in the operating area. The UUV agent accepts this request and a goal is added to the Mission model of the UUV agent's world model. To be specific, the new goal that is added is 'C2 is aware of mine location'.

The addition of the goal into the UUV agent's world model automatically triggers the core rule of attempting to achieve goals. The UUV agent identifies those capabilities (Table 6.1) defined in its world model where the preconditions and expected effects are satisfied. The preconditions pertain to the current world state (the agent is on surface, the agent does not know about any mines, etc.) and the expected effects pertain to the agent's goals (C2 is aware of the mine's location). The only capability that satisfies the preconditions and expected effects is the 'Inform_ObjectLocation' capability. Therefore, the agent decides to initialize a task implementing this capability in order to achieve the agent's goal. The process corresponding to the Inform_ObjectLocation capability is called 'Find_Mine'.

The Find_Mine process is defined as the sequence of the following capabilities (as described in Table 6.1) that happen to be all actions in this case:

Find_Mine Process:

1. Dive (100 m)
2. FindObject (MLO)

3. Surface
4. Inform (C2, MLO location)
5. Loiter

ENIAK's rules engine retrieves the process definition from the agent world model and instantiates a unique instance, or plan, of the Find_Mine process and passes the plan to the Act component.

The Act component recognizes that this is a multi-step plan (represented in BPMN 2.0 notation) and processes it accordingly with a Business Process Manager (BPM) that complies with the BPMN 2.0 specification. The BPM steps through each of the plan's sub-tasks and decision points, checking preconditions and then executing each task by passing the task and associated parameters to the associated actuator driver. The agent updates the world model accordingly as the tasks are performed.

Use Case 2: Automated Fault Detection and Isolation

Unlike the UUV agent that receives its goals from an external entity, the Power Supply agent's mission is pre-configured in its world model. Its underlying mission objective is a persistent goal to ensure that all of the monitored power supply modules are operating nominally, where the characteristics of nominal conditions are contained in its world model and derived at startup. At startup, the agent begins receiving voltage measurements from the modules and inserts them into its world model, where inference rules run over the measurements to determine if the module is not operating nominally, i.e., it is either under-voltage or over-voltage.

In the case of an under- or over-voltage situation, this fact is automatically inserted into the agent's world model and ENIAK's rules engine automatically creates a new goal to return the power supply module to nominal operating conditions. The addition of a goal triggers the core rule of attempting to achieve its goals and the agent's capabilities (Table 6.2) are examined to identify those with the expected effect of returning the power supply to nominal status and the precondition of an over-voltage power supply.

The Replace_PowerSupply_Process is defined as the sequence of the following capabilities (as shown in Table 6.2) that are all actions:

Replace_PowerSupply_Process:

1. Open_Alert(ID)
2. ShutdownPowerSupply(ID)
3. CreateReplacePowerSupplyWorkOrder(ID)
4. WaitForNominalOutput(ID)
5. Close_WorkOrder(wo_id)
6. Close_Alert(alert_id)

The Replace_PowerSupply composite capability is returned and a plan is instantiated from its associated process and inserted into the world model. The rules engine then passes the plan to ENIAK for execution and inserts a triple into the world model alerting that the over-voltage situation has been identified and it is attempting to fix the problem.

ENIAK passes the plan to the Act component, which then passes it to the BPM, as it is a multi-step plan. As with the UUV use case, ENIAK and the BPM execute the plan step-by-step, where the preconditions for each of the plan's tasks are checked immediately prior to execution. Through this basic scenario, the agent is able to automatically detect issues with a monitored component, identify the exact nature of the issue and immediately create a detailed work order, significantly reducing the time to detect and isolate problems with the system.

6.5 Collaboration and Communication

A key requirement for autonomous systems is to collaborate with other machine and human teammates to achieve mission objectives that the system cannot fulfill on their own. An autonomous system may also be required to operate under the guidance or command of a human teammate. This section outlines how the CAE framework enables collaboration, in particular how communication events (the sending and receiving of messages) as well as higher-level collaboration are represented in the world model.

Inter-Agent Communication

Given that communication, either directly via explicit exchange of messages or indirectly via changes in the environment, is with actors external to the agent, all communication must go through sensors for incoming messages and actuators for outgoing messages. In general, a sender of a message, whether initiating or responding to a conversation, has some sort of associated expectation of the behavior of the recipient of the message. For example, agents send messages with a purpose or goal in mind. Additional processing is required to represent that intent and resulting effects of sending the message in the world model. In particular, we draw upon human speech act theory [14] that also forms the basis for the communication and collaboration standards established by the Foundation for Intelligent Physical Agents (FIPA). FIPA specifies a set of 'Communicative Acts' that convey the illocutionary effect (intent) of the message content. Example communicative acts are Inform, Request, Agree (to a request), Refuse (a request), Query, Call for Proposal, Propose, Accept Proposal and Reject Proposal. Some of them are used for initiating a conversation, some are in response to received messages.

Sending Messages

As with all effects on the environment, the agent uses one or more actuators to send messages to other agents and humans. Different actuators may be used for different types of communication protocols, such as message queues, chat, email and voice. The CAE world model represents the semantics according to the communicative acts defined by FIPA with the preconditions (on the sender) for sending the message and the expected effects (of both the sender and the recipient) that arise from sending the message. Clearly, upon sending a message, the expected effects can only be added to the sender's world model. While the act of sending the message has expected effects, there are no guarantees of the desired outcome or that the recipient will actually behave according to the expected effect. For example, the message may be lost in transition, the recipient may not be able to fulfill a request, or the recipient may not trust the sender. All messages contain at least the recipient, illocutionary speech act representing the purpose of the message (such as) and the content of the message. Table 6.3 outlines the corresponding preconditions and expected effects as represented in the sending agent's, $Agent_S$, world model when sending representative messages (inform and request) to the receiving agent, $Agent_R$.

Table 6.3. Semantic representation in the world model of capabilities associated with sending messages.

Capability	**Precondition(s)**	**Expected Effect(s)**
$Agent_S$ sends "Inform Fact" to $Agent_R$	I ($Agent_S$) believe Fact; I can communicate with $Agent_R$; $Agent_R$ does not believe Fact	$Agent_R$ believes Fact
$Agent_S$ sends "Request Service" to $Agent_R$	I have goal 'Service is done'; I can communicate with $Agent_R$; $Agent_R$ can execute Service	$Agent_R$ has goal 'Service is done'; Service is done; I know Service result

Receiving Messages

Similarly, as with all data coming from the environment, an agent uses communication sensors to receive incoming messages. There may be different sensors for different communication protocols. In all cases, the corresponding sensor driver takes the output of the sensor and converts it into a form to update the world model. As described in Section 6.3, the messaging sensor driver represents the fact that the message has been received along with the timestamp that the data was received from the sensor, the confidence level associated with the sensor, as well as receipt by sensor as justification.

Furthermore, upon receipt of a message, the recipient automatically updates its world model with beliefs that correspond to receipt of the

Table 6.4. The beliefs an agent adds to its world model upon receiving messages.

Message	Belief about message receipt	Beliefs about sender	Belief about content
$Agent_R$ receives "Inform Fact" from $Agent_S$	I ($Agent_R$) believe $Agent_S$ sent me "Inform Fact" *Confidence*: C_1	$Agent_S$ has Goal 'I ($Agent_R$) believe Fact' *Confidence*: $C_2 < C_1$; $Agent_S$ believes Fact *Confidence*: $C_3 < C_2$	I believe Fact *Confidence*: $C_4 < C_3$
$Agent_R$ receives "Request Service" from $Agent_S$	I believe $Agent_S$ sent me "Request Service" *Confidence*: C_1	$Agent_S$ has goal 'Service is Done'; $Agent_S$ has goal '$Agent_S$ knows Service result' *Confidence*: $C_2 < C_1$	I have goal 'Service is Done'; I have goal '$Agent_S$ knows Service result' *Confidence*: $C_3 < C_2$

message itself, believed intent of the sender, and actual effect on the recipient. The recipient can thus reason about and perform according to the original intent of the sender. Table 6.4 shows the corresponding beliefs of $Agent_R$ when receiving some representative messages from $Agent_S$. Of course, $Agent_R$ receiving the message in and of itself has no effect on $Agent_S$'s world model.

An altruistic agent will adopt the goals of other agents, but this does not necessarily mean that $Agent_R$ will achieve the goal, or even intends to achieve the goal. In particular, upon further reasoning, the $Agent_R$ may decide that the goal creates a conflict or has a lower priority than existing goals and will send a Refuse message back to $Agent_S$ in accordance with FIPA.

Collaboration

In a system comprising heterogeneous agents from different vendors, not all agents may fully understand, be able to reason about or even follow the intent of individual messages. Thus, FIPA introduces the notion of 'Interaction Protocols' that prescribe sequences of communicative acts among agents to ensure the exchange of the proper messages to achieve a desired result. For example, the FIPA-ContractNet interaction protocol describes a sequence of messages, whereby a sender (manager) issues a 'call for proposals' to recipients (bidders) to achieve a particular task, the recipients provide 'proposals' (or bids), the sender accepts the bids by some of the bidders and the accepted bidders perform the task and inform the sender of the results. FIPA only specifies the content and sequence of exchanged messages—it does not specify what the agents actually do between messages to decide which messages to send (e.g., create bid, evaluate bid, and do the task). The CAE framework enables representation of not only the messages but also the interim tasks to derive the messages. Ultimately, these are simply plans with different roles that can be represented in BPMN with the corresponding internal execution as well as messaging tasks.

The above only addresses the cases of benevolent and altruistic agents. In the real world, a number of other factors must be taken into consideration, such as:

Authentication: Potential spoofing of a communication channel, so that senders and recipients are not actually who they say they are. While human authentication is done via two- or three-factors (something they know (password), something they have (CAC card), something they are (fingerprint, iris)), in the case of a machine system, the machine is what it knows, has, and is; therefore, standard authentication mechanisms are not applicable. For example, blockchain technology can handle the authentication of machine-to-machine communication and the application of blockchain to agent interaction is the subject of ongoing research.

Trust: Some agents (whether human or machine) may not be trustworthy, especially if they are an insider threat or have other motives. Just like for humans, a machine's trustworthiness is substantiated through experience or verified by reaching out to a trusted third party; therefore, mechanisms for establishing trust among agents must be established.

Use Case 1: Autonomous Mine Hunting

The UUV agent's communications with other agents and humans is achieved per the CAE core communication capabilities. It receives a message from Command and Control (C2) to notify C2 of the location of a mine in the area. This message is specified according to the FIPA standards so that it is interpretable by the UUV agent. To be specific, C2 and the UUV agent use the QUERY interaction protocol to collaborate. C2 sends the UUV agent a QUERY_REF message asking it the location of a mine. The UUV agent, upon accepting the request, responds with an AGREE message. Once the UUV agent has identified a mine, it sends C2 an INFORM-RESULT message containing the mine's location.

In order to further the discussion of collaboration, assume that there is a C2 agent implemented using the CAE framework that interfaces with the UUV agent. Before the UUV agent's mission begins, the C2 agent obtains a goal from the Commander to 'Clear the Area of Mines'. This process of accepting a goal and identifying a plan to achieve the goal follows the same basic process as described for the UUV agent. A precondition for clearing the area of mines is to know the location of the mines, so the C2 agent automatically requests the UUV agent to notify it of the location of a mine. The receipt of this request initiates the processing described in Section 6.4, where the UUV agent's goal is now only a subset of a larger mission. Upon receipt of the message with the location of a mine sent from the UUV agent, the C2 agent requests that a Mine Clearing agent neutralize the mine located by the UUV agent. Through collaboration schemes such as these, agents and humans can work together to achieve mission objectives in an organized, yet self-governing, manner.

Use Case 2: Automated Fault Detection and Isolation

The Power Supply agent is only one type of autonomous agent envisioned in a multi-agent system that supports the operations and maintenance of a mission-critical information system. Other agents distributed throughout the enterprise, including but not limited to monitoring agents, analytics agents, human assistant agents, resource management agents and database agents, form the multi-agent network that collaborate with each other and with human teammates in support of the mission, using standards-based collaboration and business process execution techniques such FIPA and BPMN. Collaboration in dynamic and distributed environments reduces the workforce demand, the load placed on humans and program costs while increasing system performance and real-time situational awareness for decision makers.

6.6 Summary and Future Work

In conclusion, we have outlined and demonstrated an approach to enabling autonomous behavior by creating and maintaining a world model about which the system can reason for making decisions to achieve commander's intent and mission objectives, both individually and in human/machine teams. Other than specifying the software underlying the sensor and actuator drivers, no domain-specific software or language is needed to specify agent behavior. All aspects of agent behavior, including collaboration, are specified through data, namely the ontology, processes and a small set of abstract rules. Different machine learning and automated planning and scheduling algorithms can be added to the system to enable more cognitive, or sophisticated, behaviors. The capabilities are demonstrated in two different domains, addressing 'autonomy at rest' as well as 'autonomy in motion'.

Future Research

Sharing world models Given that different agents will have different experiences and different levels of accuracy in their sensors, their world models will necessarily diverge and, worse, be inconsistent. In many cases, that is not an issue. However, there will be times when agents work together on a common task, making consistency across world models desirable. Research is required as to how to identify inconsistency and how to resolve it when necessary. For example, the world model allows for representing the following beliefs within an agent: I (self) believe:

(Fact1) with confidence 80%
(AgentX believes Fact2 with confidence 90%) with confidence 100%
(AgentX is correct) with confidence 90%
(Fact1 and Fact2 are inconsistent) with confidence 100%

A mechanism is required to reconcile these beliefs. This may require third party adjudication, further information exchanges, or simply learning from experience. This is particularly complicated in the face of the supposedly benevolent teammates who are in fact adversaries (the 'insider threat').

Enhancing cognition A common domain-independent application programming interface (API) is required to incorporate other inferencing and decision-making algorithms operating on the semantic knowledge base that constitutes the world model.

Scalability The world model can easily get extremely large with billions or trillions of beliefs. While current GPUs can process large graphs, they may not be sufficient to handle this scale. The goal of the DARPA HIVE program is to increase scalability of graph analytics by 1,000 [15]. Further mechanisms need to be introduced to trim the model without losing fidelity. This is equivalent to the hypothalamus in the mammalian brain that distinguishes between short-term and long-term memory.

Collaboration without communication A key requirement in autonomous teams is the ability to collaborate where communication is not possible (adversarial jamming) or undesirable (keep silent to avoid detection). While pre-defined plans help significantly, real life does not always go as planned. The ability to store and reason about a model of other agents (teammates) enables an agent to resolve what it believes the other agents may do and can plan accordingly.

Acknowledgement

The work described was performed under the RAPID (2016) and COUGAR (2017) Independent Research and Development projects at Northrop Grumman. The authors are profoundly grateful to Alexander Strickland, Brent Dombrowski, David Hamilton, Dorota Woodbury, Jason Kemper, Lisa Chandler, Ivan Caceres, Marilyn Jarriel and Todd Gillette for their valuable contributions in researching and implementing the approach outlined in this chapter.

This paper is dedicated to the memory of Donald Steiner, who passed away during its revision. Donald was a visionary in the AI field, leading research & development in applied cognitive systems technologies and developing one of the first cognitive cybersecurity products that transitioned fundamental research to the marketplace. He drove innovation and built teams of university researchers and small businesses. He has over 60 publications and over 10 inventions and patents in advanced analytics technologies. Donald was a truly caring and brilliant mentor and friend; he will be greatly missed.

References

[1] United States Army Training Doctrine and Command. 2017. Multi-Domain Battle: Evolution of Combined Arms for the 21st Century, Version 1.0.

[2] Defense Science Board. 2016. Summer Study on Autonomy.

[3] Kurzweil, Ray. 2005. The Singularity is Near. Viking Press.

[4] Goertzel, B. and Pennachin, C. (eds.). 2007. Artificial General Intelligence. Springer.

[5] Market Guide for Robotic Process Automation Software. https://www.gartner.com/doc/3835771/market-guide-robotic-process-automation.

[6] BPMN Specification—Business Process Model and Notation. http://www.bpmn.org/.

[7] Foundation for Intelligent Physical agents. http://www.fipa.org/.

[8] Hayes, M.P. and Gough, P.T. 2009. Synthetic aperture sonar: A review of current status. In IEEE Journal of Oceanic Engineering 34(3): 207–224.

[9] Berners-Lee, Tim et al. 2001. The semantic web. In Scientific American 284(5): 34–43. JSTOR, JSTOR, www.jstor.org/stable/26059207.

[10] OWL—Semantic Web Standards. https://www.w3.org/OWL/.

[11] Böse, B., Avasarala, B., Tirthapura, S., Chung, Y.Y. and Steiner, D. 2017. Detecting insider threats using radish: a system for real-time anomaly detection in heterogeneous data streams. In IEEE Systems Journal 11(2): 471–482.

[12] Molineaux, M., Klenk, M. and Aha, D.W. 2010. Goal-driven autonomy in a navy strategy simulation. AAAI Press.

[13] Héctor Muñoz-Avila, Ulit Jaidee, David W. Aha and Elizabeth Carter. 2010. Goal-Driven autonomy with case-based reasoning. In Proceedings of the 18th International Conference on Case-Based Reasoning Research and Development (ICCBR'10), Isabelle Bichindaritz and Stefania Montani (eds.). Springer-Verlag, Berlin, Heidelberg, 228–241.

[14] Searle, J.R. 1969. Speech Acts: An Essay in the Philosophy of Language. Cambridge University Press.

[15] https://www.darpa.mil/program/hierarchical-identify-verify-exploit.

[16] Bordini, Rafael H., Jomi Fred Hübner and Michael J. Wooldridge. 2007. Programming Multi-agent Systems in agentSpeak Using Jason. Wiley Series in agent Technology. Chichester: Wiley.

[17] Jason—a Java-Based Interpreter for an Extended Version of agentSpeak. http://jason.sourceforge.net/wp/.

[18] McCabe, Frank G. and Keith L. Clark. 1994. April—agent process interaction language. In International Workshop on agent Theories, Architectures, and Languages, 324–340. Springer.

[19] Carey, Scott, Martin Kleiner, Michael R. Hieb and Richard Brown. 2002. Standardizing battle management language–Facilitating coalition interoperability. In Proceedings of the European Simulation Interoperability Workshop.

[20] Breton, R. and Rousseau, R. 2005. The C-OODA: A cognitive version of the OODA loop to eepresent C2 activities. In Proceedings of the 10th International Command and Control Research and Technology Symposium, Washington DC, USA.

[21] FIPA Agent Communication Language Specifications. http://www.fipa.org/repository/aclspecs.html.
[22] Poslad, Stefan. 2007. Specifying protocols for multi-agent systems interaction. In ACM Transactions on Autonomous and Adaptive Systems 2, no. 4: 15–es.
[23] Bellifemine, Fabio Luigi, Giovanni Caire and Dominic Greenwood. 2007. Developing Multi-agent Systems with JADE. Hoboken, NJ: John Wiley & Sons, Ltd.
[24] Jade - Java agent DEvelopment Framework. http://jade.tilab.com/.
[25] Steiner, D., Burt, A., Kolb, M. and Lerin, C. 1995. The conceptual framework of MAI2L. *In*: From Reaction to Cognition. MAAMAW 1993. Lecture Notes in Computer Science (Lecture Notes in Artificial Intelligence), vol. 957. Springer, Berlin, Heidelberg.
[26] Wooldridge, Michael. 2009. An Introduction to Multiagent Systems. 2nd edition. Chichester, U.K: Wiley.
[27] OpenCog Foundation. https://opencog.org/.
[28] Laird, John E. 2012. The Soar Cognitive Architecture. MIT Press.
[29] Shanahan, M. 1999. The event calculus explained. *In*: Wooldridge, M. and Veloso, M. (eds.). Artificial Intelligence Today, pp. 409–430. Springer Lecture Notes in Artificial Intelligence no. 1600.
[30] Shoham, Yoav. 1987. Temporal logics in AI: Semantical and ontological considerations. Artificial Intelligence 33, no. 1: 89–104. https://doi.org/10.1016/0004-3702(87)90052-X.

Probabilistic Scene Parsing

Michael Walton,[1,*] *Doug Lange*[1] and *Song-Chun Zhu*[2]

7.1 Introduction

Command and Control (C2) systems must effectively represent information to a mission commander, leading to rapid and correct decisions. Key goals of such systems include providing an interface to force {projection, readiness, employment}, intelligence, and situational awareness. Often, the approach to these interfaces is to present information in a manner that is natural for humans to interpret, that is, visual displays of information (Bemis et al., 1988; John et al., 2004; Smallman et al., 2001). Such displays apply across a variety of domains, from mission planning to tactical command and control, because not only are they intuitive for a human to understand but also such displays easily represent spatial and temporal relationships among objects (Smallman et al., 2001). The challenge moving forward is the complexity of the modern battle space increases the difficulty of creating effective windows to the required information for a commander (John et al., 2004). Because the information density is increasing, important information and relationships hide from human view making integrating information for planning difficult. The question is, can we augment or replace human capability with computer systems that discover the relevant relationships and information a commander requires?

[1] Space and Naval Warfare Systems Center-Pacific, San Diego, CA.
[2] University of California, Los Angeles, CA.
* Corresponding author: mwalton@spawar.navy.mil

Scene understanding is a sub-field of computer vision concerned with the extraction of entities, actions and events from imagery and video. In many ways, scene understanding may be seen as a holistic or gestalt approach to computer vision which incorporates multiple vision tasks into a common pipeline. Traditional computer vision challenges include low-level feature extraction,[1] image segmentation[2] and object classification.[3] Additionally, scene understanding is concerned with the semantic relations between visual objects. At any level of this hierarchy of imagery analysis tasks, context may play a substantial role.

A particularly well-studied approach to the problem of natural scene understanding using probabilistic graphical models (PGMs) is scene parsing (Yao et al., 2010). This area of research strives to encode the semantic information of an image as a parse tree, or more generally a parse graph. Associated algorithms for manipulating and traversing these structures enable sophisticated capabilities such as causal inference (Fire and Zhu, 2015; Pei et al., 2011) natural language text generation (NTG) and answering queries about the content of an image or video (Tu et al., 2013). Working from an analogy to natural language processing in which a parse tree is a structured interpretation of a piece of text (Yao et al., 2010), image parsing computes a parse graph which represents the most likely interpretation of an image. Extending this analogy into the naval tactical domain, tactical scene parsing computes the most probable interpretation of the available Intelligence, Surveillance and Reconnaissance (ISR) information. Notionally, a tactical parse tree is a structured decomposition of the totality of a ship or battlegroup's ISR datasources[4] such that all input feeds are explained. A "tactical parse graph" is subsequently augmented with lateral edges allowed at all levels of the hierarchy to specify spatial and functional relations between nodes.

Fundamental to graphical models is the decomposition of joint distributions into factors. These factors are obtained through the identification of conditional independence relationships between variables.

---

[1] The representation of raw, high dimensional input images in a feature space via either learned or engineered features. Common engineered methods include Scale Invariant Feature Transform (SIFT) descriptors, various color and gradient histograms. Feature learning has been broadly explored in diverse research domains and a fair discussion is beyond the scope of this chapter.

[2] For example, partitioning the set of pixels which comprise an image of a bicyclist into two unlabeled groups.

[3] In which a segmented object in a scene is mapped to some high-level semantic concept. For example, consider the labeling of pixels which compose a bicycle and cyclist with distinct classifications.

[4] Generally, the reader may conceive of these data as the extracted entities and relations that may be rendered in some form of C2 display which directly interfaces with an operator or analyst.

Determining independence between random variables in a graphical model is essential as it makes inference tractable for larger models of more complex distributions. Scene parsing provides data-driven mechanisms for determining statistical dependencies of factors and provides well-defined formalisms for describing a probabilistic interpretation of context in vision and other domains.

7.2 Scene Parsing

For completeness, we briefly review relevant concepts in PGMs and approximate inference with Markov Chain Monte Carlo (MCMC) before moving on to And-Or Graphs and Data-Driven MCMC as special cases of these methods.[5] An authoritative treatment and detailed background on Graphical Models and MCMC may be found in (Koller and Friedman, 2009) and (Andrieu et al., 2003) respectively. A more detailed survey of work specific to scene understanding can be found in (Tu et al., 2014).

Probabilistic inference fundamentally depends on obtaining some posterior distribution of interest over some collection of random variables. In the simple case of discrete random variables, exact solutions may be obtained by constructing an appropriate conditional probability table (CPT). Here inference simply amounts to consulting the appropriate entries of the CPT, however, even in the simplest case of binary random variables, the table grows 2^n for n random variables. The complexity of the resulting intractable inference problem can be significantly reduced using PGMs to compactly express conditional independence relationships between random variables. Broadly, PGMs may be directed, acyclic (often referred to as Bayesian Networks or Belief Networks) or undirected (commonly known as a Markov Random Field aka MRF) (Heckerman, 2008). A crucial and useful property of PGMs is that their joint distributions reduce to a product of factors (the Markov blanket of the query variable in Bayesian networks, or cliques in the case of a MRF).

Exact inference in PGMs still requires enumeration over the appropriate factors and can still be inefficient for large or densely connected graphical models. However, sampling based approximate inference can be efficiently accomplished using a variety of Markov Chain Monte Carlo (MCMC) algorithms in which the stationary distribution of the Markov chain is the desired posterior distribution. In particular, we will discuss a data-driven variant of Metropolis-Hastings (Sidhartha and Greenberg, 1995) and its application to approximate inference in a MRF for a particular probability model known in the scene parsing literature as an And-Or Graph.

[5] Those familiar with these methods may skip the remainder of this section and move on to the discussion specific to And-Or Graphs in Section 7.2.1.

7.2.1 And-Or Graphs

In statistical natural language processing (NLP), a stochastic grammar is a linguistic framework with a probabilistic notion of grammatical wellformedness and validity (Manning and Schtze, 1999). In general, a grammar is a collection of structural rules which may generate a very large set of possible configurations from a relatively small vocabulary. The And-Or Graph (AOG) is a compact yet expressive datastructure for implementing stochastic grammars (See Figure 7.1). It is observed in (Zhu and Mumford, 2006) that a stochastic grammar in the form of an AOG is particularly well suited to scene parsing tasks. An AOG is constructed by nodes where each OR node has child nodes corresponding to alternative sub-configurations and the children of an AND node correspond to a decomposition into constituent components. Intuitively, this recursive definition allows one to merge AOGs representing a multitude of entities and objects into larger and increasingly complex scene graphs. Theoretically, all possible scene configurations could be represented by composition of all observed parse graphs in a dataset. Therefore, the AOG is a compact formalization of the

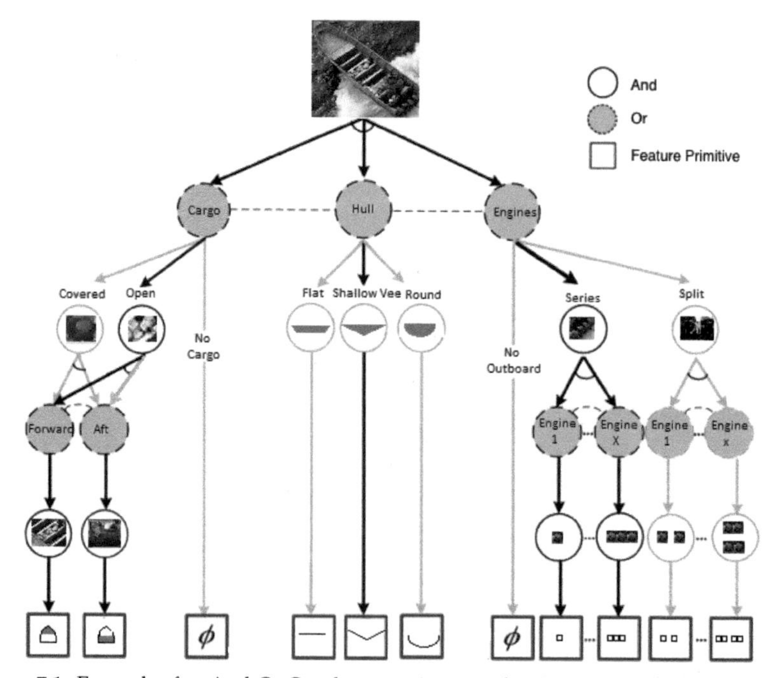

Figure 7.1. Example of an And-Or Graph expressing a stochastic grammar for boats (Walton et al., 2017). The AND nodes in the graph specify the feature dependencies to satisfy a particular sub-grammar; for example, a cargo ship must have cargo in the forward and aft sections of the boat. The OR nodes indicate alternate sub-configurations for a particular attribute, for example a boat may have a round hull or a shallow vee hull.

set of all valid parse graph configurations that may be produced by the corresponding grammar.

Example of an And-Or Graph expressing a stochastic grammar for boats. The AND nodes in the graph specify the feature dependencies to satisfy a particular sub-grammar; for example a cargo ship must have cargo in the forward and aft sections of the boat. The OR nodes indicate alternate sub-configurations for a particular attribute, for example a boat may have a round hull or a shallow vee hull.

The lateral edges in an AOG correspond to relations which allow the graph to encode contextual information between entities at all levels of the hierarchy And-Or tree subgraph. These edges form subject predicate object (SPO) triples. The relations may be distance based, geometric or semantic. Distance based relations may be of the form *A* near *B*; similarly, geometric relations may span large distances but encode complex relationships regarding the arrangement of entities in a scene (for instance *C* collinear with *D*, *E* concentric to *F*). Semantic or functional relations encode abstract information about entities in the scene. Examples from the tactical imagery domain include "boat carrying covered cargo" or "person holding firearms"; in the tactical domain these could represent essential details that could be easily lost in the face of overwhelming information density such as "ship traffics drugs" or "agent is hostile". The development of efficient means of complex inference, anomaly detection and operational summarization on these data for C2 is an open applied research question. In (Walton et al., 2017) tactical scene parsing is proposed as a promising means of addressing these challenging problem areas.

7.2.2 Parse Graphs

In the image parsing framework originally proposed in (Zhu and Mumford, 2006), AOG stochastic grammars may be learned from data represented as parse graphs. A parse graph is a labeled directed graph where each node corresponds to an entity with some semantic attribute indicating its type. In this approach, energy terms are assigned to each node of the AOG which may be drawn from expert knowledge or learned from data. Intuitively, the energy of each OR node is inversely proportional to the likelihood of each alternative configuration; similarly, the energy associated with compositional AND nodes with respect to their children captures the uncertainty associated with these relations. In addition to discussing AOGs as a compact representation of the space of valid configurations, the graphical model specifies a prior probability distribution over parse graphs. Parse graphs may then be inferred from data by maximizing the posterior probability which is proportional to the prior defined by the AOG and the uncertainty of a generated candidate scene parse (Tu et al., 2005).

To ground our discussion, up until this point we have only been explicit regarding spatial scene parsing and the corresponding Spatial AOGs applied to single images and have deliberately only discussed generalizations to this model in notional terms. In more recent literature, extensions to this framework have been proposed that incorporate multiple datasources as well as forming augmentations to the model for representing temporal and causal relations. The Temporal And-Or Graph (T-AOG) represents a stochastic event grammar in which the terminal nodes are atomic actions and the vocabulary consists of a set of grounded spatial relations such as positions of agents and interactions with the environment. The associated algorithms have been used to successfully detect actions and fluent[6] transitions (Pei et al., 2011) as well as infer causal relationships as suggested in (Fire and Zhu, 2015).

7.2.3 Causal Inference & Intent Prediction

Learning perceptual causality is a crucial capability of humans which enables us to infer hidden information from sensory inputs as well as make predictions of future events in the environment. This is further compounded in the tactical domain in which a commander must make correct, efficient C2 decisions based on the estimated tactical scene. As a motivating example, consider the case of an agent looking in a box but unable to see what is in the box. An agent may not be able to directly observe what another person has grabbed from the box (due to occlusion or low resolution) however the system may observe the agent with a drink in hand afterwards. Independently, a spatial parse of this scene may draw conclusions regarding 'person' 'box' 'drink'; similarly, a temporal parse may detect the action 'opening box'. However, from the causal joint of these perspectives, far richer inferences may be made (e.g., 'person has drink' 'person is thirsty'). Parses of this form may also be extended to inference of the entirely unobservable *intent* governing an agents' actions (Pei et al., 2011).

Spatial AOGs (S-AOG) (Zhao and Zhu, 2011) and temporal AOGs (T-AOG) model the spatial decomposition of scenes and the temporal decomposition of events into atomic actions respectively. Similarly, a causal AOG (Fire and Zhu, 2015) (C-AOG) models the causal decomposition of events and fluent changes. Correspondingly, a STC-AOG jointly models all three perspectives in an interconnected graph. In STC-AOGs, a taxonomy is formed with a universal node type as the root and all considered objects, events and fluents defined by their respective ontologies as subtrees. This shared structure is crucial for computing semantic similarity between concepts in the case of joint inference.

[6] A logic and artificial intelligence term indicating some condition or attribute of agents or environment which can change over time.

7.2.4 Joint Parse Graphs & Inference

In joint parsing tasks across parse graphs generated from multiple data sources, three types of challenges arise and criteria must be derived for resolving: coreference, deduction and rejection (Tu et al., 2014). Coreference refers to the procedure by which multiple references to a singular entity are associated across multiple parse graphs. In the case of an identified smuggler ship and a separate identification of a recreational vessel, both of these should be associated with references to the higher-level classification ships. Related semantics for singular entities are detected and resolved by their ontological similarity. For example, the Ship type is a parent of Smuggler ship and recreational vessel and entities of these types should possess strong semantic similarity; weighted similarity measures may be defined for each edge in the ontology. For real world scenarios, in which multiple text, video, and picture inputs must be incorporated, the treatment of co-reference is done post-parsing of the inputs and involves finding the nearest common parent nodes.

As we have discussed, events and actions may be inferred from a scene; further, we would like our models to be capable of predicting fluent changes and future agent actions based on our observations. Scene parsing methods make the open world assumption (Russel and Norvig, 1994); they are not constrained to inference based solely on the current state of the environment, rather these models enable complex deduction by incorporating a probabilistic notion of actions and outcomes. Concretely, deduction is accomplished by inserting candidate subgraphs into the joint parse graph created by the STC-AOG. We only consider inserting subgraphs that increase the prior probability of the joint parse graph's occurrence. Inserting subgraphs also increases the energy of the joint parse graph, by applying an energy threshold that can be added from a given deduction constrains the amount of deduction that can be performed. At times, several possible deduced parse graphs will fall within the energy threshold; in this case, the total number of deductions may be constrained by limiting the total entropy change of the parse graph deduced given an initial parse graph energy. Equation 7.1 forms the basis for constraining the iterative deduction process.

$$H(pg_{de}|pg_{jnt}) = -\Sigma_{i=1}^{N} P(pg_{de}^{i}|pg_{jnt}) log P(pg_{de}^{i}|pg_{jnt}) > \frac{logN}{c} \quad (7.1)$$

Here, the entropy of a particular deduction pg_{de} given a joint parse graph pg_{jnt} represents the parse graph before the insertion of the deduced subgraph is bounded by the number of candidate subgraphs N and a hyperparameter c. Scene parsing algorithms will continue inserting low energy deductions until this threshold is satisfied.

Information received by an operator may also be conflicting, for instance when initially classified smugglers' boats adjust course and speed toward a strike group. Revision is performed to resolve conflicts in the STC-AOG. In this example, multiple reports may categorize the same object as neutral in some parse graphs and foe in others. This violation of their tactical AOG renders this an impossible occurrence necessitating revision of one or more sub-parse graphs. Changes to each parse graph will increase its associated energy, therefore, scene parsing methods minimize the total number of revisions by setting a threshold on total parse entropy.

7.2.5 Probabilistic Modeling

The prior probability of a parse graph is inversely proportional to the energy present in that parse graph. In the following discussion, we will freely change between probability and energy to simplify the mathematical expressions; in Eq. 7.2 the probability of a parse graph and its energy can be easily interchanged using the relation:

$$P(pg) = \frac{1}{z} e^{-E_{STC}(pg)} \tag{7.2}$$

where Z is the normalization factor, and $E_{STC}(pg)$ is the energy of that parse graph in the STC-AOG. To calculate the energy of a parse graph for an STC-AOG we sum up the energy of each individual graph and the energy incurred by the number of joint interactions.

$$E_{STC}(pg) = E_S(pg) + E_T(pg) + E_C(pg) + \textstyle\sum_{r \in R^*(pg)} E_R(r) \tag{7.3}$$

Here the terms $E_S(pg)$, $E_T(pg)$ and $E_C(pg)$ are the energy terms defined by the spatial, temporal and causal AOGs, respectively. $R*$ is the set of relations across the spatial, temporal and causal domains. Each AOG has a parse graph energy defined by the sum of the energy associated with the configuration selected at the OR node $E_{or}(v)$ and the energy associated with a relation between AND nodes $E_R(r)$.

$$E(pg) = \textstyle\sum_{v \in V^{or}(pg)} E_{or}(v) + \textstyle\sum_{r \in R^*(pg)} E_R(r) \tag{7.4}$$

From this general definition of total energy for a parse, we may specialize these models by defining the energy of their constituent nodes uniquely for the spatial (Zhu and Mumford, 2006; Tu et al., 2005), temporal (Pei et al., 2011), and causal domains (Fire and Zhu, 2013).

7.2.6 Spatial Parsing

Visual spatial parsing relies on computing 3D scene layout, detecting 3D objects, detecting 2D faces, and segmenting background. For a spatial AOG, we consider the stochastic scene grammar to be defined by a 4-tuple $G = (S, V, R, P)$. S represents the start symbol or the scene node. V is the finite set of nodes that make up the scene; these nodes may be further classified as $V = V^N \cup V^T$, where V^N represents the non-terminal nodes and V^T represents the terminal nodes. R represents the production rules which allow us to traverse from a parent node α to its child nodes β, $R = \{r: \alpha \rightarrow \beta\}$. The final node P represents the expansion probability for a rule r, $P(r) = P(\beta \mid \alpha)$; this gives the likelihood of observing child β from parent α.

Given this definition of a stochasitc grammar, we may now consider the total energy of a particular parse tree pt given an image I. A single parse tree consists of OR node connections representing "type" or semantic attributes $A_T(Ch_v)$, as well as AND node connections representing "geometry" attributes $A_G(Ch_v)$; where Ch_v denotes the child nodes of a sub-tree rooted at v. Finally, the end of the parse tree will define some image features $I(\Lambda_v)$ in some image area Λ_v. With these terms, we can then define the parse tree energy for a given scene as the summation of the AND, OR, and terminal energies, E^T

$$E(pt \mid I) = \sum_{v \in V^{OR}} E^{OR}\left(A_T(Ch_v)\right) + \sum_{v \in V^{AND}} E^{AND}\left(A_G(Ch_v)\right) + \sum_{\Lambda_v \in \Lambda_I, v \in V^T} E^T\left(I(\Lambda_v)\right) \tag{7.5}$$

In the equation above we see that the energy of all three portions are calculated seperately. The energy of the OR nodes gives the likelihood that the OR node v belongs with the corresponding type attribute A_T. The probability that $v \rightarrow A_T(v)$ is found by taking the ratio of the number of ways that v can belong to $A_T(v)$ with all the ways that v can belong to alternative attributes $Ch(v)$.

$$E^{OR}\left(A_T(v)\right) = -logP\left(v \rightarrow A_T(v)\right) = -log\left\{\frac{\#\left(v \rightarrow A_T(v)\right)}{\sum_{u \in Ch(v)} \#\left(v \rightarrow u\right)}\right\} \tag{7.6}$$

In the scene parsing literature, the energy of AND node attributes may be learned from data using a particular type of constrained MCMC called Data-Driven MCMC over an AOG as a special case of a Markov Random Field (MRF). To accomplish this, we label AND nodes of the graph as cooperative, "+", as well as OR nodes labeled as competitive, "–". A + relation means that they correspond to a concurrent pattern in the scene. A – relation encodes exclusivity of patterns in a scene. To maximize the posterior of the induced MRF, we define a set of cooperative parameters, λ^+, and competitive parameters, λ^-. These parameters may be associated with sufficient statistics denoted as cooperative h^+, or competitive h^-. The energy of the AND nodes is then defined as:

$$E^{AND}\left(A_G(v)\right) = \lambda^+ h^+\left(A_G(Ch_v)\right) + \lambda^- h^-\left(A_G(Ch_v)\right) \tag{7.7}$$

An example of a sufficient statistic on a + relation for 2D faces would be the quadratic distance between connected joints: $h^+(A_G(Ch_v)) = \sum_{a,b \in Ch_v} (X(a)$

$-X(b))^2$, whereas the h^- relation would specify the overlap rate between the occluding image area $h^-(A_G(Ch_v)) = (\Lambda_a \cap \Lambda_b)/(\Lambda_a \cup \Lambda_b), a, b \in Ch_v$.

The terminal energy E^T is found using bottom-up image features $I(\Lambda_v)$ on the image area Λ_v. A variety of methods can be used to extract the terminal node energy, including 3D orientation map (Lee et al., 2009) and surface labels of geometric context (Hoiem et al., 2007).

These energy calculations need to account for every combination for configurable components on the parse tree; this creates the NP-hard problem of enumerating all parse trees in large spaces. In order to address this problem, the top-down and bottom-up approaches can be combined using hierarchical cluster sampling (Zhao and Zhu, 2011).

Hierarchical cluster sampling is a two-part process. First, we use production rules and contextual relations to find all the possible higher-level structures, called "clusters". These clusters consist of cooperative relations +. When several + relations are found they tend to bind together, these tight bonds create a regular structure that may be used to define the cluster. Clusters (CL) are formed if cooperative constraints are satisfied. The probability for each cluster is then defined as

$$P_+(Cl \mid I) = \prod_{v \in Cl^{OR}} P^{OR}(A_T(v)) \prod_{u,v \in Cl^{AND}} P_+^{AND}(A_G(u), A_G(v)) \prod_{v \in Cl^T} P^T(I(\Lambda_v)). \quad (7.8)$$

Only clusters with a likelihood above a threshold are saved; this threshold can be learned through a probably approximately admissible (PAA) bound (Zhu et al., 1997). Here the notation has transitioned from energy to probability; this equivalence relation is defined similarly in Eq. 2.

With a finite set of clusters defined, scene parsing leverages an extension of MCMC to search the combinational space of possible scene configurations. This sampling stage relies on Data Driven Markov Chain Monte Carlo (DDMCMC), which speeds the convergence of the metropolis hastings algorithm by initializing the proposal probability to be dependent on estimated cooperative and competitive relations.

The Metropolis-Hastings algorithm is a popular method for estimating unknown probability distributions using Markov chain Monte Carlo (MCMC). The goal of MCMC is to allow sampling from a distribution when the actual underlying distributions are unknown. In the case of Metropolis-Hastings, this is accomplished by sampling from some related distribution; In MCMC, we assume a known function, $f(x)$, that is at least proportional to the underlying target distribution. In the case of Metropolis-Hastings, we then perform a random walk by choosing some starting point x_0, then apply some new conditional distribution, $g(x \mid y)$. For each iteration, we start with our current value x_t then sample a known distribution, $g(x' \mid x_t)$ to create a new sample, x'. Once we have the new sample, we then calculate a threshold value, a, which determines acceptance criteria for each new sample.

$$a = \frac{f(x')}{f(x_t)} \frac{g(x_t \mid x')}{g(x' \mid x_t)} \quad (7.9)$$

The value of a defines the probability of selecting the new sample, x', versus rejecting it and keeping the old sample, x_t, by:

$$x_{t+1} = \begin{cases} x' & if\, a \geq 1 \\ x' & if\, a < 1 \text{ with probability } a \\ x_t & if\, a < 1 \text{ with probability } 1 - a \end{cases} \quad (7.10)$$

However, MCMC can still be slow to converge; the rate of convergence is largely dependent on the agreement between the assumed distribution, $g(x' \mid x_t)$, and the underlying true distribution. DDMCMC speeds up the Metropolis-Hasting by selecting a distribution representative of the underlying objects. To perform DDMCMC, we condition the initial distribution on the data $g(x_t \mid x') \rightarrow g(x_t \mid x', I)$ by taking advantage of the type of data present in the system; then spatial, causal, and temporal recognition can be completed much faster in the Metropolis-Hasting framework.

For instance, in the case of spatial parsing, we will define a proposal probability based on the relations discovered while creating clusters.

$$g(pt^* \mid pt, I) = P_+(Cl^* \mid I) \prod_{u \in Cl^{AND}, v \in pt^{AND}} P_-^{AND}(A_G(u) \mid A_G(v)) \quad (7.11)$$

This proposal density gives more support to proposals with tight relations via P_+, while punishing any proposals that would include competitive relations via P_-^{AND}. All of these probabilities can be easily calculated during the clustering stage, and then stored in look-up tables. This method satisfies the detailed balance principle[7] to ensure the convergence of MCMC.

7.2.7 Temporal Parsing

To perform temporal parsing, first the video is spatially parsed frame by frame (Pei et al., 2011). The segmentation of events is then integrated into the parsing process and each interpretation of the video is segmented into single events, I_Λ. The terminal nodes of a temporal AND/OR graph are composed of atomic actions, a. An atomic action is determined by a series of relations occurring, $a = \{r_1, \ldots, r_j\}$. Two types of relations exist in temporal parsing: the unary relation $r(A)$, or fluent, represents a relation an object/agent, A, has with itself over time. A fluent has a probability of occurring

[7] Detailed balance in this context refers to a criterion imposed by the definition of reversible Markov Chains. Briefly, the assumed Markov process must possess a stationary distribution (implying certain symmetric relations in the transition matrix) and that any closed cycle of states have equivalent probabilities in both directions.

which can be computed via foreground properties such as aspect ratio and intensity histogram of a bounding box. The spatial relationships between agents are denoted with the binary operator, $r(A, B)$, where A and B are separate objects or agents. These binary relations are computed by the relative positions of A and B, where the exact parameters of the distribution between positions of binary relations can be learned from training data. We define the probability of an atomic action as the product of the probabilities of its independent relations in frame I_t.

$$p(a \mid I_t) = \frac{1}{Z} \prod_{j=1}^{J} p(r_j) \propto e^{-E(a)} \tag{7.12}$$

The grammar for an event consists of a 5-tuple $AoG = \langle S,V,R,\Sigma,P \rangle$, where all the terms in this grammar are defined exactly as in the previous section. The additional term Σ represents the set of all valid configurations of atomic actions that led to the occurrence of an event. The determination between nodes is based on the relative temporal length. That is to say, that for a chain of events, consider sub-events A, B, and C. Given event durations τ_A, τ_B, τ_C, then we can use the relative duration of τ_B with respect to τ_A, τ_C. Using this method, we can then define temporal filters as $F = (F_1, F_2, F_3)$, and measure the duration of the event in terms of its sub-events $\tau_{E_1} = (\tau_A, \tau_B, \tau_C)$. We can then treat this problem like image filtering and take the inner product of the duration of sub-events, τ_{E_1}, and the filters, F, in order to get response of our temporal data to a filter $Tr = \langle \tau_{E_1}, F \rangle$. This response function follows a continuous distribution given by its histogram, h.

$$p(Tr) \sim h(\langle \tau_{E_1}, F^* \rangle) \tag{7.13}$$

In order to find these filters, the minimax entropy principle is used (Zhu et al., 1997). The minimum entropy principle relies on the minimax principle to find the filters which minimize the maximum possible difference between the observed histogram and the histogram developed by the current model.

$$F^* = \arg \min_{|F| = \tau} \left\{ \max_{h \in \Omega_F} entropy(h) \right\} \tag{7.14}$$

The parse graph energy, as before, will consist of the sum of its individual nodes. The terminal nodes, V^t, are represented by the atomic actions $\{a_1, \ldots, a_{n_T(pg)}\}$ found by the given parse graph. The OR nodes are represented by the probability that a $\{v_1, \ldots, v_{n_{or}(pg)}\}$ chooses its sub-nodes in pg. The AND nodes have temporal relations defined by the response functions, $\{T_{r_1}, \ldots, T_{r_n(R)}\}$. We can then say the energy of a parse graph, $\epsilon(pg)$, is given by the sum of the energy of each of these terms.

$$\epsilon(pg) = \sum_{a_i \in V^t(pg)} E(a_i) + \sum_{v_i \in V^{or}(pg)} - logp(v_i) + \sum_{Tr_i \in R(pg)} - logp(Tr_i) \tag{7.15}$$

The first and third terms have already been defined in this section, while the Or-node energy can be determined in the same manner using Eq. 7.6.

The energy in Eq. 7.6 represents a parse graph for a single event, E_i. The interest in temporal parsing is finding particular sequences of events that could explain the video I_Λ. Therefore, we define a sequence parse graphs, $P(K, pg_1,\ldots,pg_K)$, where K gives the number of parse graphs of single events we wish to consider in our sequence. We then define the energy of our parse graph *PG* as :

$$E(PG \mid I_\Lambda) = poiss(K)\ \Sigma_{k=1}^{K}\left(\epsilon(pg_k \mid I_{\Lambda pg_k}) - logp(pg_k)\right) \tag{7.16}$$

Here the term $p(k)$ is the prior that the sequence *PG* contains pg_k this term allows us to penalize unlikely associations between *PG* and pg_k. The term $\epsilon(pg_k \mid I_{\Lambda pg_k})$ gives the energy for the single event pg_k and is defined in Eq. 7.16. The item $p(K)$ allows us to penalize the energy of an event sequence for containing numerous single events. This causes the energy to prefer results with simpler explanations. The penalty is defined as a Poisson distribution, $poiss(K) = \lambda_T^K e^{-\Lambda_T}/K!$, where the term λ_T represents the number of parse graphs in the video I_Λ. Given the energy of $PG \mid I_\Lambda$, you can then find and maximize the posterior probability giving the maximum likelihood solution.

$$PG^* = \underset{PG}{\operatorname{argmax}}\, p(PG \mid I_\Lambda) \tag{7.17}$$

As each frame is segmented, a collection of AND nodes consistent to the current frame is kept by the temporal parser. The parsing algorithm is based on the Earley parser and performs three processes to complete the temporal parsing: prediction, scanning, and completion.

Prediction is performed as each frame is spatially parsed. Predictions estimate the most likely event to occur in the next frame. This is accomplished by selecting the lowest energy increase that occurred in the next frame. Scanning operations then check the detected object in the new frame and advance the pending derivation of the AND nodes from the previous frame. During scanning iterations, the early parsing algorithm prunes any derivations with high energies to limit the number of nodes in the parse graph. When the sequence terminates, the algorithm highlights the most likely atomic events to have occurred during each step in the video.

7.2.8 Causal Parsing

The causal AOG represent knowledge of the causal relationships between events and fluent changes. Causal parsing is concerned with two types of fluents. Fluents may refer to the local state of an entity or object, or used to express latent utility functions in a multi-agent setting. The And-nodes of a C-AOG represent a composition of conditions and events that can cause fluent changes. For example, to cause a door to open, it can be that the door is unlocked and someone pushes the door. An Or-node represents an alternative cause that can result in a fluent change. Alternatively, it is possible that the door is unlocked and pushed open or the door could be automatic and triggered by a proximity sensor.

Simple datastructures such as a contingency table may be used to represent the relation of each action-fluent pair. The most likely relationship is then determined by maximizing the information gain between a full contingency table and the expected contingency table predicted by the models current iteration. To maximize the information gain, we may equivalently maximize the KL divergence:

$$cr^* = \underset{Cr}{\text{argmax}} IG = \underset{cr}{\text{argmax}} KL(\mathbf{f} || \mathbf{h}) \tag{7.18}$$

Here the term $\mathbf{f} = (f_0, \ldots, f_{n \times m-1})$ represents the relative frequencies within an n × m table representing the n actions with m fluents. **h** encodes the relative frequencies of the current iteration. The causal relation is handled similarly to the temporal method using an Earley parser algorithm rooted around detected fluent changes.

7.2.9 Text Parsing

C2 decision making does not exist in a vacuum; in addition to other ISR datasources, semi-structured textual information, such as intel reports or chat logs, should also be incorporated. Luckily, these datasources may be represented as structures compatible with joint parsing using well-understood computational linguistic methods. Briefly, the procedure for text parsing consists of three stages: text filtering, speech tagging, and dependence filtering.

When receiving a communication such as "RED has operatives in the area, and has control of a nearby port", we first need to perform name-entity recognition (NER) using tools such as Stanford NER (Finkel et al., 2005). NER identifies which part of the text relates to a name, in this case "operatives" and "port"; additionally, compound noun phrases of the sentence need to be identified. In the sentence above, one must recognize the descriptor "RED" is acting on "operatives", and "control" is acting on "port". To further simplify the problem, many parsing methods condense compound noun phrases to a single object using systems such as Apache OpenNLP (OpenNLP).

To extract described events from a piece of unstructured text, a part-of-speech (POS) tagger, such as the Stanford Lexiccalized Parser (The Stanford Parser), is commonly employed. This identifies nouns, verbs, adverbs, adjectives, etc. After POS, dependence filtering is applied to map the relationships identified through POS and generalize to ontological to a constrained set of terms. This means converting "RED operative" to a general term like "enemy". At this point, a parse graph compatible with joint parsing may be generated. Using the attribute grammar of the text, dependence filtering can be accomplished with a look-up table or utilizing a tool like the WordNet dictionary (WordNet). For brevity, we have only briefly discussed text and image parsing in this chapter; our primary interest is in the algorithms that facilitate the fusion or joint parsing across multiple ISR domains. Further detail on parsing text and imagery may be found in (Tu et al., 2014).

7.2.10 Joint Graph Parsing

With the groundwork laid for single-datasource parse graphs, the scene parsing literature is extended to joint parsing across parse graphs of different datatypes in (Tu et al., 2014). Relating the multiple parse graphs from several inputs means considering the energies ranging over entities and relations, x, in a joint parse graph, pg_{jnt}. This energy is then defined with respect to which parse graph the relation belongs.

$$E(x) = \begin{cases} \alpha_1 + d(x_j, x_1) & if x \in pg_1 \notin pg_2 \\ \alpha_2 + d(x_j, x_2) & if x \in pg_2 \notin pg_1 \\ \alpha_{1,2} + d(x_j, x_1) + d(x_j, x_2) & if x \in pg_1 \cap pg_2 \\ \alpha_\phi & if x \notin pg_1 \cup pg_2 \end{cases} \tag{7.19}$$

The equation above assumes two seperate joint graph inputs, but it can be generalized to several inputs. The term α_n represents the energy that the relation x belongs to the domains defined by n and no other parse graphs. In the case of α_1, it gives the energy that x is a relation given by input 1 and not by any other input α_n; energy will be set to low values when excluded parse graphs would be unable to show the relation x. α values can be either manually inputted or learned from annotated training data (Dodge et al., 2012). The value d is a distance measure that models the difference in how relation x is annotated from multiple inputs. This distance is a linear combination of all considered distances, semantic temporal and spatial distances. The semantic distance represents the difference between the two semantic types, this can be measured in several different ways (Lee et al., 2008; Thiagarajan et al., 2008; Pesquita et al., 2009). The temporal distance is the time-based measure between the occurrence of two relations, while the spatial distance gives how far apart two events occur in their respective locations.

$$d(x, y) = \alpha_0 d_o(x, y) + \alpha_t d_t(x,y) + \alpha_s d_s(x,y) \tag{7.20}$$

By calculating these distances we can find the joint energy of our system and define a relation resulting from several inputs.

7.3 Applications

7.3.1 Answering Questions with Parse Graphs

A joint parse graph is a structured, semantic representation of the objects, events and fluents present in a data set. These joint parse graphs can be used in semantic queries to answer natural language questions about the state of a tactical scene. These questions may vary in complexity and include dependencies on scene parsing algorithms' unique capabilities for entity resolution, joint inference of partially observable information, deduction and prediction. For example:

- Is Red active in this area?
- Have there been local issues in the nearby port town?
- Where are the small vessels inbound to ship Green's location?
- Does smuggler ship *S* have armed crewmen on its deck?
- Is *S* carrying any large containers?
- Has *S* recently dropped anything overboard?
- What types of activity has *S* engaged in over the last 24 hrs?
- Where has *S* been observed over the past 2 months?
- Does *S* have fishing equipment?
- What types of activity do the agents aboard *S* appear to be engaged in?

Note that the joint parsing methods developed in (Pei et al., 2011) are capable of generating multiple parses of a single scene. The multiple parse graphs correspond to different interpretations of the same scene; therefore, to answer a question accurately, multiple interpretations may be combined to determine the probability of a particular interpretation, $P(a)$, by summing the posterior probability $P(pg)$ of each parse graph where pg implies interpretation a; this is denoted with an indicator function $\mathbb{1}(pg \models a)$

$$P(a) = \Sigma_{pg} P(pg)\mathbb{1}\,(pg \models a) \tag{7.21}$$

In (Pei et al., 2011) the authors propose and demonstrate a user-facing query engine for interacting with and extracting critical information from parse graphs. A natural language query is composed by the user and entered into a web-application GUI front-end. Text input is parsed into SPARQL queries which are compared against RDF decompositions of joint parse graphs. Responses to the queries are then presented to the user in the GUI with the associated marginal probability, interpretable as a confidence of the response.

Using this representation, simple metrics such as ROC (Receiver Operating Characteristic), and associated precision and recall measures are suitable and commonly used in measuring performance in prediction or query answering problems. Precision is proportional to the number of relevant objects and relations indicated by the scene parse with respect to ground truth. This quantity degrades when superfluous information is included in the graph. Recall is proportional to the number of objects and relations present in both the ground truth and the scene parse. Questions may be of the form of Who, What, When, Where or Why, and answers may be boolean-valued or return the empty set or elements of the STC-AOG ontology dependent on the query.

7.3.2 Generating Text Summaries

In (Yao et al., 2010; Tu et al., 2005) scene parsing methods are used to summarize and annotate natural images using natural language text generation. In this scenario, a parse graph is mapped to natural language sentences that maximally explain the input. This differs from the question answering scenario in that the models' outputs are non-interactive and express the full content of their parse graphs. Such human readable summaries are appropriate for rendering directly in C2 displays, or for reporting and automated brief generation.

It is readily apparent the traversal of an AOG from the leaves to the root correlate well with a high-level notion of semantic abstraction. An open question for more fundamental investigation is how higher-level information may be incorporated as indications and warnings into C2 interfaces. For example, it may be possible to deduce agent intent, threat assessment or other entirely un-observable information from joint parse graphs. Deductive reasoning capabilities enabled by scene parsing play a key role in these use cases, for instance (following from our previously assumed example scenario):

- Parses of text reports may indicate the exchange of funds with an individual for weapons or explosives.
- This individual may be an owner of, or in some way associated with, a small boat located nearby.
- Imagery may indicate that this vessel has large containers and armed crewman onboard.
- Various other electronic intelligence may indicate that the same vessel is inbound.

In isolation, none of this information is particularly useful. However, jointly parsing across these otherwise disjoint analytic pipelines could produce a very clear indication which could be displayed directly to an operator without explicit queries. Furthermore, using the scene parsing methods we propose, all logical premises leading to a conclusion would, by design and technical necessity, provide data provenances and explicitly represent the uncertainty associated with a particular deduction.

7.3.3 Behavioral Modeling

Finally, we briefly survey methods for learning and inferring latent aspects of a tactical scene such as the utility functions of agents and their corresponding intent. Explicit examples of this may include the affordance[8] of objects, the utility of taking courses of actions, and the policy or pattern of behavior of agents in tactical scenes. Developing models of strategic behavior is essential for both inferring the intent of potential adversaries as well as obtaining a rigorous, data-driven estimation of the correct behavior of blue-force human operators and autonomous systems.

Scene parsing with AOGs provides an ideal formal framework for representing and inferring latent relationships. A method for inferring the physical forces exerted on the bodies of human actors observed sitting in various chairs is proposed in (Zhu et al., 2016). The inference objective is to estimate the relative utility of chairs (e.g., their perceived comfort). In this work, a finite element based physics engine is used to simulate the physical interaction of RGB-D reconstructions of parametric human actors and their environment. This then serves as training data to estimate the physically constrained likelihood of human poses in a scene. Using this inference algorithm, the authors then illustrate that one may estimate the utility of a particular chair in the environment.

Ongoing and future work in Naval tactical scene understanding seeks to extend recent developments in scene parsing to estimate utility of various platforms afloat. This capability would enable improved assessment of C2 environments and a more statistically grounded method for defining measures of effectiveness. Further, assuming the rational choices of agents, it may be possible to infer their most likely courses of actions by extending the approach pioneered in (Zhu et al., 2016).

7.4 Conclusion

Methods for inferring the context of events and objects in an environment from sensor measurements are critical for the Naval tactical command and control domain. In the field today, much of this inference is left to human operators; the efficacy of which is largely dependent on operator training, experience and intuition. In future systems, we hope to provide a means of assisting and automating this process to inform operators' assessment of increasingly complex and rapidly changing battlespaces. Scene parsing

[8] The likelihood of agents taking actions on an object in their environment.

using And-Or graphs provides a promising and efficient means of representation, summarization and decision making based on complex, uncertain contextual relationships extracted from Naval tactical ISR data.

References

Andrieu, C., de Freitas, N., Doucet, A. and Jordan, M.I. 2003, Jan 01. An introduction to mcmc for machine learning. Machine Learning 50(1): 5–43.

Apache opennlp. (n.d.). Retrieved from http://opennlp.apache.org/.

Bemis, S.V., Leeds, J.L. and Winer, E.A. 1988, April. Operator performance as a function of type of display: Conventional versus perspective. Hum. Factors 30(2): 163–169.

Dodge, J., Goyal, A., Han, X., Mensch, A., Mitchell, M., Stratos, K., . . . others. 2012. Detecting visual text. pp. 762–772. In Proceedings of the 2012 Conference of the North American Chapter of the Association for Computational Linguistics: Human language technologies.

Doug Lange Michael Walton and Song-Chun Zhu. 2017. Inferring context in scene understanding. In Proceeding of 2017 AAAI Spring Symposium on Computational Context. Association for the Advancement of Artificial Intelligence (AAAI).

Finkel, J.R., Grenager, T. and Manning, C. 2005. Incorporating non-local information into information extraction systems by gibbs sampling. pp. 363–370. In Proceedings of the 43rd Annual Meeting on Association for Computational Linguistics.

Fire, A. and Zhu, S.-C. 2015, nov. Learning perceptual causality from video. TIST 7(2): 1–22.

Fire, A.S. and Zhu, S.-C. 2013. Using causal induction in humans to learn and infer causality from video. In Cogsci.

Heckerman, D. 2008. A tutorial on learning with bayesian networks. pp. 33–82. *In*: Holmes, D.E. and Jain, L.C. (eds.). Innovations in Bayesian Networks: Theory and Applications. Berlin, Heidelberg: Springer Berlin Heidelberg.

Hoiem, D., Efros, A.A. and Hebert, M. 2007. Recovering surface layout from an image. International Journal of Computer Vision 75(1): 151–172.

John, M., Manes, D.I., Smallman, H.S., Feher, B.A. and Morrison, J.G. 2004, sep. Heuristic automation for decluttering tactical displays. Proceedings of the Human Factors and Ergonomics Society Annual Meeting 48(3): 416–420.

Koller, D. and Friedman, N. 2009. Probabilistic graphical models: Principles and techniques—adaptive computation and machine learning. The MIT Press.

Lee, D.C., Hebert, M. and Kanade, T. 2009, June. Geometric reasoning for single image structure recovery. In IEEE Computer Society Conference on Computer Vision and Pattern Recognition (cvpr).

Lee, W.-N., Shah, N., Sundlass, K. and Musen, M.A. 2008. Comparison of ontology-based semantic-similarity measures. In Amia.

Manning, C.D. and Schtze, H. 1999. Foundations of statistical natural language processing. The MIT Press.

Michael Walton, D.L. and Zhu, S.-C. 2017. Inferring context in scene understanding. In Proceeding of 2017 AAAI Spring Symposium on Computational Context. Association for the Advancement of Artificial Intelligence (AAAI).

Pei, M., Jia, Y. and Zhu, S.-C. 2011, nov. Parsing video events with goal inference and intent prediction. In 2011 International Conference on Computer Vision. Institute of Electrical & Electronics Engineers (IEEE).

Pesquita, C., Faria, D., Falco, A.O., Lord, P. and Couto, F.M. 2009, 07. Semantic similarity in biomedical ontologies. PLoS Comput. Biol. 5(7): 1–12.

Russell, S. and Norvig, P. 1994. Artificial intelligence: a modern approach (2nd edition). Prentice Hall.

Siddhartha Chib and Edward Greenberg. 1995. Understanding the metropolis-hastings algorithm.

Smallman, H., John, M.S., Oonk, H. and Cowen, M. 2001. Information availability in 2d and 3d displays. IEEE Comput. Grap. Appl. 21(4): 51–57.
The stanford parser: A statistical parser. (n.d.).
Thiagarajan, R., Manjunath, G. and Stumptner, M. 2008. Computing semantic similarity using ontologies. HP Laboratories. Technical report HPL-2008-87.
Tu, K., Meng, M., Lee, M.W., Choe, T.E. and Zhu, S.C. 2013. Joint video and text parsing for understanding events and answering queries. CoRR, abs/1308.6628.
Tu, Z., Chen, X., Yuille, A.L. and Zhu, S.-C. 2005. Image parsing: Unifying segmentation, detection, and recognition. Int. J. Comput. Vision 63(2): 113–140.
Wordnet: A lexical database for english. (n.d.).
Yao, B.Z., Yang, X., Lin, L., Lee, M.W. and Zhu, S.-C. 2010, aug. I2t: Image parsing to text description. Proceedings of the IEEE 98(8): 1485–1508.
Zhao, Y. and chun Zhu, S. 2011. Image parsing with stochastic scene grammar. pp. 73–81. *In*: Shawe-Taylor, J., Zemel, R.S., Bartlett, P.L., Pereira, F. and Weinberger, K.Q. (eds.). Advances in Neural Information Processing Systems 24. Curran Associates, Inc.
Zhu, S.C., Wu, Y.N. and Mumford, D. 1997, nov. Minimax entropy principle and its application to texture modeling. Neural Computation 9(8): 1627–1660.
Zhu, S.-C. and Mumford, D. 2006. A stochastic grammar of images. Foundations and Trends in Computer Graphics and Vision 2(4): 259–362.
Zhu, Y., Jiang, C., Zhao, Y., Terzopoulos, D. and Zhu, S.-C. 2016. Inferring forces and learning human utilities from videos. In IEEE conference on Computer Vision and Pattern Recognition (cvpr).

8

Using Computational Context Models to Generate Robot Adaptive Interactions with Humans

Wayne Zachary,[1,*] *Taylor J. Carpenter*[2] and *Thomas Santarelli*[3]

8.1 Introduction

Human-human coordination and cooperation can be robust and adaptive, while human-robot and robot-robot interactions, in comparison, can be fragile. In this chapter, we argue that human-robot interactions most often break down because of the inability of the robot to understand and adapt to context, particularly in the sense that humans understand and use context. This inability is not a problem unique to robotics, but rather one that is characteristic of all automated systems, particularly those that involve "intelligent" automation and must interact with people. The modern study of this problem can be traced back at least 35 years to the work of Suchman (1987) at Xerox, who collected and described in detail

[1] Written while he was at CHI Systems, Inc., Plymouth Meeting, PA; currently, Wayne is the Managing Partner, Starship Health Technologies, LLC, Fort Washington, PA.
[2] Written while he was at CHI Systems, Inc., Plymouth Meeting, PA; currently, Taylor is a PhD student at the Univ. of Pennsylvania Dept. of Computer & Information Science.
[3] Director of Training and Simulation, CHI Systems, Inc., 2250 Hickory Road, Suite 150, Plymouth Meeting, PA 19462.
* Corresponding author

the often-insurmountable difficulties that human office workers had in getting supposedly "smart" photocopy machines to do what they (i.e., the people) wanted. Her analysis showed that the problem was not just one of the design of interfaces, but was inherently a cognitive problem, involving expectations and models that the humans held about automation, and the presumptions and predictions that the automation was (often implicitly) making about human users and their intentions. We would say that the details specific to an instance of a general task (photocopying), including the intentions and constraints of the person who brings forward that task-instance, constitute a context to which the person and the smart machine must adapt. In the case studied by Suchman, humans tried to adapt to the task in a context sensitive way, and made reasonable assumptions on how to do that, if the other party were a human and not a machine. The machine, on the other hand, had no representation of context and therefore no ability to adapt to it, with the result being frequent and unpleasant (at least for the humans) failure.

Unfortunately, as automation has evolved over the last quarter of the 20th century and now in the 21st, this problem has continued and in many cases worsened as the capabilities of intelligent machines and the frequency of their interaction with humans has increased, while the machine's ability to understand and adapt to context has not. This chapter presents a framework and a set of computational tools that were developed to help address this need.

We have already identified why the ability to understand and adapt to context is important—to enable more robust and effective interactions with humans. The question of how to do so is also framed by the above discussion—the computational device needs to understand and reason in a way that is compatible with the way that people understand and reason about context. More specifically, the computational system needs its context adaptations to be behaviorally compatible with those used and expected by people. In other words, the machine's understanding of context needs to help the robot or information system adapt to context in a way that would be intuitively understandable and explainable to a person participating in an interaction or situation.

8.1.1 Organization and Overview

Given this view, we begin with an overview of a computable representation of context that has been explicitly developed from theories and data on human cognition, called the Narratively-Integrated Multi-level Model of context, or NIM. We use NIM as a framework for modeling context-understanding and reasoning in a specific domain, and embedding that model into a computational system, such as a robot or an information system. With such an embedded context representation and reasoning capability, the computational system can detect contextual constraints on its behavior and adapt its behavior accordingly. We then introduce an application domain that serves as a source of examples used throughout the chapter. This domain involves human-robot interaction and cooperation in space exploration. The specific example is a small and constrained space-based or moon/planet-based habitat, where human astronauts and multiple robots work independently but in close proximity. This "space robots" domain has obvious analogies to more familiar human-robot interaction settings, such as warehouses or construction sites where robots and humans may soon be working together.

We next examine, in detail, a set of behavioral challenges in the space habitat domain in which a robot must adapt its behavior to the actions of the astronauts around it, by understanding the behavioral context in a similar way that the astronauts understand it. We have elsewhere termed this capability to adapt behavior according to context as "action compliance" (Zachary et al., 2015a). It is a behavioral/task analog of the "force compliance" concept that has driven prior research in robot physical/environmental interaction. Force compliance requires the robot to use information about the physical context to adjust its joint control, whereas action compliance requires a robot to use information about the shared behavioral context to adjust its behavior to the humans encountered in its (work) environment.

We specifically discuss two forms of a foundational type of action compliance—non-interference. Non-interference is simply the ability of a robot to not be what humans would call underfoot, or otherwise be inadvertently interfering with purposive human activity. Thus, non-interference requires proactive avoidance of collisions with, or unneeded interruptions of, people (or other robots) engaged in purposive tasks. Non-interference is a cognitively difficult task, because it requires the non-interfering agent to understand the context of activity of the other agents to the extent that it can reason about how the action will unfold and affect the future context, and how its possible actions (or inactions) could create collisions or interruptions in that future context. It is a skill that takes human children many years to acquire, as any parent will attest, and that many intelligent domestic animals (i.e., pets) never fully acquire. Pets, though, are forced to not interfere without engaging in communications with the humans involved, a constraint which may often be applied to robots as well. This is the non-communicative form of non-interference, or "keyhole non-interference", to conform with the use of the term in the artificial intelligence plan recognition literature (e.g., Cohen et al., 1981).

Our consideration of action-compliance thus begins with a description of a NIM model of context to provide a space robot working on solitary and independent tasks with an added ability to exhibit keyhole non-interference with regard to the astronauts moving about the space habitat. We show how NIM provides the robot with both the needed representation of context and the reasoning needed to build/maintain a context representation to generate the desired non-communicative non-interference behavior.

Using this model as the foundation, we then consider the communicative form of non-interference, in which the robot can communicate with an astronaut to take actions that make its non-interference more efficient and effective. In doing this, we show how the context model used at the simpler keyhole level must be elaborated to generate the stronger form of action compliance. This elaboration is in terms of both the representation itself and the procedural knowledge needed to produce that more sophisticated form of action compliance.

After presenting these increasingly complex context-based action compliance models, we then describe the Integrated Context Engine (ICE), a set of computational tools we created to support the implementation, testing/examination, execution, and integration of domain-specific context models for robot or other computational environments. The chapter ends with a discussion of limitations and broader possible uses of our model and future capabilities of our tools.

8.2 The NIM Computational Context Framework

The NIM computational framework, initially described in Zachary et al. (2013) and Zachary et al. (2015a), provides both a static and dynamic view of context with regard to a specific computational agent's point of view. The static view is a declarative Context Representation (CR), the contents of which at any point in time define the agent's understanding of the current context. The dynamic view consists of different kinds of procedural Knowledge Elements (KEs) that together both build and maintain the contents of the CR, and use the momentary CR contents to build and implement behaviors that adapt the activity of the agent to patterns in the CR. This overall structure and the details of each component were taken from different lines of cognitive systems research. They were integrated with an underlying architecture for computation to yield the NIM framework.

8.2.1 Context Representation

NIM represents context in seven hierarchical levels of abstraction. Each level can contain a set of Concept Elements (CEs), each an instance of a concept-class from a domain-specific ontology, and a set of semantic relationships that inter-relate concepts within and across levels of the CR. The lower three levels correspond to what cognitive science terms situation awareness (see Endsley, 1995; Endsley and Garland, 2000). These lower three levels are:

1. *Perception*, in which the agent posts CEs describing perceived items in the external environment, along with their status, attributes, and dynamics;
2. *Comprehension*, in which the agent posts and updates CEs describing or relating perceived concepts to more abstract functions or features that identify their inferred significance to the agent and/or to the agent's situational goals in the domain; and
3. *Projection*, in which the agent posts and updates CEs that define possible future behaviors or actions in the environment in local (i.e., near-term) time.

The upper three levels correspond to what cognitive science terms narrative understanding. They incorporate the idea of early narrative researchers (Bruner, 1991; Schank, 1995) as well as more recent work on narrative processing (e.g., Miller et al., 2006; Miller et al., 2010) that views narrative as structures that people use to contextualize human interactions and social behavior, in terms of pervasive motivations and situational objectives. As structures, narratives are composed of individual story-units, which are themselves composed of units of action made up of one or more events. Bridging the upper three narrative levels and the three lower situation awareness levels is the fourth level:

4. *Events*, in which the agent posts and updates CEs that posit specific future Events involving objects or relationship posted on lower levels of the CRs, representing the agent's expectations of what events should, or could happen in the future.

The notion of events (see Barwise and Perry, 1983; Zacks et al., 2007; Radvansky and Zacks, 2011), is the key bridging construct. An event is a concept about agents or objects, and relationships to other agents, objects, and temporal-spatial information (e.g., an astronaut Beta will begin moving at time 14:05 toward module Destiny). In NIM, a CE at this fourth level is typically a hypothesis about a future event (i.e., an expectation) which can be refined by future information, reasoned with, or transformed into a CE about real event. CEs that are future events allow the agent to reason about their implications, for example "would that expected movement bring the astronaut into conflict with my own movement plans or locations". When an expected event is concluded to be a real event (i.e., to have happened), then it becomes the lowest level building block of a new or existing narrative.

The highest NIM levels of the CR that involve narrative (rather than situational) understanding are:

5. *Action Units*, in which the agent posts and updates CEs representing specific Action Units from active or potential narratives that could be recognized, given the narratives that are possible or plausible for

the current context; action units are recognized (or discarded) as the specific Events that comprise them are recognized or discarded at the Expectations level;

6. *Story Units*, in which the agent posts and updates CEs representing specific units of narrative termed Story Units (see Miller et al., 2006) from active or potential narratives that could be recognized, given the narratives that are possible or plausible for the current context; action units are recognized (or discarded) as the specific Action Units that comprise them are instantiated or discarded at the Action Units level; and
7. *Plausible Narratives*, in which the agent posts and updates CEs representing specific narrative instances from the class of active or potential narratives that could be recognized and extended or discarded, given the Story Units that recognized (or discarded) as the specific Events that comprise them are instantiated or discarded at the Story Units level.

8.2.2 Context-maintenance and Context-application Knowledge

NIM is a representation-centric model in that it has a single integrated CR structure, but views context reasoning as the result of the operations of a large number of independent Knowledge Elements, each with specific functions and domain-specific procedural knowledge. NIM identifies three separate types of KEs that form different but related functions, as follows.

1. *Perceptual extractor (PE)*—a data-driven procedure that extracts a specific kind of information from the external environment and Posts it to the CR Perceptions level. The set of PEs keep feeding the context process with new/current information about the external environment.
2. *Representation builder (RB)*—a container of procedural knowledge that is self-activated by a pattern of information on the CR. Once activated, it competes for an opportunity to execute. When it gets that opportunity, it makes specific changes to the CR via specific Post, Update, and/or Delete transactions based on procedural knowledge it contains, operating on the CR contents. The reasoning process can be inductive—using lower-level information on the representation to create new pieces of more abstract information; deductive—using higher, more abstract pieces of information to create or modify lower level information; or abductive—using patterns of information to create possible explanations, courses of action, or hypotheses that may trigger other reasoning processes to validate (or invalidate) them. The reasoning processes of RBs can also be deconstructive—cleaning

up and removing left-over, no-longer useful, or now-incorrect CEs. Together the set of independent RBs act to build and maintain the CR in the face of changing environmental information imported by the set of PEs.

3. *Action-builder (AB)*—a container of procedural knowledge that is self-activated by a pattern of information on the CR. Once activated, it competes for an opportunity to execute. When it gets that opportunity, it uses the procedural knowledge and patterns of information contained on the CR to identify and configure an action to a separate physical action subsystem. In the robot example, this would mean generating a high-level action specification for a robot controller, or to a robot-human communication subsystem. In the course of its computations, an AB can also modify the CR contents via specific Post, Update, and/or Delete transactions.

Each KE, regardless of type, has a three-part structure: a triggering pattern which causes it to be activated, a priority function which specifies the priority for execution of a specific instance of the KE, and a body which computes what action is to be taken on the CR or directed to an external action subsystem.

8.2.3 Computational Architecture

The NIM framework views context as a self-organizing process, which continuously constructs, deconstructs, and maintains the content of a declarative context representation. It is an example of the influential Pandemonium reasoning architecture first suggested by Selfridge (1959). In more concrete terms, the CR is operationalized as a multi-panel blackboard structure (Nii, 1986; Corkill, 1991) termed the Whiteboard, with one required panel (i.e., the CR), and other (domain-specific) panels being optional. The set of KEs are treated as a collection of independent reasoning elements, termed the Swarm, which become activated when their trigger conditions are satisfied in the sensation stream or in the CR. As KEs become activated and compete for an opportunity to execute, their execution is governed by a policing process that enforces an explicit set of Principles of Operation or PoPs (Zachary et al., 2015a). We note that this abstract architecture can be executed in a sequential Von Neuman style machine or in various kinds of parallel execution architectures. We have created a specific set of tools, called the Integrated Context Engine (ICE), for building, running, and embedding NIM context processing into robots or other computational applications. Figure 8.1 shows conceptually how the different types of KEs work to constantly construct and deconstruct CR from the data obtained from the external environment, and how ABs work to apply context information to take (adaptive) action in the external environment.

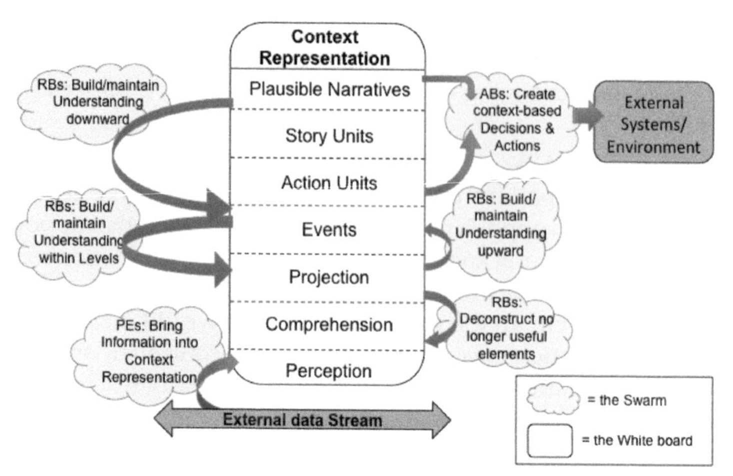

Figure 8.1. NIM framework components and function.

8.3 Example Domain – Robots in Space Exploration

We discuss the use of NIM computational context in a problem domain based on the use of mixed human and robot teams in space exploration. This domain is loosely based on the current International Space Station as it might appear in the near future, but the idea of humans and robots working together in a closed space on complex tasks is common to many other environments, ranging from undersea exploration, or work in closed or clean rooms on electronics or biological/pharmacological research to industrial settings like warehouses or factory floors. Our simple space station example has four astronauts (named Alpha, Beta, Delta, and Gamma) and one robot working in a space habitat with seven modules interconnected by hatches in three directions (see Figure 8.2). The modules bear the same name as the first seven modules of the International Space Station.

In this habitat, all astronauts work on a 24-hour schedule, with common sleep and meal times. The robot is an anthropomorphic model similar to the current Robonaut 2 robot (fitted with both arms and legs (Diftler et al., 2011)) and works in the day and recharges during the evening. Each evening, mission control publishes a daily plan, which provides a schedule for all of the astronauts' (and the robot's) tasks for the next day. The daily plan defines the intended times and places of all work tasks for the day. These typically include experiments and repair/maintenance activities that can be conducted either by a single astronaut or a team of astronauts. The daily plan also includes all times and places for meals, sleeping, exercising, attending meetings, as well as scheduled free time. Finally, in our hypothetical space habitat, the daily plan also includes the scheduled activities for the robot, which are designed to keep the robot out of the way of the astronauts. This separation is for the safety of the astronauts who fear that the large robot,

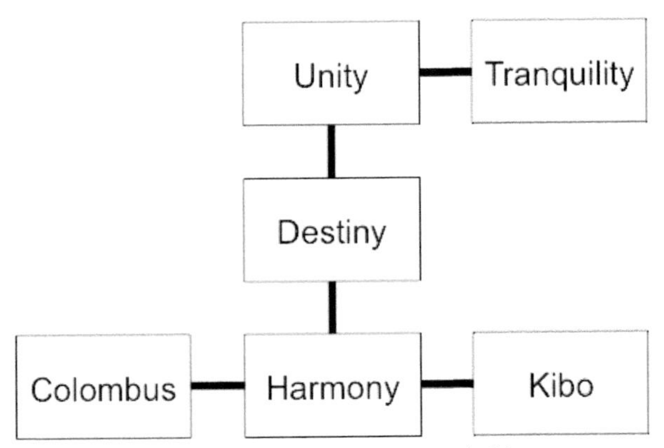

Figure 8.2. Structure of the space habitat.

lacking action compliance capabilities, may inadvertently crash into them and/or otherwise hurt them. Figure 8.3 shows a portion of a hypothetical daily activity schedule and the movements necessary to change locations between scheduled activities.

Although the daily plan is designed to preclude non-interference reasoning by keeping the robot apart from the astronauts, the schedule is always ambitious and tight. There are thus frequently situations in which astronauts may deviate in time and/or space from the plan as, for example, tasks run long, start late, or are interrupted by unexpected events. These deviations create conflicts which can deviate from the goal of keeping the robot and astronauts separate, requiring non-interferent action compliance by the robot. In the keyhole non-interference case, the robot is not allowed (or is assumed to be unable) to communicate with astronauts. In the communicative case, the robot is able and permitted to communicate with astronauts via text messages on the astronaut's personal digital assistant (PDA) devices.

In this example case, we assume that the robot has capabilities for locomotion and sensation/perception, which may vary by the type of robot. For example, an anthropomorphic robot such as Robonaut 2 would move by using its limbs to propel it through the habitat, while a flying robot such as the Sphere prototype (Fong and Micire, 2013) would use the expulsion of propellants to fly throughout a habitat, particularly a micro-gravity habitat. These locomotion, perception, and navigation capabilities are not trivial, and are in fact the subject of extensive research in robotics. However, solving those problems would still leave action compliance unsolved, and would therefore leave the robots as unaware of the context and unable to avoid conflicts and collisions that are potentially harmful to the humans around them. By assuming these capabilities, we provide points of articulation for our research with these other lines of research.

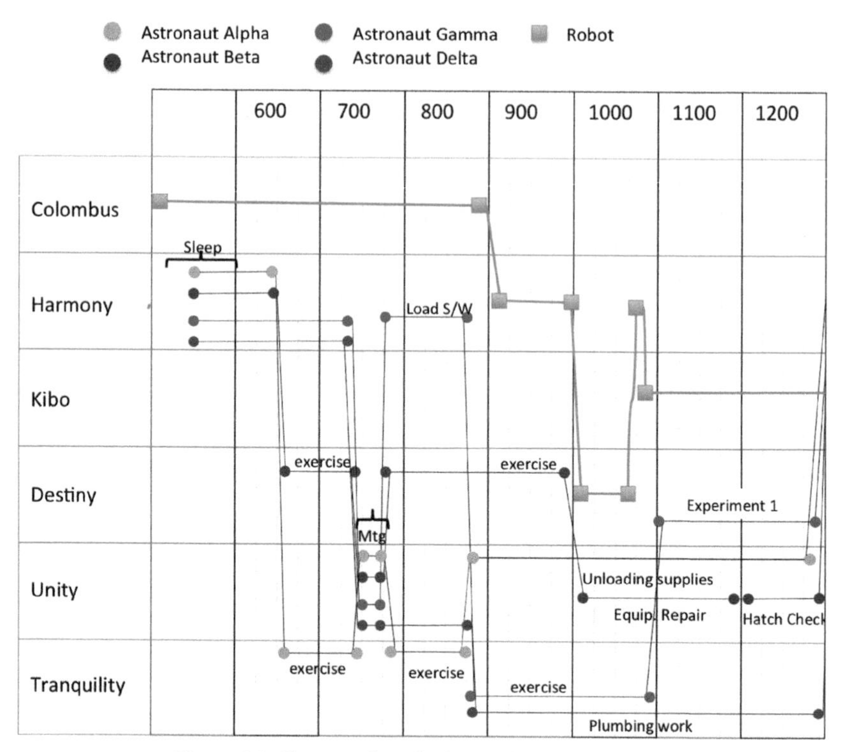

Figure 8.3. Excerpt of work plan in the space habitat.

We also assume a direct sensory mechanism to inform the robot of where it is in the habitat and an indirect sensory mechanism that allows it to perceive the location of the astronauts in the habitat. These mechanisms could be operationalized by each agent (astronaut and robot) having a positional tracking device, and the robot being given access to the data from those devices. As suggested earlier, we also assume a high-level robot behavior controller that can translate physical action intentions—such as move to workstation 2 in module Kibo in the next 5 minutes—into lower-level control operations that effect the action.

8.4 Models of Action Compliance

We used the NIM framework to develop computational context models for a (simulated) robot working in the space habitat described above. The robot then used the resulting context understanding to generate both keyhole and communicative non-interference action compliance behaviors. The keyhole model is presented first below. The communicative model is then presented in terms of the extensions and enhancements it required beyond the simpler keyhole model.

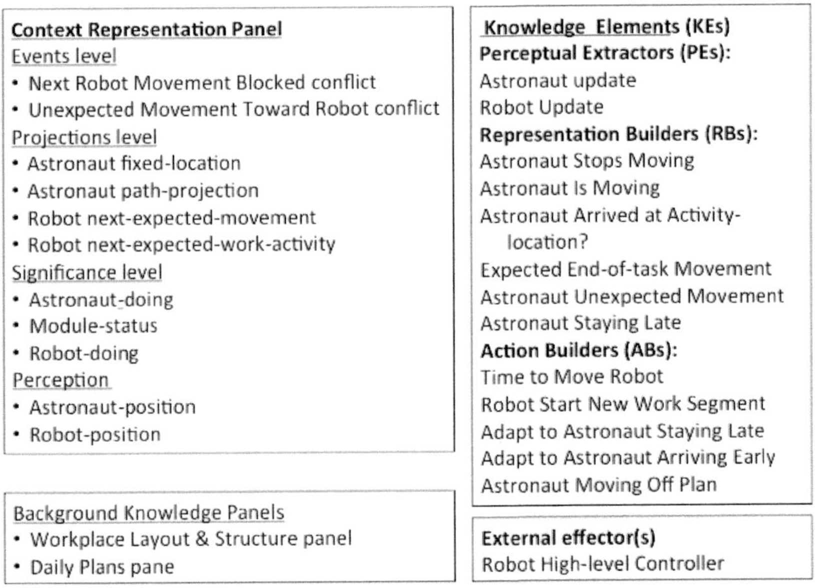

Context Representation Panel
Events level
- Next Robot Movement Blocked conflict
- Unexpected Movement Toward Robot conflict

Projections level
- Astronaut fixed-location
- Astronaut path-projection
- Robot next-expected-movement
- Robot next-expected-work-activity

Significance level
- Astronaut-doing
- Module-status
- Robot-doing

Perception
- Astronaut-position
- Robot-position

Background Knowledge Panels
- Workplace Layout & Structure panel
- Daily Plans pane

Knowledge Elements (KEs)
Perceptual Extractors (PEs):
Astronaut update
Robot Update
Representation Builders (RBs):
Astronaut Stops Moving
Astronaut Is Moving
Astronaut Arrived at Activity-location?
Expected End-of-task Movement
Astronaut Unexpected Movement
Astronaut Staying Late
Action Builders (ABs):
Time to Move Robot
Robot Start New Work Segment
Adapt to Astronaut Staying Late
Adapt to Astronaut Arriving Early
Astronaut Moving Off Plan

External effector(s)
Robot High-level Controller

Figure 8.4. Overview of the keyhole non-interference context model.

8.4.1 Non-communicative Non-interference Context Model

In the initial context model, the robot was assigned only independent tasks on its portions of the daily work plan and could not communicate with astronauts. Thus, the robot needed to understand and reason about the work context, using the daily plan and the flow of locational information on all entities in the space habitat, to generate non-interference behaviors. Figure 8.4 shows a summary of the NIM context model for this case. Details of the various components are provided below.

8.4.1.1 Keyhole Non-interference Context Model Context Representation

This CR uses only the lowest four levels of the canonical CR structure. The Perception level contains two concept-types: astronaut-position concept-types and robot-position concept-types. Each concept-instance encodes perceived information about a specific astronaut or the robot. We note that the inclusion of the robot in the context is essential (here and in other levels) so that the robot has a representation of itself as part of the context, which allows the robot to reason about how the context affects it(self), and how its behaviors alter the context. One interesting by-product of this self-representation in the robot's context model[1] is that it allows much of the

[1] We deliberately avoid the more philosophically contentious phrasing of "self-awareness", though one could argue that self-representation in contexts is a foundation for self-awareness.

same reasoning to be used to: (a) direct the robot through its work activities as specified in the daily plan; and (b) to generate its non-interference action-compliance behaviors.

As an example of the concept structure in this level, the astronaut-position concept has this structure:

Astronaut-position concept-type = (name, location trail, meta-status)

where the attributes of a concept-type instance are in the parentheses, and are defined as follows:

- 'name' is the astronaut's name (e.g., "Alpha");
- 'location trail' is a stack of five ordered pairs ($location_i$, $time_i$), in which the pair in the first position contains the latest location of Astronaut 'name' and the rest of the stack contains the next four most recent (location, time) data on that Astronaut; and
- 'meta-status' is a semaphore which signals whether an instance of this concept-type has been evaluated for significance since its last change.

The robot concept-type robot-position has analogous structure to the astronaut-position concept-type.

The Significance level contains three concept-types: astronaut-doing, robot-doing, and module-status. There are four instances of astronaut-doing (one for each astronaut), one instance of robot-doing for the robot, and one instance of module-status for each module in the habitat. For example, the astronaut-doing-type concept has the structure:

Astronaut-doing concept-type = (name, status, task, meta-status)

where the attributes of a concept-type instance are in the parentheses, and are defined as follows:

- 'name' is the astronaut's name;
- 'status' is astronaut's movement state as either 'static' or 'moving';
- 'task' is the astronaut's presumed current task, with possible values of:
    - 'transit', meaning the astronaut is presumed to be moving to somewhere;
    - 'task-name', meaning the astronaut is presumed to be busy doing the named task from the daily schedule;
    - 'unknown', meaning the astronaut's presumed task is not known; or,
    - 'extended', meaning the astronaut is presumed to be continuing a daily-schedule task beyond its scheduled completion time;

- 'meta-status' is a semaphore which signals whether an instance of this concept-type has been evaluated for projection since its last change of 'task' or 'status'.

The robot concept-type robot-doing has analogous structure to the astronaut-doing concept-type, and the concept-type module-status tracks the 'occupied'/'not-occupied' status of each module in the habitat.

The Projection level contains four concept-types: Astronaut-fixed-location-projection, Astronaut-path-projection, Robot-next-projected-movement, and Robot-next-projected-work-activity. There may be multiple concept instances of each type—one of the astronaut-type for each astronaut, and one of the robot-type for each robot. As an example, the concept type Astronaut-fixed-location-projection has the following structure:

Astronaut-fixed-location projection concept-type = (name, projection-location, projected-end)

where the attributes are defined as follows:

- 'name' is the astronaut's name;
- 'projection-location' is the module and workstation in the module in which the named astronaut is projected to remain; and
- 'projected-end' is the future time until which the named astronaut is projected to remain in the projection-location.

Astronaut-path-projection is the movement-based analog of Astronaut-fixed-location-projection. The two robot projection concept-types are the analogs of the two astronaut projection concept types. The projection concepts are important because they represent the robot's inferences about the future events representing deviations from projected behavior that could lead to potential interference situations.

The Events level is where such future events are represented. The Events level contains two concept types that correspond to the two kinds of events that can signal such deviations: Robot-next-movement-blocked and Unexpected-movement-towards-robot. The first concept-type deals with a class of conflicts in which the robot's work plan calls for it to move to a new work location in the near future. This planned robot move contains an underlying precondition that the new location must be empty, as must all intermediate modules along the path to the work destination. As soon as an astronaut stays in a module past the time the robot projects it to leave (based on the daily work plan), this sets up a condition in which the Astronaut's continued presence in that location can implicitly block the robot's ability to move into it. Thus, the concept type Robot-next-movement-blocked has the following structure:

Robot-next-movement-blocked concept-type = (meta-status, name, current-module, activity, next-move-time, next-destination, blocking-astronaut, blocking-module, time-late, time-posted)
where the attributes are defined as follows:

- 'meta-status' is as defined in other cases above;
- 'name' is the robot's name;
- 'current-module' is the robot's location when this concept was posted';
- 'activity' is the robot's current or just-completed activity when this concept was posted';
- 'next-move-time' is scheduled time for the robot's movement from 'current-module' to its next work location;
- 'next-destination' is location to where the robot is scheduled to move at 'next-move-time';
- 'blocking-astronaut' is name of the astronaut which is the source of the projected conflict event;
- 'blocking-module' is the name of the module in which the projected conflict event would occur;
- 'time-late' is how late the astronaut was with regard to the scheduled movement, at the time this conflict concept was posted; and
- 'time-posted' is the clock time at which the concept-instance was posted on the CR.

The other concept-type, Unexpected-movement-towards-robot, deals with a complementary class of conflicts to the first, in which the robot is working according to plan in a specific location and an astronaut begins to move contrary to projections, towards the robot's current location. This sets up a future situation in which the Astronaut (whose destination or intention for moving are unknown to the robot) could move into the robot's current module, and in which the robot could then be interfering.

The context process also depends on two other types of background information: the daily-schedule panel and the habitat-layout. Each is represented as a separate panel of the CR whiteboard for conceptual consistency with the CR structure. We note, however, that in principle such non-contextual information and knowledge could be represented in any appropriate knowledge-representation structure. For example, the habitat-layout background knowledge panel is represented as a panel with a single level, containing one concept type for each module, defined as:

Module-type = (name, {(connected module, connection point)})

where the attributes of a concept-type instance are in the parentheses, and are defined as follows:

- 'name' is the name of a habitat module; and
- {(connected module, connection point)}, is a list of one to six modules that directly connect to the named module, with the structure that each module, like a die, has six possible numbered connection points. Each ordered pair (connected module, connection point) identifies first the name of the adjoining module, and the number of its connection point.

This panel gives the context process knowledge of the layout of the habitat, so that movement paths can be computed and so that propositions such as "*Astronaut Alpha* moving toward *Robot*" can be evaluated.

The daily plan background knowledge panel contains the set of plans for each astronaut and the robot. Each plan for a given agent is a list of assignments with a start and end time, module-name, task-label, and activity-flexibility (which, for astronauts, can be 'fixed' or nil for unscheduled time). The plan panel gives the context process knowledge of the daily plan in a form needed for the reasoning to populate the Significance, Projection, and Events levels.

8.4.1.2 Non-interference Context Model Knowledge Elements (KEs)

The different type of NIM KEs—Perceptual Extractors (PEs), Representation Builders (RBs) and Action Builders (ABs)—provide the procedural knowledge to build and maintain the content in the CR, to use the CR contents to detect situations requiring action compliance (here, non-interference) behavior, and to generate the appropriate behavioral response. As above, each KE is an independent knowledge source with a triggering pattern of information, such that when an instance of that pattern is present in the CR, the KE becomes active and creates an instance of itself bound to the triggering pattern-instance. That KE instance then competes with other activated KE instances (including other instances of the same activated KE) for the opportunity to be executed. Figure 8.4 above listed the individual KEs in the non-interference context model.

The NIM notion of quantized time provides the basic structure for the execution process for this set of KEs. At each tick of time, new sensory data become available to the context system. Then, all PEs that are activated by the types of data that are newly available activate instances of themselves. Multiple instances may be created and activated by the new data as needed. For example, if new positional data are made available for all four Astronauts, then four instances of the *Astronaut Update* PE will be created, one bound to the new datum for each Astronaut. The PE instances are allowed to execute before any other KEs, changing the CR with new or updated perceptual concept-elements as a result. Next, the RBs and ABs activate instances of themselves based on the patterns of information that exist after the PEs have updated the CR. Each activated RB and AB instance also calculates a priority heuristic for itself. The priority value and the logic in the Principles of Operation are then used by the context system to give each RB and AB instance an opportunity to execute, changing the CR and making directives to the Robot's action systems as needed.[2]

The two PEs—*Astronaut Update* and *Robot Update*, bring the stream of updated sensory data on the astronaut and robot locations into the CR. The six RBs—1. *Astronaut Stops Moving*, 2. *Astronaut Is Moving*, 3. *Astronaut Arrived at Activity Location*, 4. *Expected End-of-task Movement*, 5. *Astronaut Unexpected-Movement*, and 6. *Astronaut Staying Late*—build and maintain the CR. The first four RBs have the effect of tracking the Astronaut movements against the daily schedule on the Significance level and forming projections about future Astronaut actions on the Projection level of the CR, using specific information from the daily work plan and space habitat layout as needed. The fifth and sixth RBs become activated when there are patterns of information that indicate a deviation from the work plan—an astronaut leaving early for a next task, staying late to complete an on-going task, or simply moving in a way that is not readily interpretable in terms of the daily plan. These two RBs use those patterns to construct future Events on the Events level that could require Robot action compliance behaviors.

The five ABs—1. *Time to Move a Robot*, 2. *Robot Start New Works Segment*, 3. *Adapt to an Astronaut Staying Late*, 4. *Adapt to an Astronaut Arriving Early*, and 5. *Adapt to an Astronaut Moving off Plan* (early or late)—both move the robot through its daily work plan and generate action compliance behaviors needed to maintain non-interference with the Astronauts.

The first two ABs move the Robot through its daily work plan, using the CR and the robot's tasking in the daily plan. Viewed as an on-going process, there will always be a concept at the Projection level of the CR that defines the Robot's next expected movement time and activity. The first AB functions as a 'clock watcher'. It activates itself when the next robot movement is scheduled to start, terminates the current work activity, and initiates the movement to the next task-location. It also creates a projection of itself (i.e., the robot) finishing that activity later. The second AB similarly activates itself when the robot's movement is finished, invokes the high-level controller to initiate the new work activity, and creates projections for future movements in the CR. In both cases, the ABs perform house-cleaning by deconstructing the old concept projections in the Projection level.

In addition to moving the robot through its daily work plan, these two ABs also provide an infrastructure for action compliance behaviors. They allow other ABs to insert nearer-term actions and movement projections, or to modify existing projections, to implement non-interference behaviors

[2] It is possible that changes made by higher-priority KEs could eliminate the triggering patterns for lower-priority KE, effectively de-activating them. These, and other conflict situations are dealt with in the Principles of Operation (Zachary et al., 2015a).

(e.g., "move out of the way", "wait until the astronaut passes", or "freeze in place until the astronaut passes").

The final three ABs are activated by patterns of information on the CR that represent different root causes of astronauts deviating from projections—an astronaut staying late in a module (presumably to complete a task), an astronaut arriving early in a module (presumably to start early or to prepare for a next-work task), or an astronaut moving off the plan (for an unknown reason). In each case, the logic of the AB implicitly assumes the preceding sentence to determine if an action compliance is necessary, and, if true, to generate non-interference behavior.

The third and fourth ABs respectively address two variants of the case in which an Astronaut's 'static' location blocks the Robot, creating an Event in which the robot must move but cannot. In the first case (AB3), the Robot's only choice is to wait for the Astronaut to leave, assuming that the Astronaut is staying a few minutes late to finish an incomplete task. In the second case (AB4), the Astronaut cannot be assumed to be moving in time for the next scheduled Robot move; in that case the Robot asks Mission Control for a new plan.

The fifth AB concerns the more complex task of the Robot wanting to stay in its current location to finish its task, but an Astronaut is unexpectedly moving (raising the possibility that the Robot will be blocking the Astronaut). This AB examines the CR to determine if an Astronaut is moving toward the Robot (making a robot-blocking conflict possible). If not, no conflict is involved, and the AB terminates. If so, then the consideration moves on to whether or not it can infer a destination for the Astronaut (e.g., to the next work location). If it can make such an inference, then a path to that destination is deduced, along with an assessment of whether or not that path crosses through the robot's current location, creating a conflict event that will require a non-interference action. The preferred non-interference strategy for such projected conflict event is to move out of the way—stop work, find an adjacent un-occupied module, and move to it until the Astronaut passes, then move back and resume work. If no destination can be inferred, then a strategy of backing-away is used. This involves moving to an adjacent unoccupied module to see what the Astronaut does. This backing-off strategy may have to be repeated (i.e., by repeated activations of AB5) if the Astronaut continues toward the Robot and no destination can be inferred. If (or when) no such adjacent unoccupied module exists, the AB applies a strategy of "hunkering down"—freezing in place before the Astronaut enters the Robot's module then unfreezing and resuming work after the Astronaut passes.

Tests of this context model, using simulated robot and astronauts, showed that this keyhole context model performed well in keeping the robot out of the astronauts' way, even when they deviated from their work plan in several different ways (Zachary and Johnson, 2015). This relatively simple keyhole context model integrated directing the robot's execution of its daily work plan with action compliance reasoning needed to avoid interference with astronauts without communicating with them. It worked well because the context model was able to use domain-specific knowledge—of the published daily schedule and the habitat's geometry—to make default assumptions that differentiated expected astronaut actions (e.g., moving on-schedule to next scheduled activities) from unexpected ones (e.g., moving off-schedule). Further, it was able to use this same domain-specific knowledge to develop assumptions about the purpose of the unexpected actions (e.g., staying late to finish, or leaving early for next activity), and to take proactive adaptive action based on those assumptions (e.g., remaining in place and waiting for an astronaut to leave a work area, or moving out of an astronaut's expected movement path).

Still, the keyhole model was inefficient in several ways. It could result in the robot stopping work to move out of an astronaut's way when it wasn't necessary, remaining in place doing nothing when it wasn't necessary, or moving very inefficiently in attempting to stay out of an astronaut's way. Such conditions occur when the default diagnosis assumptions about astronaut actions were incorrect, or when the context model could make no diagnosis about the unexpected astronaut behavior. These are the points where the efficiency of the context model and non-interference action compliance reasoning could be improved by allowing communication with an astronaut. Extending the keyhole context model to one for communicative non-interference is discussed in the next subsection.

8.4.2 Context Model for Communicative Non-interference

To enable a communicative context model, we first created a limited communication capability for the Robot, enabling it to communicate with Astronauts via text messages to/from each Astronaut's wearable personal digital assistant (PDA). The framework from which we approached robot-human communication was that of speech act and pragmatics theory from linguistics, in which communication is viewed as purposive and functional (i.e., as an action), and adaptive to the local context of the communication (Searle, 1969; Searle et al., 1980; Thomas, 2014). From this perspective, the unit of communication of interest was not the individual utterance or text (by astronaut or robot), but the dialog *initiated by the robot* to accomplish a specific purpose of function. The functional goals of a dialog in this non-interference model were:

- determining an Astronaut's intention (i.e., reason for unexpected behaviors);

- gathering information to develop an action plan for non-interference (e.g., to ask about when or where the astronaut might be leaving/ staying/going); and/or,
- obtaining or negotiating to obtain an Astronaut's concurrence with a specific non-interference action plan (e.g., asking if it could remain in location *Y* until the astronaut is finished).

Moving from the keyhole case to one that included purpose-driven robot-initiated communication with an astronaut required three broad elaborations of the keyhole non-interference context model. The first was extending the CR and set of KEs to allow the robot to construct Robot speech acts, understand Astronaut speech acts, integrate them into a dialog structure, and to reason about the dialog contents to achieve the desired disambiguation needed to derive the best possible action compliance behaviors. The second was extending the CR and set of KEs to allow the context model to represent and reason about the mechanics and pragmatics of the interaction with an Astronaut, such as timeliness of responses and adhering to general conventions of human-human conversation, so that the interaction seemed 'normal' to its human participant. The third was extending the CR and set of KEs to make more fine-grained distinctions in constructing action compliance adaptations, based on the relevant dialog, than the three simple strategies used in the keyhole case (i.e., wait, move to a specific location, or hunker-down). Each is described below. Details of specific concept-types in the CR and specific Knowledge Elements are omitted except where absolutely necessary, with the presumption that the reader can deduce them from the prior discussion. Additional specifics of the communications processing can be found in Zachary and Carpenter (2017), which discusses that aspect of the communicative non-interference model in detail.

8.4.2.1 Dialog-related Context Model Extensions

The most basic extension needed to accommodate communications was adding a new concept type 'Astronaut-communication-type' to the Perception level of the CR to represent individual astronaut text-communications. The attributes of each instance of this type included astronaut-name, meta-status, time-received (defined in other cases above), plus 'directed-to', indicating the robot communicated to, and 'communication-content' providing the literal text-string received from the astronaut's PDA.

The main enhancements, however, involved creating a new 'Dialog' panel as a separate component of the context representation, that dealt with linguistic sub-context of robot-astronaut dialogs. This Dialog panel had two levels. The lower (less-abstract) Communications level contains concepts about the texts by the two parties in a dialog, and contains two concept types, one for Robot communications and one for Astronaut communications. Instances of astronaut-communication are created on this level by the PEs and RBs activated in initial processing of an astronaut-communication instance on the Perception level of the main CR panel. Here, on the Dialog panel, the literal content of the text is replaced with a 4-tuple that specifies the:

- intent inferred from the text contents;
- destination of the current or next-movement by the astronaut;
- time at which the astronaut anticipates any next-movement to be started;
- handshake component (if any) intended to close the dialog (e.g., 'bye'); and
- time the text was sent.

The concept for each Robot-communication instance is created in the Dialog panel Communications level as part of the Action Builder that generates the robot text.

The higher (more abstract) Active-Dialog level of the Dialog panel contains concepts on each active dialog in which the robot is involved. It has only one concept-type (called 'dialog-type') whose instances contain information on specific dialogs. Each instance of the dialog concept-type defines the function or purpose of that dialog (e.g., to determine why the astronaut is staying late at a location), the astronaut and robot involved, the time the dialog was initiated, which party texted last, and the processing meta-status of the dialog. Also of critical importance, are several relationships that are instantiated between concepts on the Dialog panel:

- *Belongs to dialog*, which maps a specific robot- or astronaut-communication concept to a specific dialog concept, making the communication instance part of the specific dialog instance;
- *Starts-dialog*, which maps a specific robot-communication concept to a specific dialog concept, indicating that specific robot-communication was the start of that dialog;
- *Astronaut-responds-to*, which maps a specific astronaut-communication concept to a specific robot-communication concept, indicating that the astronaut text was the direct response to the robot-text;
- *Robot-responds-to*, which maps a specific robot-communication concept to a specific astronaut-communication concept, indicating that the robot text was the direct response to the astronaut-text;
- *Working-hypothesis (WH)-of-dialog*, which maps a specific WH from the main CR panel to a specific dialog, indicating that the WH is the

presumed astronaut intent that the dialog is intended to confirm or disconfirm. (The formation of WHs is discussed below.)

These relationships and the concepts they inter-relate represent the dialog, its structure, and its content.

8.4.2.2 Conforming Robot Communications to Human Conversational Norms

It is important that the communicative style be one that would not be immediately alarming or irritating to the human participants. For example, providing a virtually immediate response to an astronaut's text could easily be perceived as annoying or at least as 'non-human'. Thus, the Action Builders responsible for creating text responses were constructed to include reasoning about the timing of such actions, so as to not respond too soon or slowly, repeat a question too frequently, etc.

8.4.2.3 Elaborating Non-interference Reasoning

In the keyhole non-interference context model, the model made its best possible inference of the astronaut's intentions from the astronaut's actions and domain-knowledge of the daily plan and space habitat, and acted on that inference directly. It was forced to do this because the robot could not interact with the astronaut to confirm/disconfirm or otherwise disambiguate among multiple possible inferences. This resulted in the inferred potential conflict that could lead to interference being explicitly represented, on the CR's Events level, as a concept about a specific conflict that it was trying to avoid. Once it is possible to create a dialog to confirm/disconfirm and gain further information on that conflict, however, the conflict by itself became insufficient to construct the dialog. Rather, it became necessary to explicitly represent all attributes of this best inference as a working hypothesis (WH) concept. The WH attributes included the:

- inferred reason for the astronaut's behavior (e.g., leaving late);
- inferred intention associated with that behavior (e.g., trying to complete a task before leaving);
- inferred next destination of the astronaut and its current status (e.g., occupied/unoccupied); and
- time constraint that could affect the robot's flexibility to resolve the conflict without a full new plan from mission control (e.g., how long the robot could stay in the current location before a different astronaut occupied it to start a different work activity).

In addition, several Representation Builder KEs had to be added to construct, maintain, and later deconstruct the WH concept instances, and to establish relationship instances among the WH instance(s) and the conflict instances. The new relationships required were:

- *Conflict behind*, which maps a specific Robot-next-Movement-Blocked-Conflict instance or an Unexpected-Movement-toward-Robot conflict instance to a specific WH instance that explicated the reasoning behind that conflict; and
- *Working-hypothesis-of conflict*, which maps a WH instance to a specific Robot-next-Movement-Blocked-Conflict instance or an Unexpected-Movement-toward-Robot conflict instance.

Thus, these two relationships are the inverse of one-another, and allowed a KE to link from a WH to the conflict it described as well as vice versa. Finally, multiple ABs were added to respond to new information that was gleaned from a dialog with an astronaut. Each of the new ABs was triggered by either:

- a pattern of information on the CR that corresponded to new specific confirming/disconfirming information about a WH; and/or
- the appearance on the CR of a new clarifying detail on astronaut intentions and astronaut plans associated with an existing WH and underlying conflict event.

Once triggered, each of these new ABs would then construct specific interference-avoidance actions for the robot to take and direct the robot controller to execute them. The existing ABs that monitored the CR and otherwise moved the robot through its work plan were unaffected and kept in place.

Tests of this communicative non-interference context model, again using simulated robot and astronauts, showed that it also performed well in keeping the robot out of the astronauts' way. The simulations also showed that the robots were able to engage astronauts in clarifying dialogs in situations where additional information could improve the efficiency of the robots adaptive behaviors (Zachary and Johnson, 2016).

8.5 The Integrated Context Engine (ICE)

The idealized architecture of the NIM computational system was the starting point for the design and implementation of a core engine that would be performant in real-world environments, termed the Integrated Context Engine (ICE). ICE is both a *software framework* that is used to define a domain-specific NIM context model, and a *software system* used to execute the model within an interactive agent or information system. This software gives the agent (e.g., a robot or information system) the information for a context understanding to make its behavior context-sensitive in a way that is explainable to and understandable by humans with which it must interact. ICE provides tools that structure the authoring, editing, debugging, and management of a NIM application while its computational engine executes a NIM context model. This runtime executable manages a set of application program interfaces that allow the engine and model to be integrated into larger computational systems.

However, ICE on its own does not serve a purpose without additional input from an author/developer in the form of domain-specific knowledge and logic. As the name "Integrated Context Engine" implies, the ICE runtime environment combines the runtime engine and an instance of a domain-specific computational NIM model, known as an ICE Application. The domain-specific information required to create a practical ICE Application takes three different forms. First, there is declarative knowledge—the specification of the domain elements in the environment and the specific data structure in which it will be represented. The structure of the declarative knowledge stored within ICE, and the corresponding NIM model, have a significant impact on the reasoning supported for a given ICE application. The ICE runtime environment cannot reason about the information it cannot represent. Second, there is procedural knowledge—the KEs that represent the reasoning that occurs over the sensor input from the environment. Lastly, there are environmental connections linking the ICE Application to an "outside" world, including the conversion of information from their native format to one ICE understands and vice-versa. ICE is not simply a system in which a NIM model can be loaded and run, but the kernel on which a domain-specific ICE Application is built.

Figure 8.5 shows the organization of the ICE runtime environment consisting of three principal subsystems:

1. Declarative Knowledge Representation—corresponding to the NIM CR and related domain-specific declarative knowledge representations;
2. Procedural Knowledge Representation—corresponding, in part, to the set of KEs; and
3. Plugin System (including the Policing Plugin)—corresponding to the Context Processor of the NIM computational system.

These subsystems, including the implications of their design, are described in more detail below. For a more detailed description of the ICE implementation, including the software languages and libraries used, see Zachary and Carpenter (2017). Additional description of ICE can be found in Zachary and Carpenter (2017a).

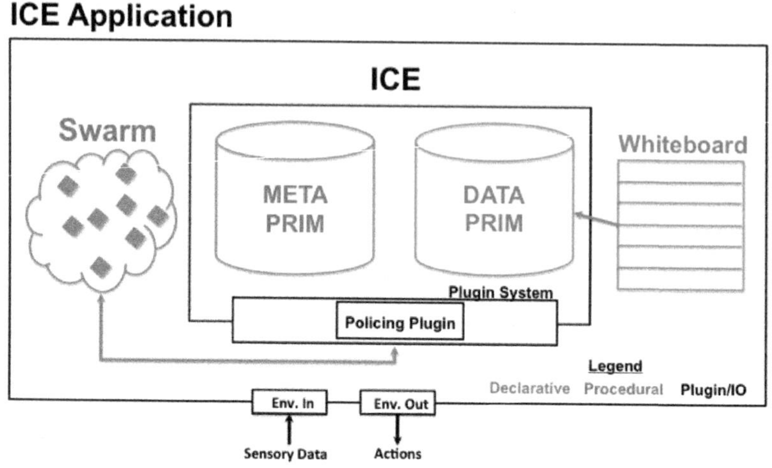

Figure 8.5. ICE application architecture, color-coded by subsystem.

8.5.1 Principles of Design

The design of the ICE runtime environment was guided by two main principles: transparency and computational performance. Transparency was required to prevent ICE from becoming a "black box", all too common with the prevalence of deep-learning architectures. Instead, the goal of the system was to create a model that is compatible with and understandable by humans. This goal could be more easily accomplished if all components of the system were accessible and visible to any interested party, be they author, supervisor, operator, or team mate. We achieved transparency by designing ICE as a stateful system, utilizing a datastore to record not only statements about the world, but also statements about system-internal information such as what actions should be taken and the current state of executing actions. In this way, ICE is a reactive, event-based, system driven by changes in its datastore. In addition to transparency, there are two main advantages to this stateful design. First, there is the capability to stop and restart the ICE system without losing data. Because all the required information is saved to the datastore, including executing actions and what they are executing on, the system can be restored from a saved state. This feature is useful in test environments because it allows for saving and restoring the system to a particular critical point in time, rather than waiting through an entire simulation run. It is also useful in production or real-world environments as it provides a fault recovery capability. The second advantage of the stateful design is the ability to conduct meta-reasoning over the environment. Since actions that are executing are also recorded in the datastore, additional actions can be created that cascade off of the activation of earlier actions.

The second guiding principle of the ICE design was computational performance. ICE was intended to build context representations and make decisions in complex, dynamic environments. To achieve that goal in a timely fashion, the system processes information at a relatively high speed. This performance is important because ICE KEs are authored as procedural swarms to react and perform minor changes to the context representation. In this way, changes perceived in the environment cause a cascade of actions that execute sequentially before a corresponding action is sent back to the environment. The ICE runtime environment was designed to achieve single-digit millisecond identification of state changes. Following this principle led to a revision from the originally prescribed Principles of Operation (PoP) related to KE update/action cycles. The theoretical PoP describe the update/action cycle as a time-sliced process where the effect of activated KEs are calculated, non-conflicting KEs are executed, the remaining activated KEs are re-evaluated, and the process repeats as needed. When implemented computationally, however, the sequential nature of the activation process was inefficient and slow, with the activations from each cycle generating more modifications to the white board, in addition to the changes caused by the environment. If the PoP were strictly followed, the ICE executor could quickly become saturated and the context representation would become delayed and incorrect, with any resulting actions unlikely to occur at the desired time. Instead, ICE is implemented in a parallelized fashion that more closely follows the self-activating nature of the Pandemonium architecture (Selfridge, 1959). Because multiple ICE KEs can be triggered simultaneously, changes are evaluated as transactional groups defined by input from the environment. Once all the activated KEs are identified for a transactional group, they are executed simultaneously. This parallelization greatly increases the speed at which KEs are activated and executed and allows the system to be performant in more realistic settings. This change, however, creates complications relating to execution control, timing, and conflicts, discussed below with policing.

These two guiding principles led to a conflict in data representation. On the one hand, transparency could be more easily achieved by handling and storing data as verbose, human-readable representations in a persistent store, yet this would have a negative impact on performance. Conversely, the most efficient way to handle data while achieving high performance would be to store everything as compact, binary representations in memory, yet this would reduce the human readability and transparency. To compromise between the two principles, a triplestore—a database for storing RDF[3]

[3] RDF, or Resource Description Framework, is a data model format where data takes the form of "subject" – "predicate" – "object" statements, or triples. It is used to capture relationships between data, or resources, primarily in the knowledge management and Semantic Web domains. (See https://www.w3.org/TR/rdf11-concepts/)

statements, as known as triples—was chosen for data storage. The format is reasonably human-readable while being reasonably efficient in terms of storage and retrieval. The verbosity of storing all actions in the datastore and functioning as a reactive system also leads to a slowdown in operations, but it greatly increases transparency.

8.5.2 Declarative Knowledge Representation

The declarative knowledge representation is a critical subsystem of the ICE runtime environment. Decisions regarding how to represent the data affect significant portions of ICE, including the efficiency and specificity of reasoning across the context representation. This section describes the format in which the context representation is stored within ICE, including the organizing structures within ICE as well as the data structures created to support ICE's functionality.

8.5.2.1 Data Representation and Storage

At the core of the ICE runtime environment is a custom triplestore we term PRIM Store. The PRIM Store is a Java-based triplestore containing information about the environment in RDF statements, using a Structured Query Language (SQL) database as the underlying storage mechanism for efficiency. Specifically, PRIM Store is a quadstore that enforces the practice that all RDF statements include a namespace. The PRIM Store supports the use of transactions, allowing multiple changes to be grouped as a single unit to be executed atomically. The major piece of functionality that was added over the storage system on which it is based is pattern recognition, or the ability to listen for specific changes in the data—both additions and removals—and react by executing a designated callback function. In simple cases, these patterns can be specified as statement patterns—an RDF quad statement that has zero or more components left unspecified to match multiple statements. For more complex situations, patterns can be specified through a SPARQL query—the triplestore equivalent of the SQL database query language. When a callback function is executed, it is provided details about what caused the execution, including the transaction in which the change occurred and peripheral changes that were made. Transactions are processed sequentially, and the transactional lock is not released until all of the patterns have been evaluated and the appropriate callback functions notified. While this procedure negatively affects the efficiency of changes to the datastore, it ensures that the resultant state from a transaction is fully evaluated before additional changes can occur. As a result, if a watched state occurs, even if only for an instant, the appropriate callback is executed. In keeping with the principle of transparency and statefulness, a pattern is installed in the PRIM Store by posting the defining information to a unique namespace. The statements are then processed to create the corresponding listener component. In this way, the PRIM store maintains a transparent definition of all of the installed patterns in a persistent format.

ICE consists of two separate PRIM stores, Data PRIM that stores the context representation, and Meta PRIM that handles information relating to the functioning of the ICE System—described further in Section 8.5.3.1 below. Data PRIM is the primary datastore of the ICE system, containing ICE's context representation and its view of the world. It is this datastore that houses data ingested from the external world with which the installed KEs interact and reason against.

8.5.2.2 Panels and Levels

Declarative knowledge within the NIM framework is typically organized into hierarchically nested groups to maximize understandability and to improve performance. Groups of related statements are termed Levels and a group of interrelated Levels is termed a Panel. These concepts are implemented in ICE as graphs, or namespaces, within the datastore, making the statements quads rather than traditional triples. While Levels are used within the NIM framework to provide additional information about posted data, the ICE system uses the graphs for search and query optimizations—the smaller the subset of data on which the system must evaluate a particular operation, the faster it will be. While by default ICE supports the functionality of the SAN Panel—situation, assessment, and narrative—and Levels internal to the ICE runtime environment, it also provides the capability for an author/developer to create and curate additional Panels and Levels specific to the domain application. For example, an ICE Application that handles communication between ICE agents and external entities may require the use of a Dialog Panel containing Levels that store information relating to communications between ICE-controlled agents and human counterparts.

8.5.3 Procedural Knowledge Representation

The procedural knowledge representation is another key component of the ICE runtime environment. While the declarative knowledge representation dictates what information can be stored within ICE, the procedural knowledge representation specifies what operations are permitted on said data and how those processes are managed and monitored. This section describes the implementation of the Knowledge Element Swarm as well as the reasoning to execute the KEs.

8.5.3.1 Meta PRIM

ICE contains a secondary PRIM Store, Meta PRIM, which stores internal information used by the ICE runtime environment for processing and managing Data PRIM. Meta PRIM stores all of the information required to maintain the KEs of the system including what KEs exist and are registered to the system, the current state of KEs (e.g., triggered, running, completed, etc.), and what caused each KE to trigger, including the pattern that was matched and the transaction that caused it. Meta PRIM is more than just an audit log, however, as the processing of KEs—their creation, triggering, dispatching, and collecting—is controlled through the posting of statements to Meta PRIM and the execution of specialized listeners waiting for the appropriate statements. There are numerous benefits to structuring the logic this way. First, maintaining all the required information in a standardized and auditable format means the system can be easily stopped or resumed to a particular state. Second, by listening and reacting to changes of KE-related statements, a level of meta-reasoning about what other actions are being taken by the system can be constructed. Included in this meta-reasoning is the ability to implement policing and other important system functionality as modular components. These components can incorporate the current state of the system, or changes to the system, as part of their reasoning. In addition to KE-related information, Meta PRIM stores a register of the components, whether internal (KEs) or external (environment), responsible for the transactions that occur. When this information is combined with the stored information of what transaction resulted in a KE triggering, a basic level of traceability becomes available. While it does not provide all of the information necessary to develop a complete picture of why a particular action was taken, it allows for statements such as "KE *X* resulted in a new Transaction *Y* being made which triggered KE *Z*". These statements proved to be vital in examining the flow of changes throughout the life-cycle of the system.

8.5.3.2 Knowledge Elements

KEs are procedural swarms that react to, and perform minor changes to, the context representation. KEs contain both a representation for their self-activating triggering conditions as well as a logic body that defines the behaviors/actions resulting from an execution. They are implemented within the ICE system as networked representations that exist as separate processes from the main ICE System, potentially even on different hosts. KEs are also implemented as "hot-swappable" in that they can be added or removed while the system is running. There are four main components to KEs: trigger conditions, locks, priority, and a body. The trigger condition is a list of one or more patterns installed in Data PRIM as listeners.

Trigger conditions are provided as a list to reduce the tedium of creating multiple KEs with the same body but different triggering conditions. The patterns within the trigger condition list can be any combination of simple patterns—RDF wildcard statements—and complex query patterns. The implementation of the locking mechanism used to handle conflicts between KEs is simplified from what is presented in the NIM framework and PoP. Rather than automatically determining which triggered KEs will result in conflicting changes—computationally expensive and inefficient—KEs are required to contain explicit locks when policing is required. These locks are specified alongside the relevant trigger condition pattern. The priority system is implemented as an integer specified at authoring time, rather than a function evaluated at runtime that factors in the conditions under which the KE was triggered. While this means priority is meaningless between instances of the same KE, it does allow, when combined with the locking mechanism, the authoring of "override" KEs meant to shadow default functionality if their trigger conditions are met. Implementation of the more sophisticated priority mechanism is currently restricted by the lack of a sufficiently expressive language to represent the logic.

The remaining, and at times most complex, component of the KE is the body. The body is currently implemented as a function in native Java source code. It is mostly unrestricted in what it is allowed to do, able to leverage the entirety of the Java language. While this code is flexible, it also means the author of a NIM model must be knowledgeable about Java or have the assistance of a developer to translate the logic into a given NIM model. It is desirable to have a simplified language that supports the manipulation of statements in ICE for the creation of an authoring tool understood and used by knowledge engineers.

8.5.4 Plugins

A useful feature of ICE that builds upon the flexibility of maintaining system and world-state information with the PRIM stores is the Plugin system. The Plugin system increases the modularity of ICE runtime by allowing reasoning elements to be installed as listeners to either or both Meta and Data PRIM. Plugins differ from Knowledge Elements in that their execution is not policed and they live closer to the underlying datastores, providing quicker access time. As all operations within the ICE system happen as a result of statements and changes within Data and Meta PRIM, Plugins can be used to substantially modify the reasoning of the ICE system. It is through this mechanism that efficient meta-reasoning, such as the policing component, is achieved. In addition to meta-reasoning components, auxiliary Plugins were created to support the debugging and analysis capabilities, including one that records changes to Meta and Data PRIM to a text file, and multiple visualization Plugins that provide different views into the current contextual state.

8.5.4.1 Policing

Policing is crucial to maintaining order within the self-activating environment of the ICE system. The policing logic is, as previously mentioned, implemented as a modular Plugin component. This has allowed two different policing mechanisms to be implemented. The first policing Plugin created was a basic pass-through component. Rather than perform any level of reasoning about the KEs that have been triggered and should proceed, the Plugin simply activated any KE that was triggered. While this does little to enforce order in the system, it proved that the policing logic could be implemented as a Plugin for progressing KEs through their states. The second policing Plugin that was created, and is currently in use, is the Explicit-Lock Plugin. This policer requires locks to be explicitly defined within the triggering condition of KEs. A lock consists of two parts: a type, which determines the KE instances taken into account when evaluating for conflicts, and an object to lock (LO). The LO allows the contents of the trigger match to be used by the conflict reasoning (i.e., only KEs with locks on the same objects will be considered to be in conflict). An LO is specified depending on the type of pattern the lock is associated with; if associated with a statement pattern, the LO is specified by a statement component (e.g., Subject, Predicate, Object, Graph), whereas if the lock is associated with a query pattern, the LO is specified by a variable name present within the query. In the event of a conflict, one KE instance is selected as a "winner" and allowed to execute, while all other instances are canceled.

A limitation of the current locking definition format is that only a single locking variable can be defined. It is not currently possible to lock on multiple objects within the matched pattern, which could be useful in some situations, as it is unclear what multiple objects would mean; would two objects mean "only conflict with other locks that also match these two objects", or would it mean "conflict with other locks that match either of these two objects". In the latter case, the potential for deadlocking[4] would need to be addressed.

There currently exist three distinct types of locks that are recognized by the Explicit-Lock policer: *Transaction*, *Transaction-Knowledge-Element*, and *Knowledge-Element*. In addition to specifying the scope for evaluated KE instance conflicts, the type of lock also determines how a "winner" is

[4] Deadlocking is a stalemate situation in concurrent computing in which competing parties are each stuck waiting for the opposing party to release a resource (see Coulouris et al., 2012).

Table 8.1. Characteristics of the lock types supported by the explicit-lock policer.

Lock Type	Conflict Scope	Winner Selection
Transaction	All KE instances within a transaction	Highest Priority
Transaction-Knowledge-Element	All KE instances of the same type, within a transaction	Random
Knowledge-Element	All KE instances of the same type, including any currently running	Already running, if present, else random

chosen from the set of conflicting KE instances, if a conflict is detected. The characteristics of the different lock types are presented in Table 8.1.

The differing characteristics of the lock types resulted in unique situations in which each type was used. For example, the *Knowledge-Element* lock type was used in situations in which a KE could be triggered in short succession, but multiple simultaneous executions would result in an invalid state, such as when the Robot's next intended action is being declared. Meanwhile, the *Transaction* lock type was commonly used in situations in which one KE was intended to "override" another activated KE—that is, one KE specializes a particular generic KE response, so if the special KEs conditions are met, it should run in place of the generic KE. While there are certainly additional lock types that could be created, primarily by modifying the conflict scope, these three lock types proved sufficient for NIM models of the Space Robot domain.

8.6 Conclusions and Future Research

The NIM framework is a general but abstract approach to developing computational implementations of context understanding and reasoning process for robot and other information system applications. The specific context model applications presented here for robot-human interaction demonstrate NIM's practical application. We believe that the core keyhole (non-communicative) non-interference model was both simple in structure and effective. We are further encouraged by the fact that the model could be easily extended—with virtually no changes to its structure or content—to include robot-human communications to deal with the limitations of the core model's default reasoning approach. Still, additional applications are needed to assess the broader utility of the NIM theory and framework and the ICE toolset.

There are several directions that future research based on the results reported here can take. One is the extension of the model discussed here beyond non-interference action compliance to active human-robot cooperation and teamwork. Such research could explore the role that convergent versus divergent context understanding can play in the pragmatic coordination of teamwork activities. Another is further exploration of the role of dialog context, including whether it should be integrated into overall context representation or whether it is best treated as a separate facet that is separate from, but interconnected with, the non-dialog content of the current context representation model.

We also envision several lines of extension and elaboration of the ICE toolset. Future iterations of ICE may focus on defining functionality necessary for authoring KE bodies in a language format that is expressive and easy to use. It has been our experience that if specialized knowledge of a system-specific language is required to author a model, it is unlikely to gain widespread support in the community. Java, while not an ideal modeling language, is a well-known, expressive language that could serve this purpose if an authoring environment was provided to guide a modeler through commonly used design patterns.

Other ICE developments are contemplated which extended run-time capabilities. While implementing different use-cases with ICE, we found that in many cases we still wanted multiple conflicting KE instances to run, just not at the same time, rather than being canceled. It would be useful to have a "wait-until completed" option, in which KE instances are held in a waiting state until the conflicting instance has been completed. This functionality is currently planned, with the main outstanding question being how to determine if a KE instance that is being resumed from a wait condition is still valid to be run. Additionally, to increase traceability of an application during execution, we are exploring additional explanation tracing logic throughout the execution engine. Ideally, this tracing capability will be combined with a visualization/exploration facility that can map execution activity to the domain-specific knowledge in the application.

Finally, we note that the Pandemonium-style architecture of NIM makes NIM-based context models inherently open to parallel computing as well as highly-distributed execution. Such implementations could set the stage for very large and high-level context representation that could integrate information across large-scale Internet of Things (IoT) systems.

Acknowledgements

This research was supported in part by Contract NNX14CJ38P from the NASA Johnson Space Center and in part by Contract N00014-14-P-1186 from the Office of Naval Research. We gratefully acknowledge their support. This paper reflects only the ideas and conclusions of the authors and not the sponsoring agencies; any errors or omissions are theirs alone. Special thanks are due to Dr. Kimberly Hambuchen from NASA and Dr. Jeffrey Morrison from ONR for thoughtful comments and insights on our research, and to Dr. William Lawless for his comments on earlier drafts of the manuscript.

References

Barwise, J. and Perry, J. 1983. Situations and Attitudes. Cambridge, MA: MIT Press.

Bruner, J. 1991. The narrative construction of reality. Critical Inquiry 18(1): 1–21.

Cohen, P.R., Perrault, R. and Allen, J. 1981. Beyond Question-Answering, Report No. 4644, BBN Inc., Cambridge, MA.

Corkill, D.D. 1991. Blackboard systems. AI Expert 6(9): 40–47.

Coulouris, G., Dollimore, J., Kindberg, T. and Blair, G. 2012. Distributed Systems Concepts and Design, 5th Edition., Pearson Education.

Diftler, M.A., Mehling, J.S., Abdallah, M.E., Radford, N.A., Bridgwater, L.B., Sanders, A.M., Askew, R.S., Linn, D.M., Yamokoski, J.D., Permenter, F.A. and Hargrave, B.K. 2011. Robonaut 2-the first humanoid robot in space. 2011 IEEE International Conference on Robotics and Automation ICRA (pp. 2178–2183). NY: IEEE.

Endsley, M.R. 1995. Toward a theory of situation awareness in dynamic systems. Human Factors 37(1): 32–64.

Endsley, M.R. and Garland, D.J. 2000. Situation awareness analysis and measurement. Boca Raton FL: CRC Press.

Fong, T. and Micire, M. 2013. Smart SPHERES: a Telerobotic Free-Flyer for Intravehicular Activities in Space. Proceedings AIAA SPACE 2013 Conference & Exposition. AIAA.

Miller, L.C., Read, S.J., Zachary, W., LeMentec, J.-C., Iordanov, V., Rosoff, A. and Eilbert, J. 2006. The PAC cognitive architecture. *In*: Gratch, J. et al. (eds.). Proceedings of Intelligent Virtual Agents 2006 (IVA 2006), Lecture Notes in Artificial Intelligence, 4133. Berlin: Springer-Verlag.

Miller, L.C., Read, S.J., Zachary, W. and Rosoff, A. 2010. Modeling the impact of motivation, personality, and emotion on social behavior. pp. 298–305. *In*: Chai, S.–K., Salerno, J.J. and Mabry, P.L. (eds.). Advances in Social Computing. Third international conference on social computing, behavioral modeling, and prediction, SBP2010. Lecture Notes in Computer Science, Springer-Verlag, Berlin Heidelberg, LNCS 6007.

Nii, H.P. 1986. Blackboard application systems, blackboard systems and a knowledge engineering perspective. AI Magazine 7(3): 82.

Radvansky, G.A. and Zacks, J.M. 2011. Event perception. Wiley Interdisciplinary Reviews: Cognitive Science 2(6): 608–620.

Schank, R.C. 1995. Tell me a story: Narrative and intelligence. TriQuarterly Books.

Searle, J.R. 1969. Speech Act Theory, Cambridge: Cambridge UP.

Searle, J.R., Kiefer, F. and Bierwisch, M. (eds.). 1980. Speech act theory and pragmatics (Vol. 10). Dordrecht: D. Reidel.

Selfridge, O.G. 1959. Pandemonium: A paradigm for learning. pp. 513–526. In: The mechanism of thought processes (Proceedings of a symposium, National Physical Laboratory, Teddington, England).

Suchman, L.A. 1987. Plans and situated action: The problem of human-machine communication, New York, NY: Cambridge University Press.

Thomas, J.A. 2014, Meaning in interaction: An introduction to pragmatics, New York, NY: Routledge.

Zachary, W., Rosoff, A., Miller, L. and Read, S. 2013. Context as cognitive process: An integrative framework for supporting decision making. pp. 48–55. *In*: Laskey, K., Emmons, I. and Costa, P. (eds.). Proceedings of 2013 Semantic Technologies in Intelligence, Defense and Security. CEUR Conf. Proceeding Vol-1097.

Zachary, W. and Johnson, M. 2015. Quarterly Demonstration Report #2: Context Augmented Robotic Interaction Layer (CARIL) PH II. CHI Systems Inc. Technical Research Report. Plymouth Meeting, PA: CHI Systems Inc.

Zachary, W., Johnson, M., Hoffman, R., Thomas, T., Rosoff, A. and Santarelli, T. 2015b. A context-based approach to robot-human interaction. Procedia Manufacturing 3: 1052–1059.

Zachary, W. and Johnson, M. 2016. Quarterly Demonstration Report #4: Context Augmented Robotic Interaction Layer (CARIL) PH II. CHI Systems Inc. Technical Research Report. Plymouth Meeting, PA: CHI Systems Inc.

Zachary, W. and Carpenter, T.J. 2017. Using context and robot-human communication to resolve unexpected situational conflicts. In 2017 IEEE Conference on Cognitive and Computational Aspects of Situation Management (CogSIMA). pp. 1–7, DOI: 10.1109/COGSIMA.2017.7929596.

Zachary, W. and Carpenter, T.J. 2017a. Cognitively-inspired computational context. *In*: Lawless, W. and Mittu, R. (eds.). Proceedings, AAAI Symposium: SS-17-03: Computational Context: Why It's Important, What It Means, and Can It Be Computed? Stanford, CA: Assn. for the Advancement of Artificial Intelligence.

Zacks, J.M., Speer, N.K., Swallow, K.M., Braver, T.S. and Reynolds, J.R. 2007. Event perception: a mind/brain perspective. Psychol. Bull. 133: 273–293.

Context-Driven Proactive Decision Support

Challenges and Applications

Manisha Mishra,[1,*] *David Sidoti,*[2] *Gopi V. Avvari,*[3] *Pujitha Mannaru,*[4] *Diego F.M. Ayala*[5] and *Krishna R. Pattipati*[6]

9.1 Introduction

In recent years, the use of unmanned aerial, ground and underwater vehicles (UxVs) has become ubiquitous in civilian and military applications, such as surveillance, search and rescue operations, counter-smuggling, counter-terrorism and cargo delivery, to name a few. The primary reason behind the pervasive use of UxVs is their ability to operate from ships/sea bases, ultra-long endurance and high-risk mission acceptance. Additionally,

[1] Algorithm Engineer, Delphi Electronics and Safety (doing business as Aptiv) Kokomo, Indiana, 46902.
[2] Department of Electrical and Computer Engineering, University of Connecticut, Storrs, Connecticut, CT, 06269; david.sidoti@uconn.edu
[3] Tracking Engineer, Delphi Electronics and Safety (doing business as Aptiv) Kokomo, Indiana, 46902; gopi.avvari@uconn.edu
[4] Candidate, Department of Electrical and Computer Engineering, University of Connecticut, Storrs, Connecticut, CT, 06269; pujitha.mannaru@uconn.edu
[5] Rokk3r Labs, Miami, FL, 33127; dfm0804@engr.uconn.edu
[6] Board of Trustees Distinguished Professor, UTC Professor in Systems Engineering, Department of Electrical and Computer Engineering, University of Connecticut, Storrs, Connecticut, CT, 06269; krishna.pattipati@uconn.edu
* Corresponding author: manisha.mishra@uconn.edu

as these unmanned vehicles are made smaller and more economical, as compared to their manned counterparts, they can be easily deployed in remote locations to collect real-time data or conduct tasks that cannot be performed by humans. However, UxVs have limited ability to react against environmental perturbations (e.g., pop-up threats, sudden changes in the weather, etc.) and require intermittent human intervention (operators supervising and interacting with multiple UxVs in a number of control modes). Although the human-UxV collective has the potential to improve mission performance, as compared to teams composed of only humans or only UxVs, it is imperative to provide the operators/decision makers (DMs) with relevant information in a timely manner in order to make effective decisions, even under dynamic, uncertain and challenging mission conditions, without overloading the operators. It has been analyzed that the volume, velocity and variety of collected data can actually decrease the effectiveness of the decision makers by creating cognitive overload, especially in fast-paced decision making environments (Malik et al., 2014). As an example, operators flying an MQ1B Predator in Afghanistan in 2009 were so focused on a fierce firefight that they failed to notice that the unmanned aircraft was headed toward a mountain; the aircraft was destroyed on impact and damage was estimated to be $3.9 million USD. Moreover, eight of the soldiers who were to be provided air support by the Predator were killed (Drennan, 2010). If alerts and tasking had been appropriately allocated to the right operators on the team, those lives may have been saved. Harnessing the right information from the vast volumes of data using appropriate decision support tools is necessary to transform the current decision making process from a reactive one to a proactive and predictive process so that it is robust under unforeseen circumstances (Malik et al., 2014). Thus, research is needed to develop proactive decision support systems (DSSs) that would facilitate rapid mission planning/re-planning in a dynamic, asymmetric, and unpredictable mission environment, while considering the status of the human operators.

A key step towards development of a DSS is to understand the components of planning. A comprehensive mission planning problem involves Boethius' *who, what, why, how, where, when, with what* (Han et al., 2013), implying *who* has the expertise to make the plan (DMs who may be humans or autonomous agents), *what* needs to be planned (tasks, jobs, and actions to be executed), *why* make the plan, *how* to achieve the expected outcome (the assignment of assets to tasks), *where* the plan is executed, *when* the plan is executed including for how long, and *with what* facilities to make the plan (the DSS), as shown in Figure 9.1. Although the existing DSSs process data to decisions, they are inundated with too much data, resulting in cognitive overload of DMs and increased probability of mission failure. This requires the DMs to spend a large amount of time in identifying and

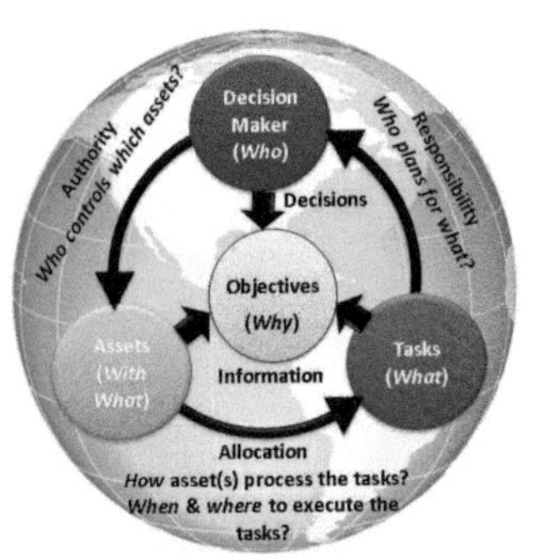

Figure 9.1. A canonical mission planning problem describing the key mission planning queries and elements.

interpreting the mission context, which is often cumbersome and error-prone. Motivated by the need to provide proactive decision support (PDS) to aid the modern war-fighters, in this chapter, we focus on two things: (a) identification of the key challenges in the development of PDS within a dynamic and uncertain mission environment; and (b) a brief discussion of solutions to address some of the PDS challenges within the maritime domain.

9.2 Proactive Decision Support

The objective of PDS is to facilitate the delivery of the *right* data/information/knowledge from the *right* sources in the *right* mission context to the *right* DM at the *right* time for the *right* purpose—a concept which has been termed 6R in the literature (Smirnov, 2006). Delivering the 6R to the DMs allows them to understand the current mission context and its projected impact on their operations, which enables them to ensure the safety of the fleet, dynamically adjust target selection, change mission timing, and tailor assets (e.g., sensors, weapons, people) and tactics (e.g., search patterns, sensor placement, route planning) to the mission needs. The fundamental challenge in achieving 6R is to conceive a generic framework that encompasses the dynamics of the relevant contextual elements, their interdependence, and their correlation to the current and evolving situation all the while taking into account the cognitive status of the DMs. This requires identifying, processing and integrating decision-relevant information from structured, semi-structured and unstructured data representing dynamic interactions between the key mission planning elements, as shown in Figure 9.2. However, these challenges are further exacerbated when the mission operates in disconnected, intermittent and low bandwidth (DIL) network-centric environment in addition to being uncertain and dynamic, where the information may be either lost or not up-to-date. Additional challenges include: (i) How to *define* the components (elements) of mission context

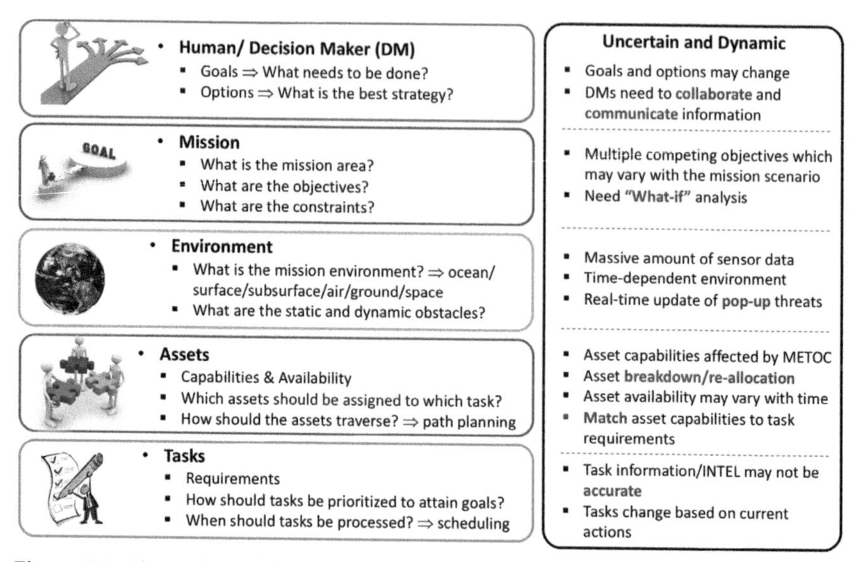

Figure 9.2. Abstraction of the mission planning elements includes the decision makers, mission, environment, assets and the tasks (or threats). The uncertain and dynamic mission environment makes the mission planning problem challenging.

and methods to represent them? (ii) How to *infer* context from operational data of dynamic interactions of contextual elements? (iii) How to *detect* anomalous situations and project potential paths (*predict the future context*) that the mission can take given an inferred context? (iv) How to *identify* the relevant information based on the source, destination, time, evolving context and information demands of DMs? and, (v) How to *disseminate* the contextual information within a DIL environment? In this chapter, we focus mainly on the first three challenges and their solution approaches.

The first step towards proactive decision making is to define the mission context in order to accurately anticipate its impact on mission performance, so that alternative courses of action (COAs) can be taken in a timely manner. The word "context" has Latin roots consisting of "con" meaning "to join together" or "to weave together" and "texere" meaning "to weave" or "to make"—thus implying weaving together the circumstances that form the setting of an event or scenario. In the past few decades various attempts have been made to elucidate context, but most of them define it based on a particular scenario. For example, in (Schilit and Theimer, 1994; Brown, 1995; Brown et al., 1997), context is defined as objects, location and identities of nearby people, etc. In most of the literature, the authors either consider context as the physical environment (real world), computing environment or the user environment (Dey et al., 2001; Brezillon, 1999). Here, we define context as a multi-dimensional feature space consisting of mission (goals and purpose), environment, assets, threats/tasks (as external context) and

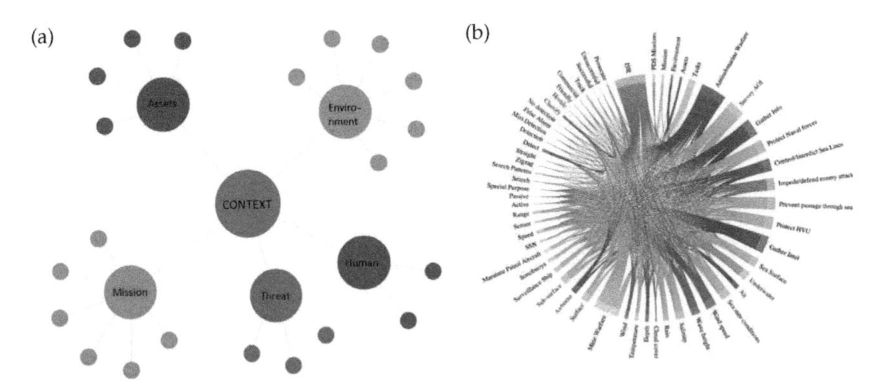

Figure 9.3. (a) Defining context as an evolving multi-dimensional feature space consisting of Mission, Environment, Assets, Threats and Humans (MEAT-H). (b) Representing the MEAT-H elements via dependency wheel (Gaspar et al., 2014).

the human/DM (as internal context), which dynamically evolves with time, as shown in Figure 9.3(a). Defining and representing mission context enables contextual information to be attached to data for later retrieval and decision making. Some software tools to represent context include, but are not limited to, Bayesian network software, Dependency wheel (shown in Figure 9.3(b)), Protégé (Bayes Server, 2017; Gaspar et al., 2014; Noy et al., 2003), etc. Other representative context modeling frameworks include action-goal attainment graphs, factor graphs, multi-functional flow graphs, hidden Markov models, dynamic hierarchical Bayesian networks (DHBN), to name a few (Murphy, 2012; Koller and Friedman, 2009; Bishop, 2006; Deb et al., 1995; Jensen, 1987; Zhang et al., 2016).

Once context is represented, the next logical step is to *detect* changes in the mission by monitoring the variables and comparing them to their expected values in order to identify the root cause of an anomalous scenario. Since context is inherently uncertain, some of the approaches to detect context include statistical hypothesis testing and information prioritization (for active learning). The impact of the identified root cause can be analyzed by projecting the context into the future via Kalman filters, particle filters or DHBN (Bar-Shalom et al., 2004; Arulampalam et al., 2002; Murphy, 2012). Malik et al. (2014) discuss a dynamic covariance kernel density estimation technique, in which the prediction process is influenced by the spatial correlation of recent changes in context. Depending on the projected context type, i.e., anticipated, unfolding or unforeseen, the DMs are accordingly informed to pre-plan, adapt or re-plan COAs, respectively, as shown in Figures 9.4 and 9.5. Our proposed PDS framework facilitates the *detection* of changes in mission context via utilization of up-to-date data sources, *diagnosis* and *prediction* of context to develop "what-if" analysis,

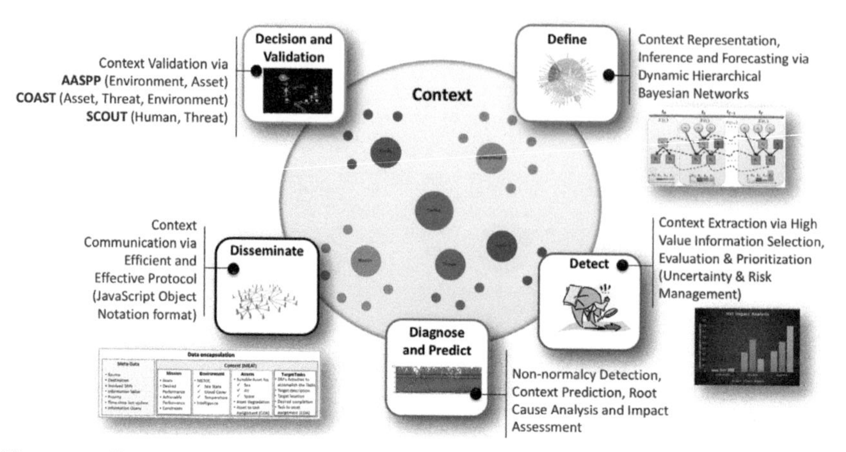

Figure 9.4. Proactive decision support requires delivering the 6R via context definition, context detection, context diagnosis and prognosis, context dissemination and context validation.

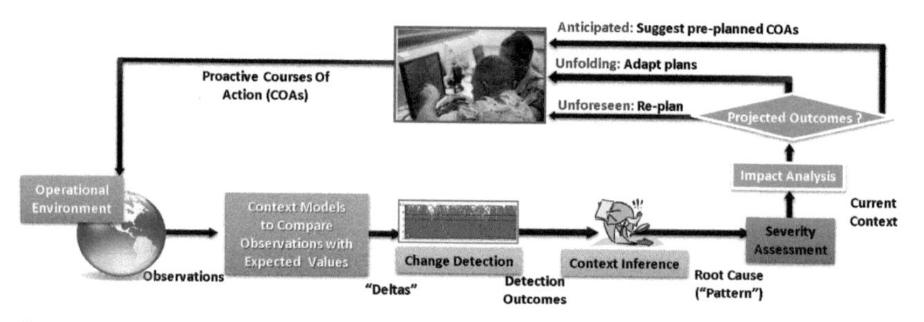

Figure 9.5. Proactive decision support framework for context detection, inference, severity assessment, impact analysis and projected outcomes.

and provision of relevant recommendations, given the DM's workload, time pressure and expertise.

9.3 PDS Applications

In this section, we demonstrate the application of our generic proactive decision making framework using three maritime domains: (i) anti-submarine warfare (ASW) mission planning where the weather impacts the asset capabilities; (ii) counter-smuggling operations in the face of weather changes, i.e., external environmental context; and (iii) dynamic unmanned aerial system (UAS) operations with environment and DM's cognitive workload state as the external and internal context variables, respectively.

9.3.1 Asset Allocation for ASW Mission Planning

The ASW mission involves allocating surface, airborne and sub-surface ASW platforms to search for, detect, classify, track and prosecute hostile submarines within a dynamic and uncertain oceanographic environment affected by meteorological and oceanographic (METOC) conditions. The Naval Meteorology and Oceanography Command's (NMOC's) mission is to provide the commanders with the *right* METOC information, delivered at the *right* time, in the *right* format to give them the decisive edge in military operations. Due to the dynamic nature of the environment, METOC data is continuously collected, analyzed and disseminated in order to develop weather predictions. As the environmental-context evolves, the collected data is used for assessing the impact of the current and forecasted environment on individual sensors and weapon platforms, as well as tactics in the form of performance surfaces (e.g., probability of detection surfaces). The DM is updated with the performance surfaces for sensors and weapons, with uncertainty associated with them, to generate plans or evaluate COAs before assets conduct actual ASW search activities. When commanders understand the forecasted evolution of the environment, its impact on their operations, and have confidence in that knowledge, it allows them to ensure the safety of the fleet, dynamically adjust target selection, change mission timing, and tailor assets (sensors, weapons and people) and tactics (e.g., search patterns, sensor placement, route planning) to the mission needs and the forecasted environment state. Currently, the existing Battlespace on Demand (BonD) framework provides a systematic approach to convert knowledge of the forecasted oceanographic environment into war-fighting and shaping decisions (Price, 2007; Sidoti et al., 2017a; Mishra et al., 2014b). However, the Decision layer in the BonD architecture is primarily a manual capability, requiring experienced, highly-trained personnel for superior mission performance. Therefore, developing proactive decision support capability is the key objective for enhancing ASW mission planning and effectiveness.

A MATLAB®-based Asset Allocation and Search Path Planning (AASPP) tool with the different MEAT (Mission, Environment, Assets and Tasks/Threats) elements of the ASW problem is illustrated in Figure 9.6, where the mission objective is to maximize the detection probability of hostile submarines over a finite time horizon. A pattern of observations obtained by the sensors and its dynamic evolution over time provides the context-relevant information to infer the target's existence and its location. The AASPP tool mimics the layers of the BonD framework and includes a realistic model of METOC effects on sensor performance and search planning, while providing context-specific COAs. The recommended COAs include the search area partitioning, allocation of assets to these search regions and search path planning for each asset to detect the moving target. The environment model allows the commanders to load the environmental data (e.g., salinity, temperature, wind-speed and wave height) within a search region, which is represented using a color coded map denoting the severity of weather parameters, as shown in Figure 9.6. These evolving environmental parameters (i.e., changing context) impacts the detection capabilities of the search assets. Therefore, the tool provides appropriate recommendations to DMs as to how often they should revise their plan to improve the search effectiveness. In the what-if analysis mode, the software allows the DMs to choose the number of assets and their ranges. We invoke the software with different evolving contexts, e.g., weather conditions, asset locations, target dynamics, and analyze the results for various scenarios. In calm weather scenarios, the target is easily detected by the assets; however, in the case of bad weather conditions, a miss-detection is highly likely as the detection probabilities of the assets are severely compromised (Mishra et al., 2017a). An interesting scenario would be to extend the existing PDS framework for theater ASW mission planning in littoral environments via incorporation of multistage pursuit-evasion games (e.g., Stackelberg security games) for intelligent and opportunistic targets.

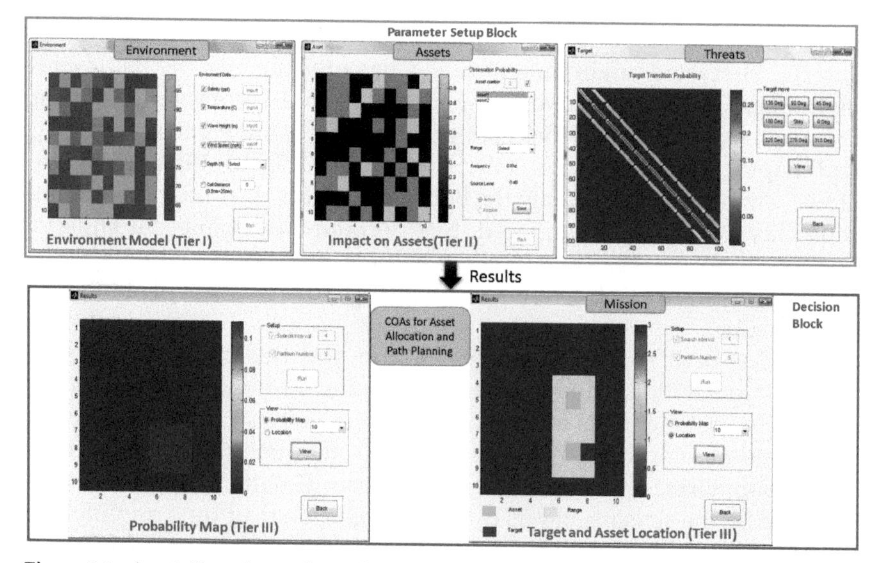

Figure 9.6. Asset allocation and search path planning decision support tool. The tool mimics the three layers (Environment, Performance and Decision) of the BonD (Battlespace on Demand) framework. We ran the software under different sea state conditions (a measure of wind speed and wave height also known as Beaufort scale). During calm weather conditions, the target (denoted by the red squares in Tier III) is detected by the asset as it lies within the detection range of the asset (denoted by the yellow squares) (Mishra et al., 2014a; 2017a).

9.3.2 Counter-Smuggling Operations

The counter-smuggling problem involves surveillance and interdiction operations as sub-goals/tasks to be executed using surveillance and interdiction assets, respectively, to best thwart potential smuggling activities, while simultaneously increasing the situational awareness. Probabilistic information on smuggling activity is generated in the form of color coded heat maps based on intelligence and meteorological and oceanographic information (e.g., wind speed and direction, wave heights, and ocean currents). This information is interpreted as probability of activity (POA) surfaces, which constitute the sufficient statistics for decision making (Bertsekas et al., 1995). Given these POA surfaces, objective of the counter-smuggling problem is to efficiently allocate a set of heterogeneous sensing and interdiction assets to maximize the probability of smuggler detection and interdiction, subject to mission constraints, such as asset availability, asset capabilities (e.g., range, speed) and sensor coordination.

The dynamic allocation of surveillance and interdiction assets under uncertainty may be viewed as a moving horizon stochastic control problem, as shown in Figure 9.7. The DMs assign the surveillance assets (e.g., P-3, E-2C surveillance aircraft) to search regions with a high probability of

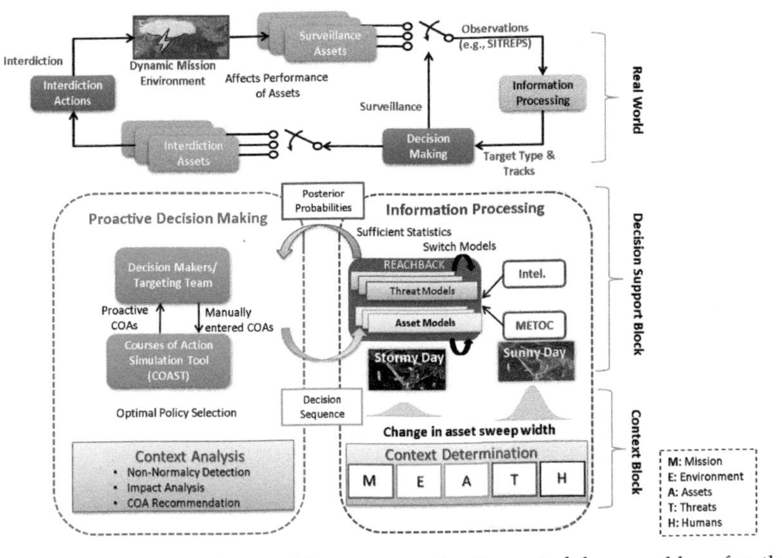

Figure 9.7. Counter-smuggling problem as a stochastic control loop problem for the allocation of surveillance and interdiction assets within a dynamic and uncertain mission environment (An et al., 2012). The decision support block consists of information processing and proactive decision making. The context block is shown at the bottom. When the environmental conditions change (i.e., a change in one of the MEAT parameters, e.g., a change in weather or asset capabilities changes the context), it affects the performance of assets and, consequently, the mission performance.

smuggling activity, and the observations from these assets are processed to characterize the target types and their trajectories. The newly collected information (e.g., intelligence, detections, interdictions, weather data, etc.) serves as a stimulus for identification of context change, and is then relayed back to the reachback cell in the form of situational reports. The reports are then extracted, processed and aggregated to generate new POA maps for the next planning cycle. The context-relevant information gathered by the surveillance assets is communicated to the interdiction assets to adapt their COAs to the new context.

The coordinated surveillance and interdiction problem is NP-hard. A branch and cut algorithm is used to solve the surveillance asset allocation problem. Sidoti et al. (2017b) use approximate dynamic programming, coupled with rollout and Gauss Seidel techniques, to solve the interdiction problem. These algorithms are embedded in the Courses of Action Simulation Tool (COAST) as shown in Figure 9.8, an optimizer in widget format that is integrated with Google Earth.

COAST automatically incorporates context, as it has access to different asset, threat, environment and DM behavior models, operational data and the inferred context. As an illustration, Figure 9.9(a) demonstrates

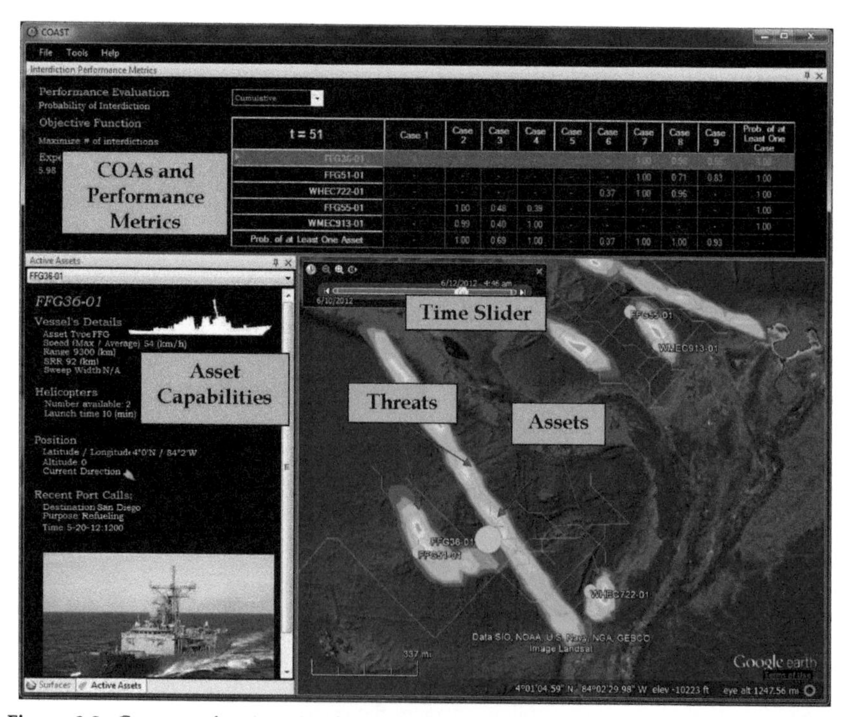

Figure 9.8. Courses of action simulation tool: a proactive decision support for counter-smuggling operations (Mishra et al., 2017b; Sidoti et al., 2017b).

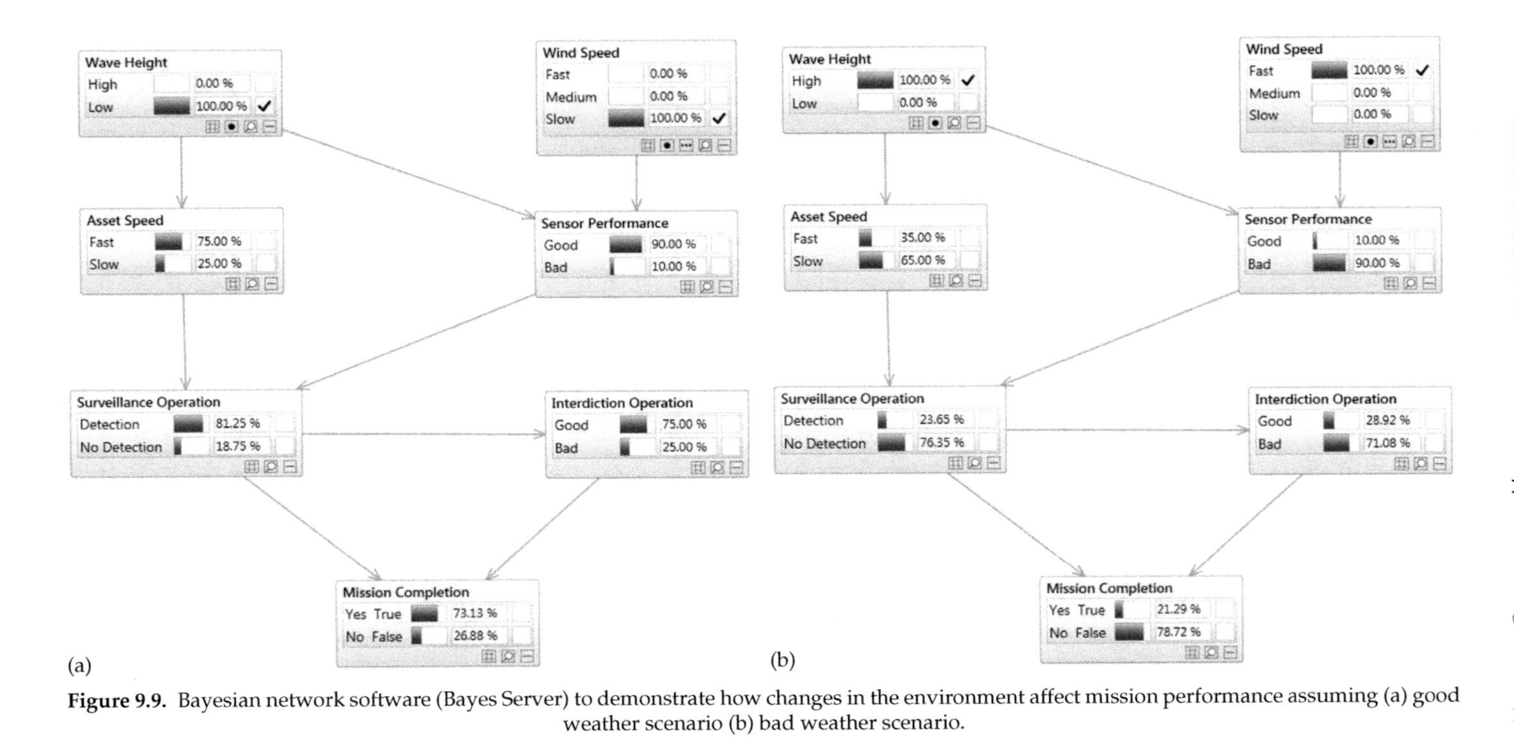

Figure 9.9. Bayesian network software (Bayes Server) to demonstrate how changes in the environment affect mission performance assuming (a) good weather scenario (b) bad weather scenario.

a good weather scenario, with low wave height and slow wind speeds, which allows the assets and sensors to function normally and, therefore, the mission completion probability is high. However, in Figure 9.9(b), the asset speed and sensor performance are degraded due to bad weather conditions and result in a low probability of mission completion. As the weather degrades (i.e., change in environmental context), the asset models are switched (sunny to stormy) and the search regions are reduced in size and shifted spatially. Additional features of proactive COAST include automatic selection of objectives (e.g., maximizing the number of interdictions for the given time horizon, maximizing the contraband interdicted, or maximizing the probability of at least one interdiction) depending on the DM's past interactions/behavior, asset parameters and constraints. The tool also provides the ability to output DM-specific *m*-best solutions for the corresponding objective function along with valid reasoning, allowing the DM to choose the best objective, number of days to solve for that objective, and the different recommendations to output with respect to that objective function. The dynamic coordination between the surveillance and interdiction assets is established via prompt (re)allocation of interdiction assets in the region, where a smuggler has been identified by a surveillance asset. The current mission scenario is projected and its impact is represented in the form of risk surfaces to the DMs as contour plots. The algorithms in COAST use these context-specific surfaces to find multiple COAs for both interdiction and surveillance problems. In the "what-if" analysis mode, the DM can utilize COAST to explore a variety of mission scenarios by considering different asset types, their capabilities, mission constraints, objectives, and algorithmic options. In the case of DIL environments, historic (or flow) POA surfaces are used as inputs for asset allocation and scheduling.

9.3.3 Dynamic UAS Operations

Current UAS operations are conducted by teams of operators with highly specialized training and roles, where the task demands for each operator are mostly independent, often resulting in sub-optimal tasking and mission performance. A 2012 report to the US Congress by the Department of Defense asserts that 68% of UAS mishaps are attributable to human error (Gertler, 2012). In order to ensure that operators are attending to the right task at the right time and that the task load is balanced across team members, Naval Research Laboratory-DC developed the Supervisory Control Operations User Testbed (SCOUT), a highly flexible simulation/testbed environment created for the purposes of exploring operator tasking and performance during unmanned system missions (Sibley et al., 2016). SCOUT is a single user platform that allows an operator to control multiple simulated unmanned assets. It has: (1) a mission planning mode, where the user must determine the best plan for sending multiple unmanned vehicles to search for multiple targets, with varying priority levels and constraints, and (2) a mission execution mode, where the user monitors vehicle status, responds to communication requests, and makes updates to the overall plan as new targets arrive or intelligence reports are received. This testbed is utilized to collect physiological data (e.g., pupillary and gaze data for human workload detection) and to iteratively develop and assess optimization algorithms (e.g., dynamic scheduling of a UAS within an uncertain environment) for proactive decision making.

Statistical machine learning and classification techniques are utilized to detect the cognitive context of DMs by analyzing the eye-tracking data collected from SCOUT. Fusing several types of features, such as pupil dilation features and gaze metrics, can improve the robustness and accuracy of cognitive context detection. In (Mannaru et al., 2016a; 2016b; 2016c), several machine learning approaches (e.g., principal component analysis, support vector machine, *k* nearest neighbor classifier, etc.) are employed to classify the operator workload where the results indicate that high classification accuracies are possible (e.g., greater than 91% accuracy in distinguishing between easy/medium workload levels and hard workload levels versus 75% accuracy between easy and medium only; these results are desirable for adaptive automation). The graphical user interface (GUI) of SCOUT is shown in Figure 9.10, where the left screen shows the mission environment (including the UAS and target positions) in Google Earth, while the right screen has the sensor feed, UAS information (speed and altitude), and chat window for intelligence updates and commands. The GUI enables scheduling of events at specific times, such as where and when various pop-up targets appear and types of unmanned assets the user is controlling. In scenarios where the mission performance is trending negatively (which may be due to operator overload), the cognitive context analysis of the operator is utilized to develop proactive recommendations

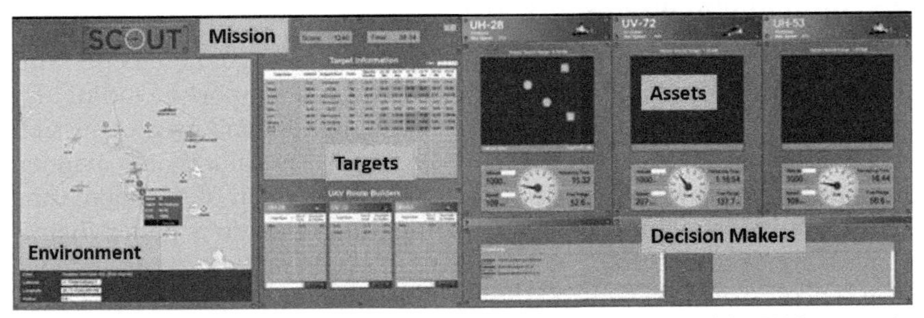

Figure 9.10. SCOUT, where the left screen (left) shows the position of the UAS, targets in Google Maps, and information about the UAS targets; the right screen (right) shows the sensor feed, UAS information (speed, altitude), and two chat windows for intelligence updates and commands (Nadella et al., 2016; Sibley et al., 2016).

for dynamic scheduling of a UAS in order to assist the operators in efficiently managing their workloads. During the planning phase, the operators must determine the best route for sending multiple UASs to search for multiple targets of interest. Each target has a specific deadline, reward associated with finding it (i.e., priority, value), and varying levels of uncertainty about its location. This planning problem is actually non-trivial since participants have a number of heterogeneous UASs with different capabilities and several targets at varied distances from a UAS, with uncertain search times and different rewards.

The dynamic UAS scheduling problem is modeled in a manner similar to the open vehicle routing problem (Brandão, 2004), which is NP-hard. However, with a single UAS or multiple UASs and a small number of targets, it is possible to find an optimal solution in a reasonable amount of time. This is sufficient, as solutions will need to be generated as the scenario evolves and as new targets appear, intelligence is updated with respect to target locations (e.g., the search radius of a target is only 0.2 km, not an originally reported 3 km) and participants are forced to reconsider their current plan. The dynamic scheduling algorithms (optimal algorithms exemplified by branch and bound, and heuristic approaches such as path time equalization, pairwise exchange combined with rollout) embedded in SCOUT make it proactive as it provides rank-ordered options, and adaptively updates plans based upon user workload, risk propensity, etc. Variables, such as target priorities, environmental factors, intelligence uncertainty, etc., are accounted for in the planning tool. UAS operators would benefit greatly from using the proactive SCOUT as it efficiently anticipates and predicts the future environmental conditions, task needs and necessary resource requirements to enable timely evaluation of COAs, while assuring operators are attending to the right task at the right time and that task demands do not exceed operator's cognitive capabilities (Nadella et al., 2016).

9.4 Conclusion and Future Work

In this chapter, we briefly discussed the challenges, concepts and applications of context-driven proactive decision support including: (1) a general PDS framework to define, detect, diagnose and predict mission context to provide proactive COA recommendations in a timely manner, and (2) application of the PDS concepts in the domain of antisubmarine warfare mission planning, counter-smuggling operations and dynamic scheduling of UAS within dynamic and uncertain mission environments. The concomitant PDS software tools, viz., AASPP tool, proactive COAST and SCOUT, were also discussed. The key advantage of using PDS is that it enables DMs to promptly understand and envision the current and projected mission context, allowing them ample time to make appropriate decisions via consideration of the uncertainty, and unknown risks stemming from the specific context via dynamic resource allocation, routing and workload balance within the mission environment.

Since PDS facilitates faster planning, especially in the readjustment of plans to meet contingencies, and provides quick responses to dynamically changing mission environments, reducing operator overload, it brings us a step closer towards proactive autonomy. However, the key technical challenge for proactive autonomy is a cooperative effort allocation scheme that balances autonomy with human interaction, sans displacement of the humans from their supervisory role of ensuring that the mission goals are achieved. Our ongoing research is addressing a number of challenges associated with human-machine agent teams tasked with conducting multiple missions. Our approach employs a unified graph-theoretic framework that brings together concepts from free energy (Helmholtz, Gibbs) optimization used in thermodynamics and information theory, approximate dynamic programming from operations research and stochastic control, active inference-based perception and action selection from neuroscience, graphical model inference and bounded rationality from probabilistic inference and cognitive science, and Feynman-Kac path costs in physics to mathematically represent, evaluate, and design hybrid team structures (Bertsekas and Tsitsiklis, 1995; Wehrl, 1991; Ortega and Braun, 2013; Kappen et al., 2012; Todorov, 2009; Aoki, 1961). The key research questions we are trying to address include: hybrid human-machine team composition, dynamic coordination, what tasks to automate, dynamic selection of the level of automation, and so on. Some viable approaches, which we plan to explore, include dynamic function allocation and inverse reinforcement learning (Sutton and Barto, 1998; Garc´ıa and Fernandez, 2015; Russell and Norvig, 2002; Subramanian et al., 2016).

Acknowledgments

The authors would like to thank NRL-MRY for their valuable contributions in designing and detailing the drug trafficking scenarios and the concept of operations embedded within our decision support software tool. The authors would also like to thank NRL-DC for providing the SCOUT software as a testbed for validating our algorithms. This work was supported by the U.S. Office of Naval Research under contract #N00014-12-1-0238, #N00014-16-1-2036 and #N00014-18-1-2838; by the Naval Research Laboratory under contract #N0017316-1-G905; and by the Department of Defense High Performance Computing Modernization Program under subproject contract #HPCM034125HQU.

References

An, W., Ayala, D.F.M., Sidoti, D., Mishra, M., Han, X., Pattipati, K.R., Regnier, E.D., Kleinman, D.L. and Hansen, J.A. 2012. Dynamic asset allocation approaches for counter-piracy operations. In Information Fusion (FUSION), 2012 15th International Conference on, 1284–1291. IEEE.

Aoki, M. 1961. Stochastic-time optimal-control systems. Transactions of the American Institute of Electrical Engineers, Part II: Applications and Industry 80(2): 41–46.

Arulampalam, M.S., Maskell, S., Gordon, N. and Clapp, T. 2002. A tutorial on particle filters for online nonlinear/non-Gaussian Bayesian tracking. IEEE Transactions on Signal Processing 50(2): 174–188.

Bar-Shalom, Y., Li, X.R. and Kirubarajan, T. 2004. Estimation with applications to tracking and navigation: theory algorithms and software. John Wiley & Sons.

Bayes Server. 2017. Advanced Bayesian network software. [Available online: https://www.bayesserver.com/].

Bertsekas, D.P. 1995a. Dynamic programming and optimal control, vol. 1, Athena Scientific Belmont, MA.

Bertsekas, D.P. and Tsitsiklis, J.N. 1995b. Neuro-dynamic programming: An overview. Proceedings of the 1995, 34th IEEE Conference on Decision and Control 1: 560–564.

Bertsekas, D.P., Bertsekas, D.P., Bertsekas, D.P. and Bertsekas, D.P. 1995. Dynamic programming and optimal control, volume 1. Athena Scientific Belmont, MA.

Bishop, C.M. 2006. Pattern recognition. Machine Learning 128.

Brandão, J. 2004. A tabu search algorithm for the open vehicle routing problem. European Journal of Operational Research 157(3): 552–564.

Brezillon, P. 1999. Context in problem solving: a survey. The Knowledge Engineering Review 14(1): 47–80.

Brown, P.J. 1995. The stick-e document: a framework for creating context-aware applications. Electronic Publishing-Chichester 8: 259–272.

Brown, P.J., Bovey, J.D. and Chen, X. 1997. Context aware applications: from the laboratory to the marketplace. IEEE Personal Communications 4(5): 58–64.

Deb, S., Pattipati, K.R., Raghavan, V., Shakeri, M. and Shrestha, R. 1995. Multi-signal flow graphs: a novel approach for system testability analysis and fault diagnosis. IEEE Aerospace and Electronic Systems Magazine 10(5): 14–25.

Dey, A.K., Abowd, G.D. and Salber, D. 2001. A conceptual framework and a toolkit for supporting the rapid prototyping of context-aware applications. Human-computer Interaction 16(2): 97–166.

Drennan, J.E. 2010. How to Fight Unmanned War. U.S. Naval Institute Proceedings Magazine, Vol. 136/11/1293, November [Online: https://www.usni.org/magazines/proceedings/2010-11/how-fight-unmanned-war].

Garc´ıa, J. and Fernandez, F. 2015. A comprehensive survey on safe reinforcement learning. Journal of Machine Learning Research 16: 1437–1480.

Gaspar, H.M., Brett, P.O., Ebrahim, A. and Keane, A. 2014. Data-driven documents (d3) applied to conceptual ship design knowledge. Proceedings COMPIT 2014.

Gertler, J. 2012. U.S. Unmanned Aerial Systems. Congressional Research Service, 7-5700, www.crs.gov, R42136, Defense Technical Information Center Report.

Han, X., Bui, H., Mandal, S., Pattipati, K.R. and Kleinman, D.L. 2013. Optimization-based decision support software for a team-in-the-loop experiment: Asset package selection and planning. IEEE Transactions on Systems, Man, and Cybernetics: Systems 43(2): 237–251.

Jensen, K. 1987. Colored petri nets. In Petri nets: central models and their properties. Springer. 248–299.

Kappen, H.J., Gomez, V. and Opper, M. 2012. Optimal control as a graphical model inference problem. Machine Learning 87(2): 159–182.

Koller, D. and Friedman, N. 2009. Probabilistic graphical models: principles and techniques. MIT press.

Malik, A., Maciejewski, R., Towers, S., McCullough, S. and Ebert, D.S. 2014. Proactive spatiotemporal resource allocation and predictive visual analytics for community policing and law enforcement. IEEE Transactions on Visualization and Computer Graphics 20(12): 1863–1872.

Mannaru, P., Balasingam, B., Pattipati, K., Sibley, C. and Coyne, J. 2016a. Cognitive context detection in UAS operators using eye-gaze patterns on computer screens. In SPIE Defense+ Security, 98510F–98510F. International Society for Optics and Photonics.

Mannaru, P., Balasingam, B., Pattipati, K., Sibley, C. and Coyne, J. 2016b. Cognitive context detection in UAS operators using pupillary measurements. In SPIE Defense+ Security, 98510Q–98510Q. International Society for Optics and Photonics.

Mannaru, P., Balasingam, B., Pattipati, K., Sibley, C. and Coyne, J. 2016c. On the use of hidden Markov models for eye-gaze pattern modeling. In SPIE Defense+ Security, 98510R–98510R. International Society for Optics and Photonics.

Mishra, M., An, W., Han, X., Sidoti, D., Ayala, D. and Pattipati, K. 2014a. Decision support software for antisubmarine warfare mission planning within a dynamic environmental context. In Systems, Man and Cybernetics (SMC), 2014 IEEE International Conference on, 3390–3393.

Mishra, M., An, W., Han, X., Sidoti, D., Ayala, D. and Pattipati, K. 2014b. Decision support software for antisubmarine warfare mission planning within a dynamic environmental context. In Systems, Man and Cybernetics (SMC), 2014 IEEE International Conference on, 3390–3393.

Mishra, M., An, W., Sidoti, D., Han, X., Ayala, D.F.M., Hansen, J.A., Pattipati, K.R. and Kleinman, D.L. 2017a. Context-aware decision support for anti-submarine warfare mission planning within a dynamic environment. IEEE Transactions on Systems, Man, and Cybernetics: Systems pp. (99): 1–18, DOI: 10.1109/TSMC.2017.2731957.

Mishra, M., Sidoti, D., Avvari, G.V., Mannaru, P., Ayala, D.F.M., Pattipati, K.R. and Kleinman, D.L. 2017b. A context-driven framework for proactive decision support with applications. IEEE Access 5: 12475–12495.

Murphy, K.P. 2012. Machine learning: a probabilistic perspective. MIT press.

Nadella, B.K., Avvari, G.V., Kumar, A., Mishra, M., Sidoti, D., Pattipati, K.R., Sibley, C., Coyne, J. and Monfort, S.S. 2016. Proactive decision support for dynamic assignment and routing of unmanned aerial systems. In Aerospace Conference, 2016 IEEE, 1–11. IEEE.

Noy, N.F., Crubezy, M., Fergerson, R.W., Knublauch, H., Tu, S.W., Vendetti, J., Musen, M.A. et al. 2003. Protégé-2000: An open-source ontology-development and knowledge-acquisition environment. In AMIA Annual Symposium Proceedings, pp. 953.

Ortega, P.A. and Braun, D.A. 2013. Thermodynamics as a theory of decision-making with information-processing costs. In Proc. R. Soc. A, volume 469, 20120683. The Royal Society.

Price, F.D. 2007. Estimations of Atmospheric Conditions for Input to the Radar Performance Surface. Master's Thesis, Naval Postgraduate School Monterey, CA.

Russell, S.J. and Norvig, P. 2002. Artificial Intelligence: A Modern Approach (International Edition). Pearson US Imports & PHIPEs.

Schilit, B.N. and Theimer, M.M. 1994. Disseminating active map information to mobile hosts. IEEE Network 8(5): 22–32.

Sibley, C., Coyne, J. and Thomas, J. 2016. Demonstrating the supervisory control operations user testbed (scout). In Proceedings of the Human Factors and Ergonomics Society Annual Meeting 60: 1324–1328. SAGE Publications.

Sidoti, D., Avvari, G.V., Mishra, M., Zhang, L., Nadella, B.K., Peak, J.E., Hansen, J.A. and Pattipati, K.R. 2017a. A multi-objective path-planning algorithm with time windows for asset routing in a dynamic weather-impacted environment. IEEE Transactions on Systems, Man, and Cybernetics: Systems 47(12): 3256–3271, December 2017.

Sidoti, D., Han, X., Zhang, L., Avvari, G.V., Ayala, D.F.M., Mishra, M., Nadella, B.K., Sankavaram, S., Kellmeyer, D., Hansen, J.A. and Pattipati, K.R. 2017b. Context-aware dynamic asset allocation for maritime interdiction operations. IEEE Transactions on Systems, Man, and Cybernetics: Systems pp. (99): 1–19, DOI: 10.1109/TSMC.2017.2767568.

Smirnov, A. 2006. Context-driven decision making in network-centric operations: Agent-based intelligent support. Technical report, Defense Technical Information Center Document.
Subramanian, K., Isbell, Jr., C.L. and Thomaz, A.L. 2016. Exploration from demonstration for interactive reinforcement learning. In Proceedings of the 2016 International Conference on Autonomous Agents and Multi-agent Systems, AAMAS '16, 447–456. Richland, SC: International Foundation for Autonomous Agents and Multi-agent Systems.
Sutton, R.S. and Barto, A.G. 1998. Reinforcement learning: An introduction, volume 1. MIT press Cambridge.
Todorov, E. 2009. Efficient computation of optimal actions. Proceedings of the National Academy of Sciences 106(28): 11478–11483.
Wehrl, A. 1991. The many facets of entropy. Reports on Mathematical Physics 30(1): 119–129.
Zhang, C., Gholami, S., Kar, D., Sinha, A., Jain, M., Goyal, R. and Tambe, M. 2016. Keeping pace with criminals: An extended study of designing patrol allocation against adaptive opportunistic criminals. Games 7(3).

The Shared Story

Narrative Principles for Innovative Collaboration

Beth Cardier

10.1 A Common Frame of Reference

How can a team arrive at decisions that are more collectively intelligent than a single person? When a group fails to come together productively, the results can be catastrophic (Tredici, 1987). William Lawless relates a true case in which a thousand people with doctorates made a very stupid collective decision on behalf of the Department of Energy: they decided to bury nuclear waste in cardboard boxes in the ground (Lawless, 1985; Ledford, 1984; Mayell, 2002). This is a negative example of collaboration, in which a poor synthesis of team expertise, management and goals led to compromised results that were dumber than most individuals could manage.

Narrative enables us to zoom into a particular aspect of this problem. In a productive collaboration, individual expertise pools so the potential of the group can exceed the capabilities of its individuals. Together, the team members contribute relevant knowledge to build a *shared story*, one that is specifically geared towards a communal goal. An unproductive team lacks this inner coherence and instead comes to be dominated by factors that are non-intrinsic to its goal, such as peer or societal status (Cohen and Lotan, 2014). This is not a productive foundation for a project—not only

Sirius-Beta, Virginia Beach, VA; bethcardier@sirius-beta.com does it fail to harness the innovative potential of a group but at a more basic level, "status-driven hierarchical processes undermine analysis and problem-solving activities in teams" (Feiger and Schmidt as discussed in Page, 2004, p. 342). Instead, the members of a team need to recast their knowledge towards a common target, a process that requires the facilitation of individual expertise into a new communal arrangement.

This chapter examines how narrative builds a shared frame of reference among its informational components. A story is a distributed system of intelligence: a sprawling network of inferences that connects diverse contexts, perspectives and forms of information (Herman, 2006). To synthesize these into a unified fabric, narrative operations connect and modify these elements in ways that are usually invisible to an audience. These processes produce a collective 'interpretive frame' that is accessible to all of its informational components, yet can also change as circumstances unfold. In narrative, the reader understands this shared frame as the emerging 'theme' or 'point'. In collaboration, this emerges as the communal understanding of a problem that knits parts of each individual's knowledge into a solution. Key features of this process are described here to provide a better understanding of how this can be applied to a heterogeneous group or collection of knowledge sources.

The value of using narrative skills for productive collaboration is well-recognized. Story-related methods enable groups to form clear narratives about their goals and research (Niki, 2017; Dreyer-Lude, 2016), develop a project culture and vocabulary (Hunicke et al., 2004; Cohen, 2006) or use storytelling techniques to transform foundational assumptions (Scharmer and Kaufer, 2013; Gray, 2016). This chapter is concerned with collaboration that requires a foundational synthesis at a theoretical level, where differences between forms and ways of knowing constitute a barrier to achieving research goals. This chapter focuses on a specific need within this endeavor: the invisible work of establishing a common frame of reference within a heterogeneous team, which is flexible enough to achieve a leap of novelty, or change according to evolving circumstances.

A diagrammatic modeling grammar is used to demonstrate that process, which I refer to as a Dynamic Story Model. It records mechanisms and vocabulary from integrative narrative processes in graphical and descriptive form. This systematic method has become the basis of formal techniques elsewhere: it was first supported by a Navy-funded research project to develop new foundations for contextual integration among ontologies (Cardier, 2013; 2015) and has since been implemented piecemeal as a type system (Goranson et al., 2015; Goranson et al., 2012), an approach that enables logic to handle situations (Goranson and Cardier, 2013; 2007; Goranson et al., 2015), a foundation for visual analytics and systems modeling (Cardier et al., 2017a; Cardier et al., 2017b), user interface design

(Goranson, 2014; 2015; Goranson and Cardier, 2014) and as a source of new ideas for games and playable media practitioners (Cardier, 2007; 2014a; 2015). Here it is offered as a precise description of a process that can be applied to autonomous systems or human teams.

A current grant from the National Academies Keck Futures Initiative supports the maturing of this research, in which it is applied to ontological interaction in biomedicine. In the process of executing this grant, five remote fields were brought together: neuroscience, cognitive narratology, computer science, design and biology (Cardier et al., 2017b; Cardier et al., 2017a). Observations about human collaboration are drawn from this project and other examples. This examination of contextual dynamics thus spans three different systems—narrative, team collaboration and ontological knowledge representation—to identify common tools that can support collaborative, intelligent systems.

10.2 Ontological Foundations

Narrative is concerned with diverting general knowledge away from expected norms. In this sense, a story is a machine for conceptual change, with easy-to-follow transitions between physical and conceptual states. One function of these transitions is to tailor general information towards a specific circumstance, and then adjust it when the next information appears. This fundamental quality has been noted in fields such as linguistic psychology (Bruner, 1991), cognitive narratology (Herman, 2009), discourse analysis (Graesser and Wiemer-Hastings, 1999) and sociolinguistics (Labov and Waletzky, 1967; Labov, 1972), which commonly note that such deviation from everyday routine is the reason for the story being told, the aspect that gives it "narrativity" (Ryan, 2005).

The mechanisms that enable this transformation are the same as those needed to build a cohering synthesis from a collaborative team. In narrative, pieces of information from an array of different sources, media and times are brought together into a coherent whole, and an emergent 'theme' or 'point' is extracted from it. This cohering idea is referred to here as a *shared story*. To describe and implement this aspect of the narrative process, a method was needed that would allow it to be tracked in detail. Techniques to represent this can be found in knowledge representation, in methods of diagramming *ontologies*.

The term *ontology* originated in philosophy but is now used in computer science to refer to a core reference framework. A computer system's ontology acts as a kind of dictionary—it is a network of conceptual structure that is general enough to interpret any incoming information for which it has been designed. A commonly accepted definition of *ontology* is that it records the "objects or entities that are assumed to exist in [a] domain of interest as well as the relationships that hold between them" (Gruber, 1993). A simple graphical example of one can be seen below in Figure 10.1:

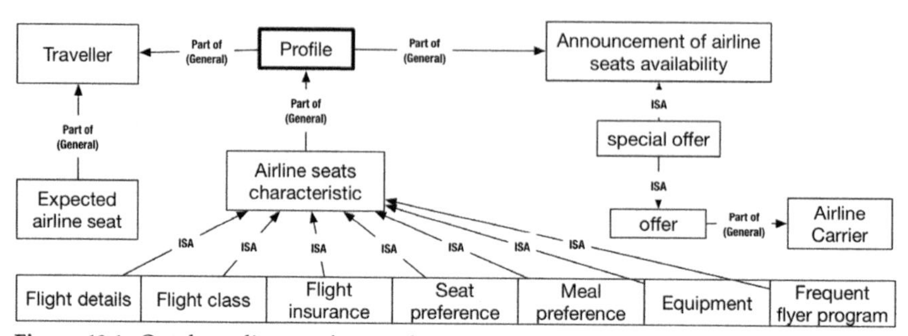

Figure 10.1. Ontology diagram for matching passengers with plane seats (Kanellopolous, 2008).

An ontology like the one depicted above, in Figure 10.1, differs from a narrative-based architecture in an important respect. In narrative, the reference framework must evolve beyond its original structure. A story is a limited representation that must account for an open world, and so with every sentence, it adapts to information that comes from outside its initial starting state. To accommodate that characteristic, this research expands the techniques of ontological diagramming to include structural transformation.

An example of a narrative version of an ontology can be seen below, in Figure 10.2. It is drawn in a manner similar to the layout of a conventional ontology—a network of conceptual nodes and relations. In this case, the network is based on the opening sentence of the title *Red Riding Hood as a Dictator Would Tell It*, a story by H.I. Phillips (Phillips, 1993). Figure 10.2 represents solely the starting state of this story (and a simplified version at that). It concerns the first chunk of text in the title sentence, 'Red Riding Hood…'. This phrase has a matchable contextual meaning, in the form of the *Red Riding Hood* story and character. Its representative conceptual network is composed of the general ideas associated with the original *Red Riding Hood* story, which are inferred by the reader: a girl, a wolf, a medieval village, a simple moral code and some other rules of the fairytale genre.[1]

In the below diagram, the commonly-known relationships of the story are depicted as a collective definition of the phrase "Red Riding Hood". This conceptual network indicates that it is possible for the child to be trusting, the wolf to be a dangerous stranger and the grandmother to be dinner.

[1] These associations were drawn from Wikipedia (Various, 2017). Wikipedia is not usually a scholarly reference but for the purpose of understanding what the ordinary public collectively believes are the key elements of Red Riding Hood, most scholarly references would not serve as well as Wikipedia.

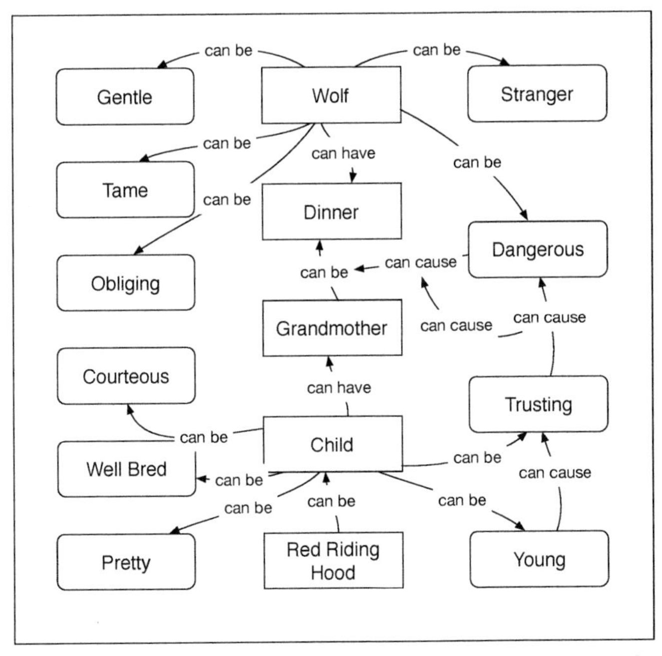

Figure 10.2. Conceptual network for the story *Red Riding Hood* using diagramming techniques from ontological knowledge representation.

This first text acts as a starting foundation for the interpretation of the rest of the title sentence, "...as a dictator would tell it". With the completion of that sentence, the overall conceptual network shifts dramatically, to become Figure 10.3, below.

In this new state, the identity and definition of the term 'wolf' in Figure 10.3 is different from the 'wolf' in Figure 10.2. Figure 10.2, which indicated that the 'wolf' was a forest predator who can be dangerous to a child. However, in Figure 10.3, the wolf is now a character in a propaganda story told by a dictator who is being devious in the service of self-interest. With the addition of six extra words, the point of the story has thus changed considerably, from a cautionary fairytale about traveling beyond safe boundaries to a warning about political duplicity. This narrative-based method identifies the mechanisms needed for human reasoning to shift easily from the first state to the second, at the same time altering the identity of its terms, such as 'wolf'.

In interdisciplinary collaboration, experts must similarly recast their knowledge. They bring a domain of expertise that was developed in relation to a particular set of problems; now they are faced with a new challenge in a different domain and their knowledge only contributes part of its solution. Even the scope of innovation changes as you move between fields. For example, in Cardier et al. (2017a), neuroscientific information about signal

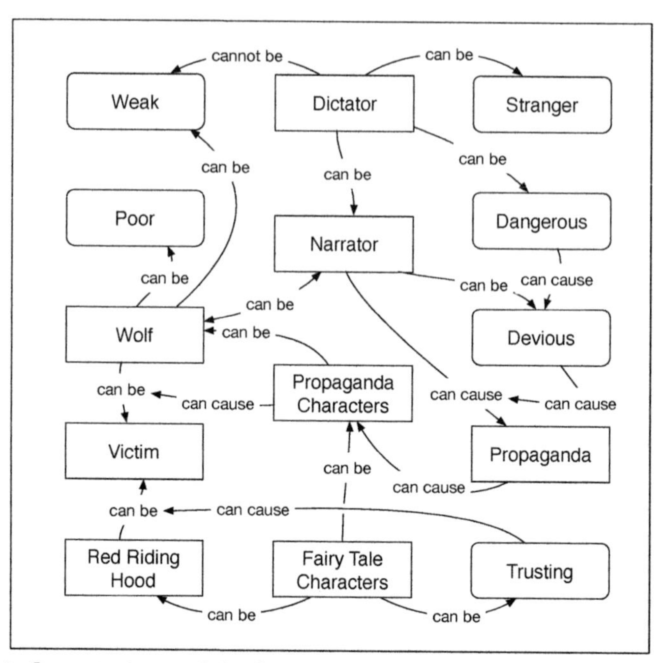

Figure 10.3. Conceptual network for the entire story title *Red Riding Hood as a Dictator Would Tell It* using diagramming techniques from ontological knowledge representation.

circuits becomes an issue of representational granularity for the designer who is developing its visualization. Before creative exchange is possible, a group needs to understand what problem they are jointly solving and develop a shared vocabulary to discuss it. Even deciding on these core terms can be tricky. This mutual adaptation of fundamental knowledge is incremental, in the manner of a developing story, so that it is organic to the project and the team members.

Let us now zoom into the integration process itself. The example story integrates the contexts 'Red Riding Hood' and 'Wolf'. Going beyond the above simple illustrative diagrams, we turn to a method that demonstrates how narrative painstakingly shifts between them. This model and method indicates the stepwise quality of information integration in storytelling. To gain an overall impression of how this technique unfolds, see Figure 10.4, which is a snapshot of the animation; the full animation can be seen at (Cardier, 2014b).

The figure uses nodes and relations, just as established knowledge representation techniques do. Nodes and relations are organized into limited representational networks in the manner of an ontology. To be clear, the formal rules of ontologies are not used, only their conceptual networks and the lessons concerning their integration. Full details of the operations and

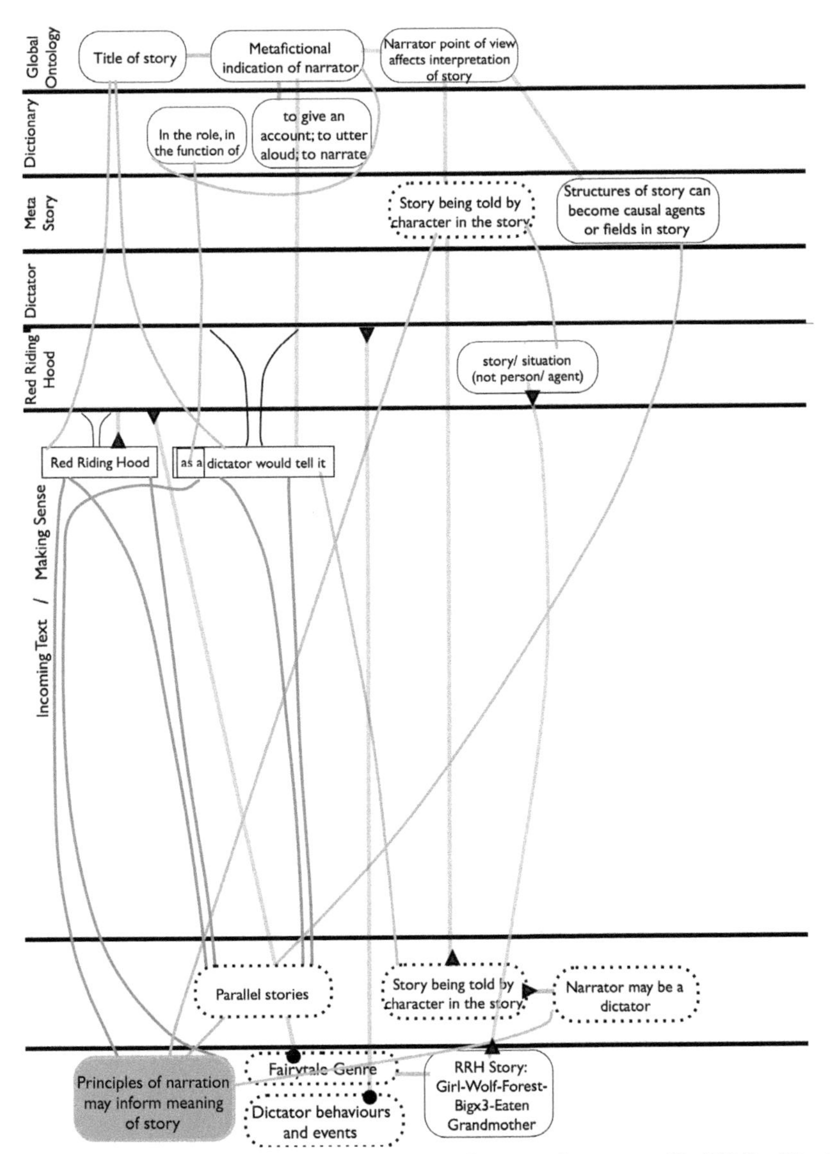

Figure 10.4. Snapshot of the inference connections between the contexts 'Red Riding Hood' and 'Dictator' in the story title *Red Riding Hood as a Dictator Would Tell It.*

its method are provided elsewhere (Cardier, 2015; Goranson et al., 2015). For this discussion, the salient feature is that the solution was to add an additional 'tier' of modeling: *situations*. By adding situations, and its formal expression in situation theory (Devlin, 1995), the system-level is included as a first-class citizen in the logical foundations. Key elements of this concept will now be explained, in relation to the problem of collaborative integration.

10.3 Component versus System

In narrative, when fragments of information aggregate into an organized system, its communal dynamics are distinct from its composite elements. This work zooms into operations that are particular to a system as a *system*. These devices include analogy, system-level identity and governance (which will be explained). Representing this activity requires an additional tier of modeling, which we include using the notion of a *situation*. Full details of how situation theory formally support this method are provided elsewhere (Cardier et al., 2017b; Goranson et al., 2015); the important detail is that it provides a way for the systems-level entities to be formally recorded.

Although the real-life process of narrative interpretation is complex, the graphical method indicates which aspects are central to this particular problem.

First, the layout distinguishes between **general** and **context-specific** knowledge, as seen in the snapshot shown in Figure 10.4. In a knowledge system, general knowledge derives from the kinds of general ontologies that can be found in biomedicine (Arp et al., 2015) or common knowledge (Reed and Lenat, 2002). In narrative, this general knowledge is drawn from semantic memory, a form of general world knowledge that people use "daily to recognize entities and objects in their environment, generate expectancies for upcoming events, and interpret language" (McRae and Jones, 2013). This general knowledge is contrasted with context-specific knowledge, which can only come from a real circumstance in the open world; in narrative, it comes from the text of a story.

This context-specific knowledge is represented in the graphical layout as two kinds of information placed in two separate areas. The first can be found in the upper area, as the weave of new structure that connects different subsets of general knowledge—e.g., the 'Riding Hood' context, the 'dictator' context (see Figure 10.4). At the bottom of the page, new interpretive structures are assembled from these contexts, guided by the story. In collaboration, this bottom area would carry information about the shared goal.

Second, **new situations** are built. This occurs when structure of any kind is aggregated and grouped. To support interpretation during this process, each newly created cluster of ideas is fenced in by a discreet situation (represented by orange boxes or separate bands across the page). This new boundary preserves the original contextual structure, indicating that one network of knowledge is different from another.

This points to another important aspect of collaborative interaction: sometimes expertise needs to be preserved in functional units. For example, members who share a disciplinary language could be asked to form sub-groups. In our NAKFI-supported project, team members from very different disciplines gave short presentations about their field and cited examples from this field when offering knowledge to the team, to situate information in its original context. The members who spoke Italian were encouraged to converse in their native language when trying to figure out a complex problem, before sharing their ideas with the English-speaking group.

Third, the **emergence** of connective structure occurs across all situations and sources. In narrative, the sum of this weaving forms the central cohering theme of the text. In collaboration, this is the emerging, shared story. It is both the details of the goal and the path to reaching it.

Fourth, connective structure across fields will sometimes be **analogical**. Analogy is a system-level feature, in which structural pattern can be aligned without having to be identical. Analogy links otherwise dissimilar situations and enables a common foundation based on structure. The means by which this occurs will be explained in a moment.

Due to the novelty and importance of these system-level features, we now turn to a closer examination of them.

10.4 Representing the Two Levels

Principles from the arts and humanities indicate how systems emerge and interact at the level of *systems* (Cardier et al., 2017a; Cardier et al., 2017b). One observation comes from the philosophy of language, first noted by Searle (1979) but developed through theories by Lakoff (1993) and Winner and Gardner (1993). They observe that in system-level devices such as metaphor and irony, interpretation occurs at least two levels: (1) individual elements, and (2) how these elements behave as a group. This is understood as two tiers of meaning: "…what is said and what is meant", in which "what is said" occurs at the level of syntactic expression and "what is meant" concerns the system-level mapping between domains. Gentner (1983) notes that the systemic, system-level mapping is key to linking domains, while Lakoff (1993) goes further to say that it is the primary influence on interpretation, because the system level records the collected rules of a system.

An example of how system-level identity operates in practice can be seen in the television series *Game of Thrones* in the episode 'And Now His Watch Has Ended' (Graves, 2013; season 3, episode 4). At the beginning of one scene, the audience expects usurped Princess Daenerys Targaryan to make a trade deal, and is judging her potential as a future ruler according to how well it goes. Daenerys offers one of her dragons in exchange for a slave army, which her supporters believe is a poor bargain. However, during the scene, Daenerys uses the dragon to incinerate the owner of the slave army and seize control, keeping both the army and the dragon, pleasantly surprising her supporters. Now the scene has changed into one of *conquest*, with its associated implications.

As a result of this shift, the audience adjusts their measure for judging whether Daenerys would be a good ruler. Even though she undermined the terms on which the last scene seemed to depend, the scene casts her as even more fit for the role of ruler, for the way she manipulated the situation to her advantage. The anticipated outcome of the story therefore also changes.

Below, the fundamental nature of this change is indicated at the level of semantic association—what-can-follow-what. This adjustment of identity and causal framework can be seen in the two diagrams in Figures 10.5–10.6, below.

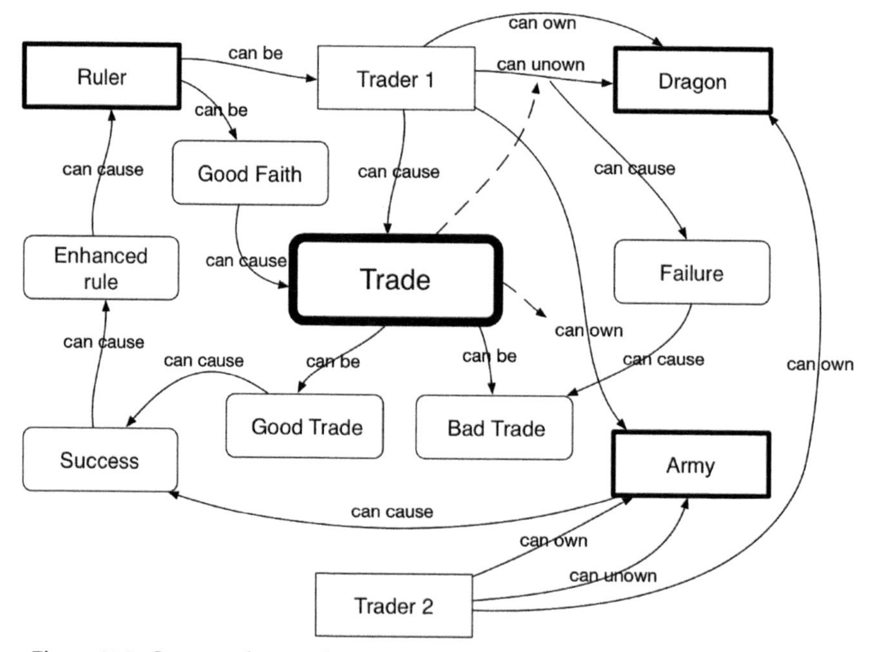

Figure 10.5. Conceptual network structure at the beginning of a scene of *Game of Thrones*.

In the example, this new arrangement positions Daenerys as more likely to be successful in her quest to regain the throne. The dragon likewise becomes redefined, starting as an object of trade and becoming a creature of power and destiny. The shift in identity of the overall situation from trade to conquest thus changes the meaning of all the terms within. That change also

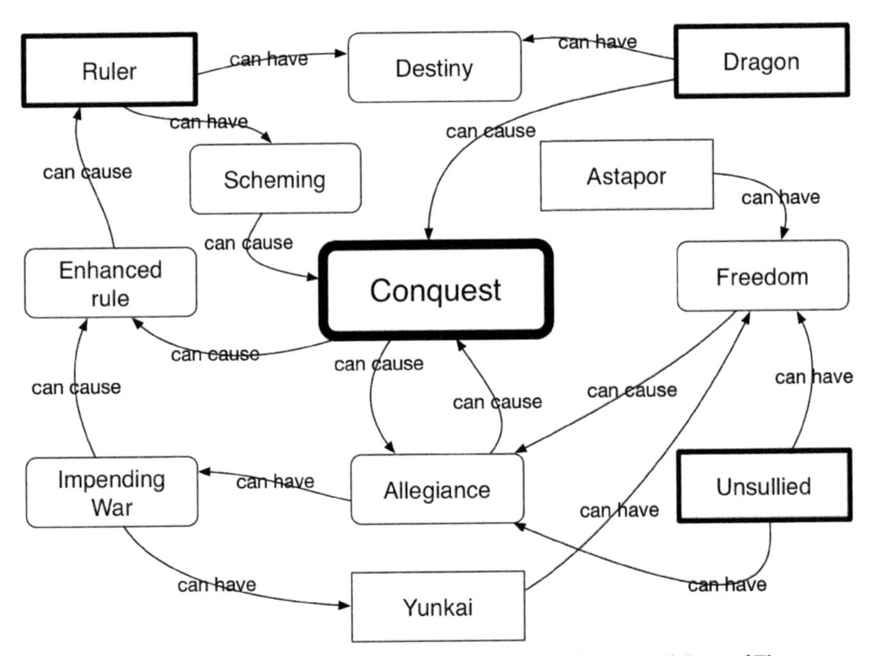

Figure 10.6. Conceptual network structure at the end of a scene of *Game of Thrones*.

brings a new causal framework (what-follows-what) and as a consequence, new anticipated outcomes. This fundamental shift in conceptual structure is similar to the paradigmatic shift observed in theories of conceptual change and scientific revolution (Thagard, 1992; Kuhn, 1973).

This example highlights the difference between identity at the level of component versus system. The 'dragon' is named at the level of individual nodes; the context in which it resides can also be named, whether 'trade' or 'conquest'. In doing so, the identity of the system contributes important structure to the component, through its relationship with similar situations in other domains. These two levels can be seen in Figure 10.7 below.

We can see this across all our examples: in *Red Riding Hood as a Dictator Would Tell It*, the definition of the 'wolf' adjusts from 'forest predator' to 'dictator's foil' along with the shift in identity of the system in which it resides, from 'fairy tale' to 'political satire'. In the *Game of Thrones* example, the first state is identified as a 'trade' while the second is 'conquest'. This defining identity is a classification based on key features, except the features are based on overall communal structure. It allows matching across systems, which is critical for overcoming barriers of context. We now explore how this occurs, through the mechanics of *analogy* as it appears in narrative.

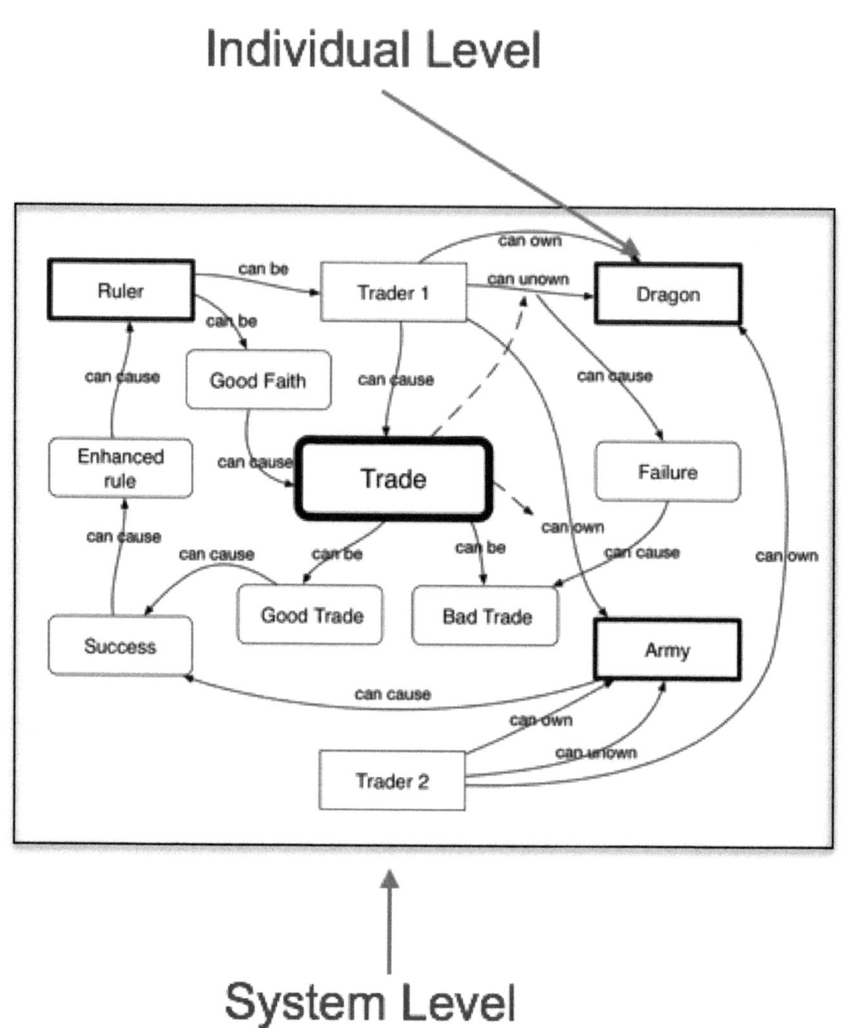

Figure 10.7. Identity can be found in at least two levels of a system.

10.4 Analogy: Connecting Heterogeneous Contexts

Analogy can bridge gaps of time, context, and scale across dissimilar situations. This is possible because analogy bridges structure at the level of systems, where local differences are subsumed to whole systems of relations (Gentner, 1983; Gentner et al., 1993; Nersessian and Chandrasekharan, 2009). With this in mind, Gentner characterizes analogy as the comparative "mapping of relations between objects" while Harrell (2005) further articulates this structure as "skeletal patterns" that recur. When these skeletal patterns align, they draw attention to and reinforce each other (Markman and Gentner, 1993). However, the likeness is not exact, and real-life comprehension of analogy accounts for this. The asymmetry of analogical connection has thus also been observed by cognitive linguists and psychologists (Nersessian and Chandrasekharan, 2009; Fauconnier and Turner, 2008; Lakoff and Turner, 1989; Markman and Gentner, 1993), but not in terms of its ability to transform meaning in an unfolding story.

This draws attention to an important feature of this work: narrative structure shows how analogy is chained for overall coherence. This takes analysis beyond the observation that analogy aligns the structure between two situations. It is also key to the production of the higher-level combined structure—the shared story. In narrative, this process produces the emergent umbrella structure of a 'theme'. A clear example follows, in analysis of the film *Guardians of the Galaxy* (Gunn, 2014).

In the opening scenes of *Guardians of the Galaxy*, two situations are linked even though they occur in different times, emotional states and planets. In the first scene, a boy tries to escape the sadness of his mother's impending death by listening to his Walkman tape player in the hospital. In the second scene, a grown man on a distant planet puts on the earphones of a Walkman tape player before clubbing alien monsters as they attack him.

The Dynamic Story model was applied to this example, and something interesting was observed, as seen in Figure 10.7. Analogical formations are lynchpins, linking situations. In this capacity, they also support the higher-level connective structure. The analogical similarities connect these scenes—for example, the common device of a tape being played in a Walkman and its music becoming the soundtrack for the subsequent filmic action is a core motif in both scenes. This motif is non-general enough to be matched across scenes even though there is no other link between them, they are set ten years apart and on different planets. Through that core cue, other analogical links form, giving them a sense of narrative succession. A diagram of this activity can be seen in Figure 10.8.

In the figure, these higher-level analogical connections are graphically represented as 'cogs' that emerge to connect the two scenes. They show how analogical structures perform the role of hinges throughout the system, allowing linked situations to build on each other. At the very top of this formation are the two cogs with red nodes at their center—these are colored to indicate dominance. These analogical structures are the sum of the other analogies in the system, a summary of a summary. They are the crystallized structure of this system, and in story terms, its core theme can be found there, accumulating as the scenes unfold. Following across the page, they eventually combine into the idea: *dorky music protects its listener from a scary and serious situation.*

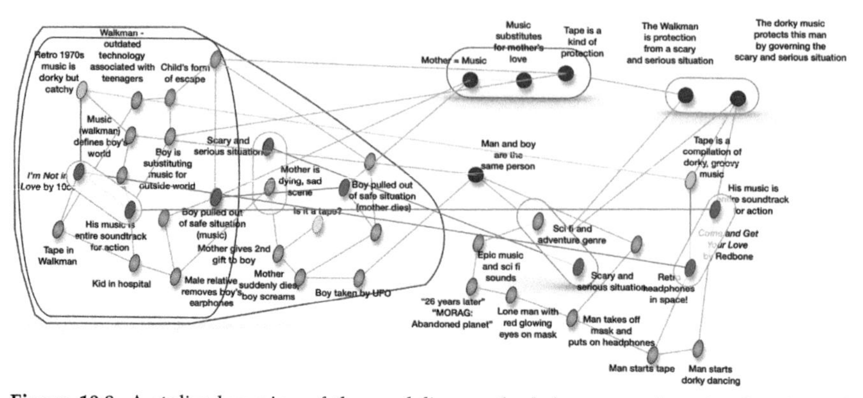

Figure 10.8. A stylized version of the modeling method draws out the role of analogical connections (seen here as 'cogs').

In this manner, the very different scenes featuring boy and man are linked. In literary terms, the linking structure would be understood as a theme—in this case the theme is: *dorky music protects its listener from a scary and serious situation*. There is no equivalent commonsense knowledge in ordinary life—this is derived from the story. This overall theme has emerged from information previous established by the story, concerning the mother, her desire to protect her son, and her need to do this via the object of the Walkman due to her impending death. These informational elements are assembled and cohered in the single idea: *dorky music protects its listener from a scary and serious situation*. This cluster of ideas is dominant over others in the system, modifying other incoming information towards its own structure, including the redefinition of terms such as 'protection' to indicate the cozy obliviousness afforded by a cocoon of music. This higher-level structure coheres two very different scenes with a very particular fabric of causal associations. It is referred to as the dominant *agent* in this system.

In collaboration, one way to replicate this linking function is by focusing on a common example. This example should characterize the target problem in a way that distinguishes it from other similar-sounding problems. Through exploration of the target problem, each participant comes to describe it using the discourse of their home field. Common properties or structures will eventually emerge across each field's description, in the sense of *analogy*, in spite of the different disciplinary languages. This example thus provides a structural focus—a Rosetta stone—over which dissimilar disciplinary languages can be overlaid. A new, project-specific vocabulary emerges during this process, one that entirely depends on the ability of collaborators to find common understanding between their fields. An example of this process can be seen in text form, in Cardier et al. (2017a), where five fields each describe the common structure of self-referential emergent organizations. In this manner, analogical similarity can overcome different lexicons and disciplinary assumptions, if it is part of a narrative-style integration process.

In both narrative and collaboration, a linking analogy can also serve as a point of reference going forward, a common touchstone from which to develop a discourse specific to the project. Forming connections across different system identities requires a means of bridging or reconciling their structures, however. The same is true of collaboration, in which researchers from different domains often do not use the same terms in the same way. These differences of context must be factored in before joint innovation can occur. These connections are managed through a process of *governance*.

10.5 Governance

Governance indicates the degree that one network imposes its structure on another. Our example focuses on the interaction between two semantic networks—'Red Riding Hood' and 'dictator'—as found in the first sentences of Phillips (1993). As generally known ideas, these two contexts have very different underlying conceptual arrangements, using different terms and sequences of cause-and-effect. The unfolding story endows one cluster with dominance, as I will explain. *Governance* is the way in which the structures of the other networks or situations adjust towards this dominant architecture. In human terms, *governance* can be thought of as the influence a strong personality has over a group, and the way other members of that group adjust to the dominant person's mode of relating. A similar modification occurs when two networks of semantics come together in narrative and interact.

Let us return to the title sentence: 'Red Riding Hood as a Dictator Would Tell It'. It indicates that the *dictator* context will likely constrain information from the *Red Riding Hood* context. This is because 'Red Riding Hood' is semantically nested in the dictator context, so we know the fairy tale will be told using many of its terms and associations. This has the effect of altering the initial 'Red Riding Hood' state in Figure 10.2 so that terms like 'wolf' becoming nested in a 'dictator' network of values, changing their associations. This is the means by which a forest predator transforms into a foil for a dictator with the additional of six words.

The dominance of the 'dictator' context carries into an interpretation of the next sentence: *Once upon a time there was a poor, weak wolf*. The phrase "poor, weak wolf" inverts the commonly known fairy tale role of the wolf, from predator to victim (Cardier, 2013; 2015). To correctly interpret this inversion, the reader combines it with knowledge of the previously established governing context: that of the 'dictator'. When the *poor, weak wolf* inversion is read in terms of the governing dictator context, a particular interpretation emerges. There is an analogical connection between the act of inverting a story and a dictator lying about his deeds. This similarity correlates those two activities; the characters 'dictator' and 'wolf' are also aligned. The resulting interpretation becomes: *the dictator is changing the traditional Red Riding Hood story in a way makes the wolf (who is also himself) look good.*

Governance thus informs and modifies meaning. This occurs on a spectrum of influence: the structure of one system can **replace** the other, **modify** the other or **collaboratively negotiate** with the other (Cardier, 2013; Cardier et al., 2017b). This explains how different system structures can find initial points of connection; for example, when one is dominant, it adjusts a small part of the structure of another towards itself. The terms of that participant yields, adapting. It can also take a turn at dominance if its information is the best fit for the next situation. Through a relationship between local and global levels, connections between heterogeneous contexts are woven, a tapestry of evolving definitions.

In this way, governance can flexibly move between networks, switching their dominant status. The network with dominant governance is described as an *agent*. In narrative, agency is primarily identifiable through the degree to which an entity's structure is able to bridge incompatible structures in a particular situation (Cardier, 2013; 2015). Looking more closely, the agent does this by donating many of its structures to the construction of a new cluster—a new situation—which can form a bridge. In this manner, it can federate nuanced knowledge from a range of fields, allowing them to come together. The new situation it produces can be a new interpretive context or the next temporal event. This is the shared story, and its emergence brings new, innovative ideas. As Boden observes: "In such cases of creative merging or transformation, two concepts or complex mental structures are somehow overlaid to produce a new structure, with its own new unity, but showing the influence of both" (Boden, 1990). This creative transformation is illustrated through processes of narrative inference, in which components come together through the formation of new system-level structure.

In collaboration, there are agents at two levels: human agents, and the terms that emerge as carrying the most load in communications between them. When trying to discuss ideas between fields, the discussion will eventually alight on an idea that resonates across fields, which is then repeatedly used as a touchstone when building further. The definition of an agent as the 'connector of remote situations' can also identify which human participants are central to a project. Unlike the status-driven activity, the human agents are not the participants with the loudest personalities or demands. They are the members who form the most connections between remote contexts in any dimension.

For the final principle from the arts, we shift from connection to its opposite. Motivating individuals to shift their interests from themselves to a collective is sometimes difficult (Lawless et al., 2017). Encouraging them to do so can involve a careful encouragement of tension.

10.6 Creative Tension

A story *deliberately* provokes dissonance between its states. Labov and Waletzky (1967) observed that a fundamental narrative quality was the complicating action, and this has been further defined by Thursby (2006) as the inciting incident that causes the event of the story to unfold. In narratology, this dissonance has been variously described as a 'breach' (Bruner, 1991), the 'unexpected' (Toolan, 2009) or a 'deviation' (Graesser and Wiemer-Hastings, 1999). By employing different kinds of complication, the urge to cohere the story can grow stronger. It also generates anticipation and speculation about what the resolution will be.

For example, in the title of the story *Red Riding Hood as a Dictator Would Tell It*, two very different contexts are brought together for compelling effect. The first phrase 'Red Riding Hood' depends on inferences to a fairytale genre. By contrast, 'dictator' draws from a network that might include Hitler, world war II politics and the moralities of human control and survival. As such, the associative priorities in each of these situations also differ. These contexts cannot be easily connected due to differences in both their terms and structure. There is also nothing in general knowledge to indicate how to link them. In order to discover how the story unfolds, the reader must consume the entire story (Cardier, 2013). The compelling drive of narrative can also be harnessed in collaborative teams.

It can be difficult to persuade people to 'buy into' the mission of a collaborative project. Each has their own problems and reasons for pursing them. Handled well, however, this incompatibility can feed to formation of a shared story. A method initiated by Dreyer-Lude (2017) harnessed this dissonance: participants with very different disciplinary problems are encouraged to share these differences with the group. They are asked: why do they care about this topic in their own research? What problem are they trying to solve? Every answer will be different.

As noted in the section on analogy, the natural inclination is to identify common themes, however. As the group discusses their professional issues, empathy between members facilitates structural alignment. Common patterns will likely emerge. This can encourage a connection between the participant's personal challenges and those of the project, establishing a mapping that transfers motivation across both. It also provides the first common touchstone and definition of the shared challenge faced by the group. A tolerance for ambiguity is thus required, along with an ability to cohere some of the information shared towards the goal.

This process is not achieved all at once. Instead, like storytelling, pieces are progressively stitched together, node by node, as the project progresses.

These new relationships adjust the information from separate sources of knowledge towards a demonstration of how contexts are connected *in this instance*. This coherence is how a motivated team emerges in conjunction with its shared story.

Teams in which members have distinctly different knowledge, especially different disciplines, have great potential for the generation of novel ideas. In connecting unlike situations or information networks, associations are formed that by design cannot be found in general knowledge. Combining multiple contexts is a rich bed for the emergence of novel and innovative ideas, including patents, which insist on novelty. In bringing interdisciplinary teams together, it is thus worth developing channels by which nuanced, domain-specific information can be accommodated.

As a point of interest, let us now turn to the way in which these system-level interactions are being currently implemented in neurobiological models.

10.7 Implementing and Visualizing System-Level Interactions

Computational knowledge models struggle to account for system level interactions and emergence—in fact, the focus on reductionist knowledge means that it is sometimes difficult to even draw attention to their existence. This lack has been lamented as a barrier to progress in systems biology and machine reasoning (Denis Noble, 2008; Sowa, 2011). Existing knowledge representation systems are famously non-agile because the bridging methods in ordinary cases are static and include exceptions (Sowa, 2016). One goal of this work has been to graphically represent the system-level behaviors described here, making visible the dynamics between whole networks of information. Another has been to develop foundations for application in knowledge systems.

Although principles of analogy have been used in numerous systems, such as Winston's (1980) system for learning and reasoning by analogy, MIDAS by Martin (1990), TACITUS by Hobbs et al. (1993) or Forbus et al. SME (2017) these approaches have been primarily concerned with natural language processing, and within that, on the lexical constructs of metaphor. This lexical focus prevents the overall structure-mapping that would enable unexpected similarities to be detected across domains. A semantic approach, such as the ACME system by Holyoak and Thagard (1989) or ATT-Meta by Barnden (2001) gives the possibility of using such structures to reason using common logical reference. However, these approaches do not use reasoning systems in such a way that unexpected systems-level effects can emerge and be aligned. In other words, their analogical mapping does not reach deeply enough into situational structure to enable the derivation of meaning towards specific contexts to occur, in the narrative manner described here.

To build foundations for the implementation of these ideas, we supplement ordinary machine logic with a second structure that could support the kinds of information found in narrative transitions. Situation theory is thus the formalizable basis for this work. Originally developed to address contextual reasoning at Stanford during the 1980s–1990s (Barwise and Perry, 1983; Devlin, 1995), situation theory is distinctive for being a two-sorted logic. In this approach, two reasoning frameworks are linked in a formal relationship. One of these, known as the 'right hand side' (RHS), is the ordinary, logical system that handles individual facts. On the 'left hand side' (LHS), is a new framework to handle contexts and their transitions, made implementable by Goranson, Devlin and Cardier (Goranson and Cardier, 2013; Goranson et al., 2015). This 'second sort', when adjusted to include narrative mechanisms, can handle the grouping and stepwise operations required to identify, bridge and modify states. This adjustment to situation theory resolves several challenges to its implementation, the result being its first practical computerized implementation.

Our narrative-based approach is similar to solutions that employ modal or higher order logics for similar purposes. What differentiates the second reasoning system is that it is not logical, nor even set-theoretic. It cannot be, because the primary challenge is that it supports reasoning over open sets (that is, most situations have unknown defining facts). It also supports reasoning over transitions between these states; narrative structure also differs from that of logic in this key respect (Bruner, 1986). This approach can thus manage situations at the system-level: what they are, how governance is arranged and what meanings they affect.

This second reasoning system is implemented using category theory. This supports the analogical and governance aspects of the process, providing a way to contextualize every fact in a vast network of situations. It works like this: when a new fact arrives, it changes the array of supporting situations that saliently support it. This change is reflected in the categoric side (LHS), where a new arrangement is found that reflects that prescribed by the text. The governance relationships between situations also usually shift. In turn, this change then also modifies the relevant ontologies on the RHS that determine the 'interpretation' (usually several) of the overall story. These principles are modeled to supplement reductionist models of information. They do not replace reductionist methods; instead they incorporate and extend them (Goranson and Cardier, 2013; Goranson et al., 2015). Operations in one sort affect the other.

This framework has been published as a methodology and primitives for a type system that satisfies implementability of the formalism outlined here (Goranson et al., 2015). The underlying situation theoretic structure also informs the visualization method. Each fact is represented by an infon, the unit of situation theory. Infons are composed into lattices—for an example, see Figure 10.9, below.

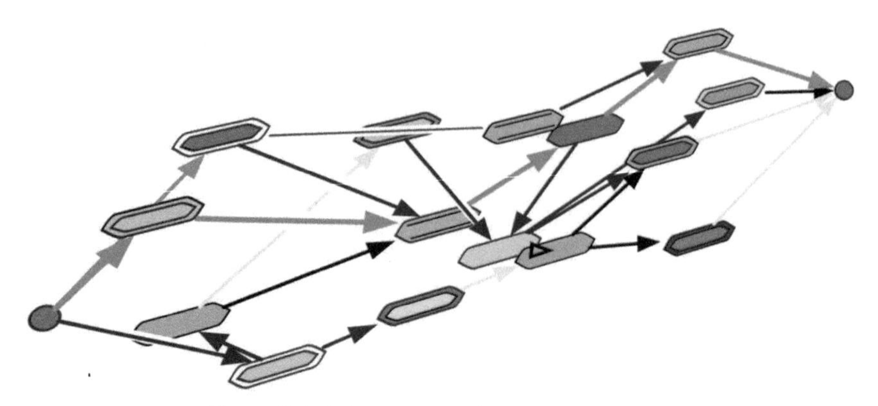

Figure 10.9. Infons structured into a causal lattice.

Each node in the network is a statement of facts that one can equate to actual data from a neurobiological system or the literal story chunks of the *Red Riding Hood*. Logically, each link is an 'and-then' relation (Lehmann, 2008) that in part conveys temporal causality by the node and of what has occurred before. Nodes are typically mappable to natural language concepts so that they can be 'read' and zoomed into for constituents. Situations are depicted as distinct interacting networks of shape or color. A user is able to move back and forth in time to see change.

This combined, collaborative structure forms into a *causal lattice*. A zoomed-out visualization of a causal lattice by Goranson (Cardier, 2017) can be seen in Figure 10.10.

A zoomed-out *causal lattice* represents the 'shared story' mentioned in the title. This umbrella structure is used by all semantic structures as a common reference point, to understand their roles *in this circumstance*. Below is another version created by Alessio Erioli and Niccolo Casas for the NAKFI project (Cardier, 2017); see Figure 10.11.

Notice how the structure in Figures 10.10 and 10.11 are essentially a macro version of the narrative version of the network in Figure 10.4. This overview enables a user to see how the shape of a system is changing as it travels from left to right. Chains of morphisms track how the overall coherence evolves as each new piece of information adjusts the system. As a user interface, the network can be scrolled, zoomed or adjusted. This 'shell' flows from left to right over time, as does the lattice, which is nominally inside as a three-dimensional object. A trained analyst will be able to read the system influences from the colors and forms to get a feel for what system governs that particular point in the event stream. In this study, developing a vocabulary of transitions becomes a focus.

The causal lattice models the 'shared story' communally created by all narrative elements. When applied to narrative, it is also a visual

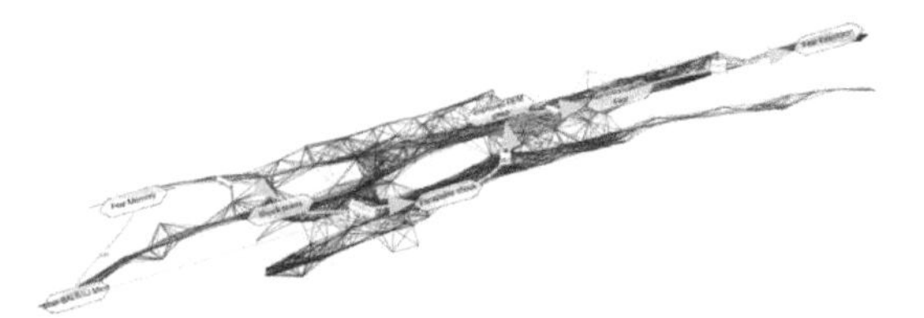

Figure 10.10. Connective structure in Figure 10.4 matured with the above causal lattice (from the NAKFI project). Here, contexts are represented as event flows composed of infons, which are semantic statements from situation theory (superimposed).

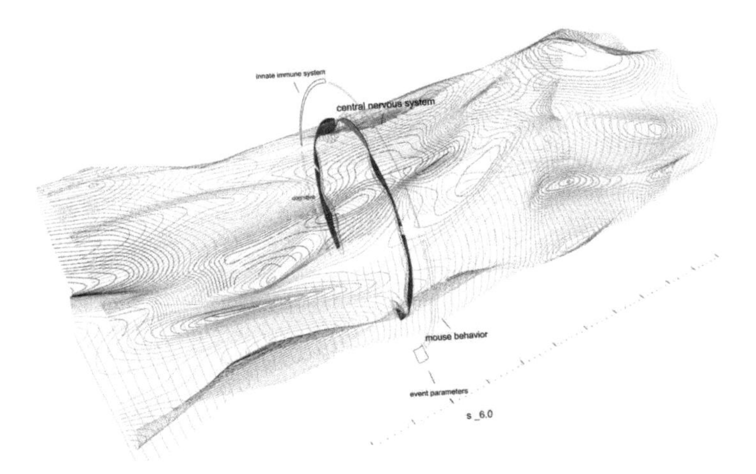

Figure 10.11. Visualization studies from the NAKFI project. The challenge is to display contextual interaction. Here, fear memory is represented by a flow of interacting processes through undulations in the 'skin' of a conceptual network. Relations are lines and interacting contexts are colored zones.

representation of the way a reader keeps track of the changing nature of context when interpreting it. In the visualization, this whole structure is animated to capture the way the causal lattice evolves.

The NAKFI study attempts to impose an explicit display of situation influence to annotate this lattice. More experiments are shown in Figures 10.12–10.13.

Our new model is thus developed to capture the phenomenon of contextual interaction, using a two-sorted situation-theoretic implementation extended by narrative-based mechanisms. One goal of this work is to understand how **connections between multiple contexts** are built. Another is to track how this affects the **governing agency** of participating elements. A third objective is to identify and record the dynamics that emerge at the **system-level** of contextual interaction. A fourth

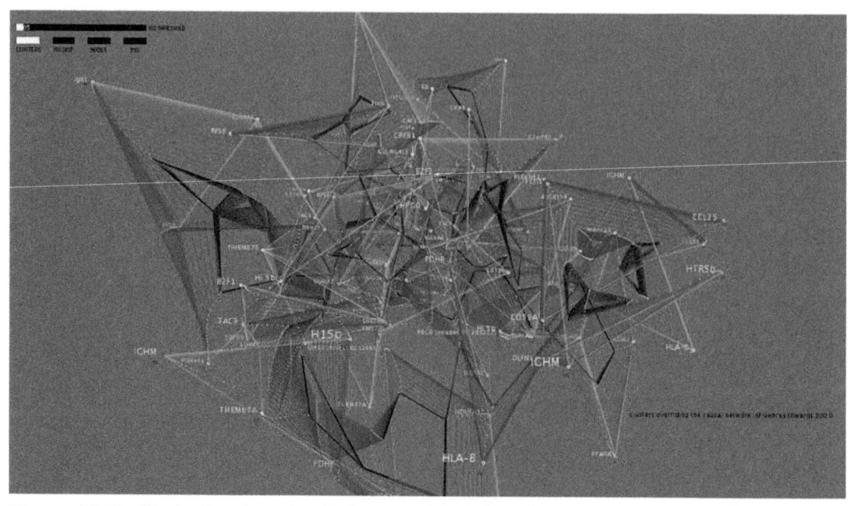

Figure 10.12. Study for the rate of change of structure between systems so that bursts of change can be seen.

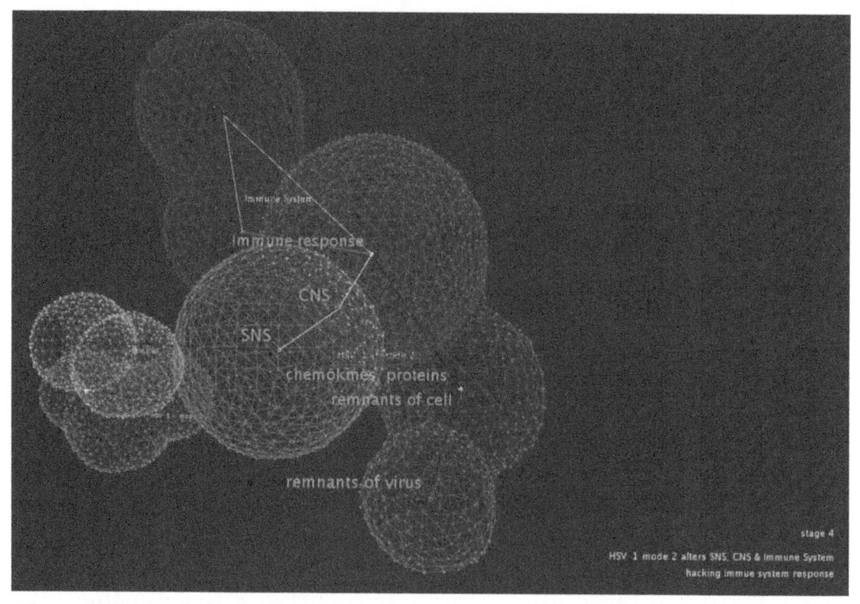

Figure 10.13. Study for the depiction and interaction between entire conceptual networks.

motivation is to better understand the role of **analogy** in this process. Finally, these observations are applied to the process of developing a **shared story among heterogeneous teams**.

10.8 Conclusion

This chapter presents a narrative-based model of contextual interaction. It shows how coherence can be derived across multiple situations, even if their information structures are heterogeneous. Techniques for this purpose are drawn from narrative processes of inference and also from the experience of facilitating interdisciplinary teams. Together these examples illustrate how individual elements can interact to build a common frame of reference. Separate contexts are connected through system-level mechanisms, such as analogy, system-level identity and governance. A communal structure emerges: a *shared story*. In narrative or an open-world situation, this coherence is continually altered, in response to each incoming piece of information.

This work is currently being used to model narrative, human collaboration and ontological interaction in neurobiological systems. In these pages, this work is presented as a practical approach, a model of narrative, an implementable method and also a visualization. Implementation is based on an extension of situation theory, which enables transitions between logical states. These different perspectives are offered their own system of distributed 'intelligence', to provide a revolving lens of insight into collaborative, heterogeneous teams.

Acknowledgment

This work is partially supported by the National Academies Keck Futures Initiative (grant NAKFI ADSEM2).

References

Arp, R., Smith, B. and Spear, A.D. 2015. Building Ontologies with Basic Formal Ontology. MIT Press.

Barnden, J. 2001. Uncertainty and conflict handling in the ATT-Meta context-based system for metaphorical reasoning. pp. 15–29. *In*: Akman, V., Bouquet, P., Thomason, R. and Young, R. (eds.). Proceedings of the Third International Conference on Modeling and Using Context. Springer, Berlin.

Barwise, J. and Perry, J. 1983. Situations and Attitudes. MIT Press, Cambridge, Massachusetts.

Boden, M. 1990. The Creative Mind: Myths and Mechanisms. HarperCollins Publishers, New York.

Bruner, J. 1986. Actual minds, possible worlds. Harvard University Press, Cambridge, MA.

Bruner, J. 1991. The narrative construction of reality. Crit. Inq. 18: 1–21.

Cardier, B. 2007. The story molecule: narrative as information. Presented at the AAAI Fall Symposium: Narrative, AAAI Press.

Cardier, B. 2013. Unputdownable: How the Agencies of Compelling Story Assembly Can Be Modelled Using Formalisable Methods From Knowledge Representation, and in a Fictional Tale About Seduction. University of Melbourne, Melbourne, Australia.

Cardier, B. 2014a. Narrative causal impetus: situational governance in game of thrones. pp. 2–8. *In*: Zhu, J., Horswil, I. and Wardrip-fruin, N. (eds.). Intelligent Narrative

Technologies 7. Presented at the Intelligent Narrative Technologies, AAAI Press, Palo Alto, CA.

Cardier, B. 2014b. Model of first paragraph of Little Red Riding Hood as a Dictator Would Tell It by H.I Phillips. http://topoiesis.org.s3.amazonaws.com/Animation%20A.mp4.

Cardier, B. 2015. The evolution of interpretive contexts in stories. *In*: Finlayson, M., Miller, B., Lieto, A. and Ronfard, R. (eds.). Sixth International Workshop on Computational Models of Narrative. Presented at the Workshop on Computational Models of Narrative, Cognitive Systems Foundation, OASICS.

Cardier, B. 2017. The moving lens: Coherence across heterogeneous contexts in narrative and biology. Presented at the AAAI Spring Symposia Series: Computational Context: Why It's Important, What It Means, and Can It Be Computed?

Cardier, B., Goranson, H.T., Casas, N., Lundberg, P.S., Erioli, A., Takaki, R., Nagy, D., Ciavarra, R. and Larry, D.S. 2017a. Modeling the peak of emergence in systems: design and katachi. Prog. Biophys. Mol. Biol. Spec. Issue Integral Biomathics Necessary Conjunction West. East. Thought Tradit. Explor. Nat. Mind Life 131c: 213–241.

Cardier, B., Sanford, L.D., Goranson, H.T., Devlin, K., Lundberg, P.S., Ciavarra, R., Casas, N. and Erioli, A. 2017b. Modeling the resituation of memory in neurobiology and narrative. Presented at the AAAI Spring Symposium on Science of Intelligence: Computational Principles of Natural and Artificial Intelligence, AAAI.

Cohen, D.B. 2006. Family constellations: An innovative systemic phenomenological group process from Germany. The Family Journal 14(3): 226–33.

Cohen, E. and Lotan, R. 2014. Designing Groupwork: Strategies for the Heterogeneous Classroom. Teacher's College Press, New York.

Dan Booth Cohen. 2006. Family constellations: An innovative systemic phenomenological group process from Germany. Fam. J. 14: 226–233.

Denis Noble. 2008. Genes and causation. Philos. Trans. R. Soc. A 366: 3001–3015.

Devlin, K.J. 1995. Logic and Information. Cambridge University Press.

Dreyer-Lude, M. 2016. Finding Your Scientific Voice.

Dreyer-Lude, M. 2017. Beyond Boundaries. Arnold and Mabel Beckman Center, Irvine CA.

Fauconnier, G. and Turner, M. 2008. Rethinking Metaphor. pp. 53–66. *In*: Cambridge Handbook of Metaphor and Thought. Cambridge University Press.

Feiger, Sheila and Madeline Schmidtt. 1979. Collegiality in Interdisciplinary Health Teams: Its Measurement and Its Effects 13A: 217–29.

Forbus, K., Ferguson, R., Lovett, A. and Gentner, D. 2017. Extending SME to handle large-scale cognitive modeling. Cogn. Sci. 41: 1152–1201.

Gentner, D. 1983. Structure-mapping: a theoretical framework for analogy. Cogn. Sci. 7: 155–170.

Gentner, D., Ratterman, M. and Forbus, K. 1993. The roles of similarity in transfer: Separating retrievability from inferential soundness. Cognit. Psychol. 25: 524–575.

George Lakoff. 1993. The contemporary theory of metaphor. *In*: Metaphor and Thought. Cambridge University Press, Cambridge.

Goranson, H.T. and Cardier, B. 2007. Scheherazade's Will: Quantum Narrative Agency. Presented at the AAAI Spring Series Workshop on Quantum Interaction, AAAI Press.

Goranson, H.T., Cardier, B. and Garcia, M. 2012. Topoiesis: A System for Reasoning over Uncompromised Reality.

Goranson, H.T. and Cardier, B. 2013. A Two-sorted logic for structurally modeling systems. Prog. Biophys. Mol. Biol. 113: 141–178. https://doi.org/10.1016/j.pbiomolbio.2013.03.015.

Goranson, H.T. 2014. System and Method for Space-Time, Annotation Capable Media Scrubbing USPTO 14/740,528 (20150279425).

Goranson, H.T. and Cardier, B. 2014. System and Method for Ontology Derivation USPTO 14/093,229 (20140164298).

Goranson, H.T. 2015. Opportunistic Layered Hypernarrative.

Goranson, H.T., Cardier, B. and Devlin, K.J. 2015. Pragmatic phenomenological types. Prog. Biophys. Mol. Biol. 119: 420–436. https://doi.org/10.1016/j.pbiomolbio.2015.07.006.

Graesser, A.C. and Wiemer-Hastings, K. 1999. Situational models and concepts in story comprehension. pp. 77–92. *In*: Goldman, S.R., Graesser, A.C. and Van den Broek, P.W. (eds.). Narrative Comprehension, Causality, and Coherence: Essays in Honor of Tom Trabasso. Lawrence Erlbaum, Mahwah, NJ.
Graves, A. 2013. And Now His Watch Is Ended. Game Thrones.
Gray, D. 2016. Liminal Thinking: Create the Change You Want by Changing the Way You Think. Two Waves Books, New York.
Gruber, T.R. 1993. A translational approach of portable ontology specification. Knowl. Acquis. 5: 1992–220. https://doi.org/10.1006/knac.1993.1008.
Gunn, J. 2014. Guardians of the Galaxy. Walt Disney Studios.
Harrell, D.F. 2005. Shades of computational evocation and meaning: the GRIOT system and improvisational poetry generation. Proc. 6th Digit. Arts Cult. Conf. DAC, 133–143.
Herman, D. 2006. Genette meets vygotsky: narrative embedding and distributed intelligence. Lang. Lit. 15: 357–380.
Herman, D. 2009. Basic Elements of Narrative, 1st ed. Wiley.
Hobbs, J., Stickel, M. and Appelt, D. 1993. Interpretation as Abduction. Artif. Intell. 63: 69–142.
Holyoak, K. and Thagard, P. 1989. Analogical mapping by constraint satisfaction. Cogn. Sci. 13: 295–355.
Hunicke, R., LeBlanc, M. and Zubek, R. 2004. MDA: A formal approach to game design and game research. pp. 1–5. *In*: In Proceedings of the AAAI-04 Workshop on Challenges in Game AI. Presented at the In Proceedings of the AAAI-04 Workshop on Challenges in Game AI.
Kanellopolous, D.N. 2008. An ontology-based system for intelligent matching of traveller's needs for airline seats. Int. J. Comput. Appl. Technol. 32: 194–205.
Kuhn, T. 1973. The Structure of Scientific Revolutions. University of Chicago Press, Chicago.
Labov, W. and Waletzky, J. 1967. Narrative analysis. *In*: Essays on the Verbal and Visual Arts. University of Washington Press, Seattle, WA.
Labov, W. 1972. The transformation of experience in narrative syntax. pp. 354–397. *In*: Language in the Inner City: Studies in the Black English Vernacular. University of Philadelphia Press, Philadelphia.
Lakoff, G. and Turner, M. 1989. More than cool reason. University of Chicago Press, Chicago.
Lakoff, G. 1993. The contemporary theory of metaphor. pp. 202–251. *In*: Ortony, A. (ed.). Metaphor and Thought. Cambridge University Press, Cambridge, MA.
Lawless, W.F. 1985. Problems with military nuclear waste. Bull. At. Sci. 41: 38–42.
Lawless, W.F., Mittu, R. and Sofge, D.A. 2017. (Computational) Context: Why it's important, what it means, can it be computed? pp. 309–314. *In*: Computational Context Technical Report SS-17-03. AAAI Press, Palo Alto, CA.
Ledford, J. 1984. Engineer claims DOE has problems with radioactive waste. United Press Int.
Lehmann, D. 2008. A presentation of quantum logic based on an and then connective. J. Log. Comput. 18: 59–76.
Markman, A. and Gentner, D. 1993. Structural alignment during similarity comparisons. Cognit. Psychol. 25: 431–467.
Martin, J. 1990. A Computational Model of Metaphor Interpretation. Academic Press, San Diego, CA.
Mayell, H. 2002. Idaho, U.S. Battle Over Nuclear Waste Dump. Natl. Geogr. News.
McRae, K. and Jones, M. 2013. Semantic memory. *In*: The Oxford Handbook of Cognitive Psychology. Oxford University Press, Oxford.
Nersessian, N. and Chandrasekharan, S. 2009. Hybrid analogies in conceptual innovation in science. Cogn. Syst. Res. Spec. Issue Analog. - Integrating Cogn. Abil. 10: 178–188.
Niki, A. 2017. Why I Launched A Story Consultancy.
Page, A. 2004. Keeping Patients Safe: Transforming the Work Environment of Nurses. The National Academies Press, Washington DC.
Phillips, H.I. 1993. Little red riding hood as a dictator would tell it. pp. 230–233. *In*: The Trials and Tribulations of Little Red Riding Hood. Routledge, New York.

Reed, S. and Lenat, D.B. 2002. Mapping Ontologies into Cyc. Presented at the AAAI 2002 Conference Workshop on Ontologies For The Semantic Web.

Ryan, M.-L. 2005. On the theoretical foundations of transmedial narratology. *In*: Meister, J. (ed.). Narratology beyond Literary Criticism: Mediality, Disciplinarity. de Gruyter, Berlin.

Scharmer, O. and Kaufer, K. 2013. Leading From the Emerging Future. Berret-Koehler Publishers Inc, San Francisco.

Searle, J. 1979. Expression and Meaning: Studies in the Theory of Speech Acts. Cambridge University Press, Cambridge, MA.

Sowa, J.F. 2011. Future Directions for Semantic Systems. Intell.-Based Syst. Eng. 23–47. https://doi.org/10.1007/978-3-642-17931-0_2.

Sowa, J.F. 2016. Semantics for Interoperability. Ontol. Summit.

Thagard, P. 1992. Conceptual Revolutions. Princeton University Press, Princeton, New Jersey.

Thursby, J. 2006. Story: A Handbook. Greenwood Publishing Group.

Toolan, M. 2009. Narrative Progression in the Short Story. John Benjamins Publishing Company, Amsterdam, Philadelphia.

Tredici, R. 1987. At Work in the Fields of the Bomb. Harper Row, New York.

Various, 2017. Little Red Riding Hood. Wikipedia.

Winner, E. and Gardner, H. 1993. Metaphor and irony: Two levels of understanding. pp. 425–433. *In*: Ortony, A. (ed.). Metaphor and Thought. Cambridge University Press, Cambridge, MA.

Winston, P. 1980. Learning and reasoning by analogy. Commun. ACM 23(12): 698–703.

11

Algebraic Modeling of the Causal Break and Representation of the Decision Process in Contextual Structures

Olivier Bartheye[1,*] and *Laurent Chaudron*[2]

11.1 Introduction

A very important point of this chapter is the following: a decision process fills a causality break. Roughly speaking, a split-up holds between decision and causality. Apparently, this observation is not so impressive but its interpretation leads to tackle one of the most difficult paradoxes modern science is faced with. In effect, rather than using solely objective arguments, this topic requires to use subjectivity because, seriously, nothing better than a human agent is able to take decisions in that so uncomfortable situation in which causal laws do not provide a suitable model to be used. In other words, what are the properties extensively studied in the frame of the anthropological paradigm able to classify the crucial differences between the human and the machine (Bartheye and Chaudron, 2015b) according to

[1] CREC-St-Cyr, Military School of Coëtquidan, Guer Cedex, France.
[2] Director, ONERA Provence Research Center (the French Air Force Academy), BA 701, 13661 Salon Air, France; laurent.chaudron@polytechnique.org
* Corresponding author: olivier.bartheye@st-cyr.terre-net.defense.gouv.fr

the Artificial Intelligence program started in the Dartmouth conference in Summer 1956.

This chapter emphasizes also the following point: contexts are crucial for defining coordinated behaviors between intelligent entities. The difficulty arises when computation contexts are required; that is, what kind of implementation should one expect from understandable context definitions? A gap holds between the conceptual definition of a context and its somewhat mechanical implementation because of a huge scale difference. In other words a computational context requires two sides: the conceptual side and the methodological side which cannot be unified. Please, don't consider that according to suitable conceptual restrictions and to suitable methodological restrictions, one should find a suitable result, since the "hole" between these two sides is comparable to the Heisenberg uncertainty principle.[1]

Recall that a context explains how the environment influences human perception, cognition and action. The relationship between perception and action is extremely complicated to represent except using duality since these two notions are actually orthogonal but this is undoubtedly the key of the computational context issue presented here. We propose to use algebraic definitions in order to formally define this relationship as a phase shift between perception and action. In effect, according to the standard notion of signal denoting what perception is, the integration of any AI action has as an immediate consequence to delocalize computation from the center of the confident signal domain where a phase holds towards its border introducing at the same time uncertainty which is represented in this study by entropy.

Once action is sufficiently dissociated from perception, action can be moved as a decision process. The expected gain is important: *consistency*. Unfortunately, the integrity of the signal collapses. In effect, integrity and acceptable universality is supported by Fourier signal analysis according to integral operators. That means actually that, in that case, perception matches with action since action is implemented as linear operators in the vector space of signals. That is, a harmonic representation holds which stands for universality. Unfortunately consistency rarely coincides with universality because of its partial nature; for instance according to model theory, several partial models hold which are mutually exclusive to partial counter-models. Once the elasticity of causality is associated with context definitions, one can set as our main thread to recover action universality according to the causal break highlighting. That is, the causal break, if any, coincides with context

[1] Heisenberg's uncertainty principle or Heisenberg's indeterminacy principle, is any of a variety of mathematical inequalities asserting a fundamental limit to the precision with which certain pairs of physical properties of a particle, known as complementary variables, such as position x and momentum p, can be known.

invalidity. Our study shows that any context is expected to be invalid, and leads that way to a causality break, according to double S-curves defined later on. In other words, causality coincides exactly with context validity. That means that the management of the causality break means to qualify what is a context, what is context identification and what is a context change according to contextual structures.

11.2 Contextual Structures and Coalgebras

11.2.1 The Context Validity Theorem

Because of the universal feature of a context, one should ask the following non trivial question: *"In which context should we use a context?"* It turns out that contexts are required

1. when *high entropy* holds (i.e., non manageable entropy),
2. when some *common stable representation* has to be defined,
3. when *valid* behaviors are expected.

Therefore a context which is expected to be computable has to integrate these *incompatible* features. In particular, one has to choose the implementation of context management between

1. a classification issue,
2. a knowledge representation issue (delimitating clear contexts or unclear contexts),
3. an action planning issue,
4. or even a proof theory issue.

If the main objective is to define interactivity, a context can be viewed as an interface. In other words, a context is a kind of middleware (see Figure 11.1).

In a computable context, particularly using Artificial Intelligence, contexts can be used to specify an entity behavior by replacing the human in Figure 11.1 with any other AI entity (a robot, a virtual computer, an IoT

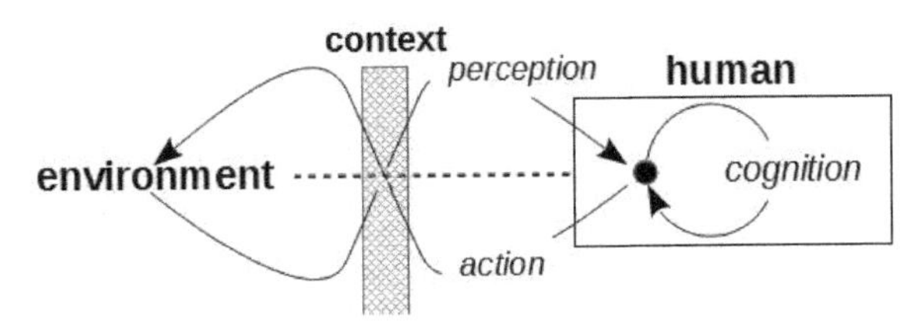

Figure 11.1. Context as a middleware.

device,...). A context is therefore a common frame well-defined for hybrid teams composed with isolated entities (see Figure 11.2).

Furthermore, a context agrees with some conservation law (*stability*), some expressive power (*understandability*), some common interpretation property and some transposition property (*universality*).

Definition 11.1 (Context). *A context is an associated set of rules and definitions with certain reproducibility properties.*

In order to tackle context management issues, in an algebraic fashion, one should use the context validity theorem. It consists to interpret the diagrams in Figure 11.1 according to category theory[2] using objects and arrows. This theorem is very important. It claims that the initiator of any context definition is actually the environment. That is, the basis of the context is the embedding of any AI entity inside a given environment. Since this embedding is the generator of any signal, it is important to classify the knowledge basis definable from the environment as the empirical basis according to next section.

Theorem 11.1 (Context validity theorem). *The following statements are equivalent:*

1. *the 3 arrows (perception, action, cognition) are subordinated to current context rules and definitions,*
2. *an order relation holds (perception precedes necessary action).*

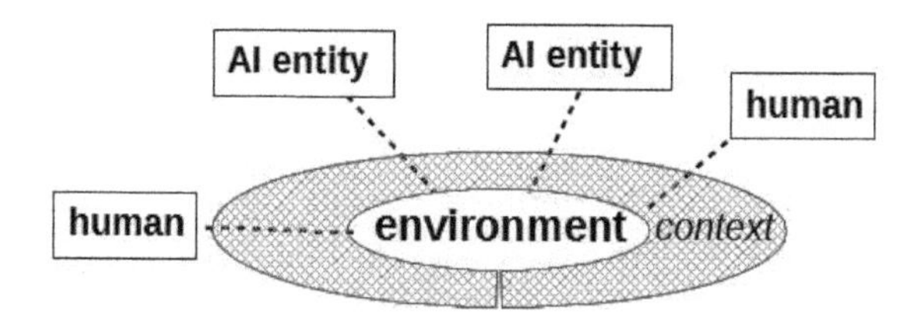

Figure 11.2. Context as a common frame.

[2] A category C consists of (i) a class **Obj**(C) of objects (ii) a class **Hom**(C) of morphisms, or arrows, or maps, between the objects. Each morphism f has a unique source object a and target object b where $a,b \in \mathbf{Obj}(C)^2$. We write $f: a \to b$, and we say "f is a morphism from a to b". We write $\mathbf{Hom}_C(a,c)$ to denote the **Hom**-class of all morphisms from a to b (iii) for every three objects a, b and c, a binary operation $\mathbf{Hom}(a,b) \times \mathbf{Hom}(b,c) \to \mathbf{Hom}(a,c)$ called composition of morphisms; the composition of $f: a \to b$ and $g: b \to c$ is written as $g \circ f$ or gf. such that the following axioms hold:
(*associativity*) if $f: a \to b$, $g: b \to c$ and $h: c \to d$ then $h \circ (g \circ f) = (h \circ g) \circ f$,
(*identity*) for every object x, there exists a morphism $\mathbf{1}_x : x \to x$ (some authors write $\mathbf{id}_x$) called the identity morphism for x, such that for every morphism $f: a \to x$ and every morphism $g: x \to b$, we have $\mathbf{1}_x \circ f = f$ and $g \circ \mathbf{1}_x = g$. From these axioms, one can prove that there is exactly one identity morphism for every object. Some authors use a slight variation of the definition in which each object is identified with the corresponding identity morphism.

As we'll see later on, if *perception* and *action* are *simultaneous*, signal analysis or harmonic analysis (e.g., Fourier analysis) can be used but there is no real context for AI action models. If any AI action model is defined, a significant phase shift holds between perception and action; in other words, AI action representation is anharmonic and performs a deviation of a system as a harmonic oscillator; as such it can be considered as *unsafe* when viewed from harmonic valuation tools. Intuitively, that means that, in such a situation, the signal domain is *not regular* or equivalently is *singular* (i.e., contains holes or singularities). The degree of anharmonicity (the degree of singularity) is represented by entropy.[3] The context validity theorem asserts that entropy is to be controlled by the order relation "*perception* ***precedes*** *action*". The present chapter tries to set the entropy level as maximal in order to perform an effective context change which nevertheless can be qualified as regular. Apparently, due to the context validity theorem, the requirement is quite simple: a context change is required when a context becomes invalid. Very surprisingly, due to regular assumption for a context change, maximal entropy or *full entropy* is expected to be an accumulation point or an adherence. It could act as a universal physical constant as for instance the speed of light in vacuum and should provide some stable condition for a context change. In other words, thanks to computer science, we should set *full entropy* as a natural framework.

11.2.2 Localization of the Causal Break and Coalgebras

If full entropy holds, one can propose to use exactly the reversal of the context validity theorem: *action* ***precedes*** *perception*. We have that way, the generic point where the causal break occurs: it occurs exactly where the context validity theorem is reversed. It is important, in that frame, to characterize what is context identification and what is a context change in an algebraic point of view. It turns out that the right term for such a characterization is actually a *coalgebra* rather than an *algebra*. The notion of coalgebra is very important in many domains: group representation, quantum groups, proof theory. Let's associate coalgebras and discrete decomposition with qualitatives. Qualitative analysis of the phase shift between perception and action requires coalgebraic tools as for instance proof theory, and in particular the sequent calculus for which an algebraic semantic can be defined using category theory. Coalgebras exist in the mathematical arsenal but the challenge here is to be able to provide a context change after the coalgebraic process. That is, the result of the coalgebra is a new algebra. The

[3] Entropy is associated with the amount of order, disorder, or chaos in a thermodynamic system; in information theory, entropy is the uncertainty of an outcome.

aim is to recover quantitatives once the phase shift between perception and action is set to be maximal according to full entropy. That way, the decision process has a clear semantic: to provide a regular transition starting from the signal algebra; that means that causality is necessary the generic point where the transition is to be performed. Since the coalgebraic process is both necessary and sufficient, that means that it is also the annihilator of the signal algebra by relaxing context signal dependency.

11.2.3 Semantic Interpretation of Coalgebras

It is not difficult to provide a semantic interpretation of algebras in the context framework using signal analysis. The context validity theorem asserts that interactions between context users are definable according to algebras. One can claim that any valid context C requires a significant domain with objects and interaction between these objects. An interactive domain of that kind is represented by an algebra (A, η, μ) where η is the A-identity operator and μ the A-multiplication. The best example is Fourier algebras; one obtains a powerful signal domain but in that case, action and perception are simultaneous. When the precedence increases, action plays a greater role and a given signal algebra A cannot suitably denote the full set of interactions. One requires a critical analysis of these interactions according to an abstract model; since this model is not perfect, it is no longer universal, in other words the delimited domain including AI macro-actions is not regular.

Coalgebras are defined in order to deal with irregularities. In the context framework, coalgebras permit to select regular subdomains (models) and to reject non regular subdomains (counter-models). Special cases of coalgebras are proof systems, case based analysis, symbolic interpretation or any discrete decomposition in order to separate "good" subdomains (i.e., regular) and "bad" ones (i.e., singular). In effect, any decomposition process which can be also considered as an unfolding process can be represented by a coalgebra, not by an algebra. That is, arrows are reversed: $x \times y \rightarrow z$ becomes $z \rightarrow x \times y$.

Another important property for coalgebras is their ability to represent proofs or computations as formal objects not solely as morphisms (or arrows) on objects. That way, the duality state/transition or formulas/proofs can be managed according to that abstraction step. It is clear that coalgebras are suitable algebraic tools to model the conceptual process. An interesting example is the carrying process in elementary arithmetic using 1-co-cycles (Isaksen, 2002).

11.2.4 Hopf Algebras

The duality between coalgebras and algebras is very important but it is not sufficient. In effect, the reversal of the context validity theorem requires much more: in particular some finite property for domains are required plus some quaternary correspondence between algebraic and coalgebraic structures. The right candidate is a very special bialgebra: a Hopf algebra[4] H (that is, both a coalgebra (Δ, ε) and an algebra (μ, η) plus a crucial arrow: the antipode S). The unfolding process is represented by the coalgebraic part (Δ, ε); resp. the coproduct and the counity are able to decompose some element $a \in A$ according to two operands $a_1, a_2 \in A^2$ such that $a = a_1 . a_2$. This process is non deterministic and must be confirmed by the algebraic part (μ, η) resp. the product and the unity recomputing the product $a = a_1 . a_2$ according to the antipode bridge S which is able to connect the coalgebraic and the algebraic components.

Recall that the idea is to find the right algebra in which the decision process can take place in order to manage the causal break. Since the decision process is a mental process, one can depict the required algebraic structure in which a decision can be computed. Therefore, according to full entropy, the causal break can be characterized by the right unfolding operator until a generic point from which one can compute exactly the reversal of the context validity theorem, i.e., a context change.

According to Figure 11.1, an important step is to qualify the cognition arrow whose input is the perception arrow and whose output is the action arrow according to the context validity theorem in order to determine how cognition can be implemented inside a *decision* Hopf algebra H. The main idea is to classify knowledge bases as it will be shown in next section. Provided that knowledge bases are classified, so should be contexts since the context definition 11.1 can be very general.

11.3 Implementation of the Context Validity Theorem

The context validity theorem must be set according to knowledge classification. The idea is to determine what a context stands for. It turns out that a context holds in a contextual basis, according to a suitable abstraction effort.

---

[4] A vector space H is a Hopf algebra if it is equipped with morphisms of the following form:

$$\text{bialgebra} \begin{cases} \text{coalgebra} \begin{cases} \Delta : H \to H \otimes H \\ \varepsilon : H \to k \end{cases} \\ \\ \text{algebra} \begin{cases} \mu : H \otimes H \to H \\ \eta : k \to H \end{cases} \end{cases} \quad \text{antipode} \quad S : H \to H$$

11.3.1 Knowledge Basis Classification and Context Localization

Knowledge classification is to be performed according to the epistemological[5] quadryptich (Chaudron, 2005; see the graphical representation Figure 11.3). According to that quadryptich, theories can be classified in four knowledge basis: the outer basis named the empirical basis provides measures and as such can, in principle, validate any knowledge; since the main method associated to this basis is experience (see Table 11.1), one might consider that the limit is set by Kant's principle *"reason cannot legislate beyond the experience"*. Since this basis is the initiator of any signal and is, at the same time, a way to validate a model, one can consider that this basis is an open embedding for any other basis. Since a context is necessary built on partial knowledge, one should find another basis to support contexts. It is equivalent to say that harmonic analysis is not a natural model for representing contexts since it is clearly insufficient in most of AI paradigms, namely AI action for which other knowledge basis are required. In terms of classification, it turns out that the conceptual basis provides concepts; the formal basis provides theories and the methodological basis provides programs.

In effect, at the conceptual level, contexts can be defined since the conceptual level is stable, understandable and somewhat universal. In fact, context definition is always conceptual. Intuitively, this is equivalent to shape recognition. In particular, if a clear context holds, if it is possible to associate a precise dedicated shape to a given picture. This is equivalent

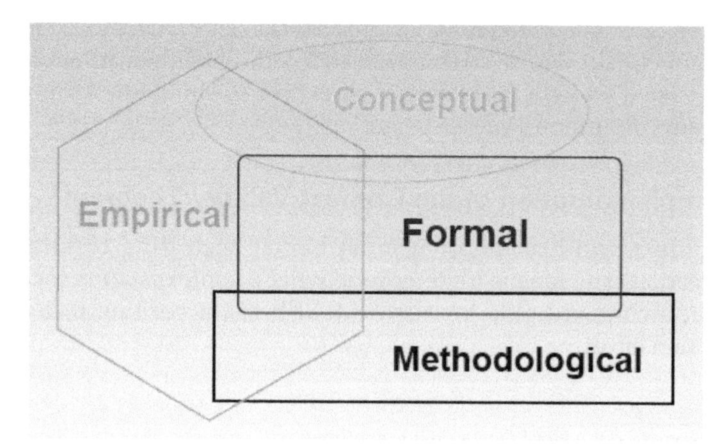

Figure 11.3. The epistemological quadryptich.

[5] Epistemology concerns the classification of scientific methods, logical models and inference mechanisms as principles, designs and theories such that their scope and their relevance can be discussed.

Table 11.1. The quaternary representation of knowledge bases.

Remainder mod 4	Sine	Knowledge Basis	Objects	Methods	Criteria
0 ≅ 4	− 2	empirical	"real world", datasets, ...	experiences, statistics	experimental
3	+ 1	conceptual	concepts, natural language, structures, boxology, …	abstract analysis	semantic, consistency
2	± 0	formal	mathematics, theory, ...	rational	proofs
1	− 1	methodological	coded data, algorithms, programs, machines, ...	sequences of actions	convergence efficiency
0 ≅ 4	− 2	empirical	"real world", datasets, ...	experiences, statistics	experimental

1. According to a problem to solve,
2. *experimental data* from protocols, processes are collected,
3. then a descriptive *conceptual model* is proposed,
4. a *formal model* is built in order to define formally objects and properties from concepts and to prove the validity of the model,
5. finally a computer program is *implemented* validating the solution and whose specification agrees with formal properties,
6. iterate at step 1.

Figure 11.4. The problem solving loop sequence.

to find saliencies of a given signal compared to the background noise. Consequently, it is clear that the methodological level is not the best one to propose a semantic denotation of a picture according to a given shape since at this level the main criteria are convergence and efficiency (see Table 11.1). At this level, contexts are necessary uncertain; this is an important qualifier since uncertainty means a high entropy level. Computation requires methodology and is at the same time the most powerful tool we have to model dynamicity and the most difficult to handle because of high entropy. In other words, high entropy holds for sake of implementation.

In the ideal case, the usability of these bases in AI is ordered using the problem solving loop sequence in Figure 11.4. This ordered sequence is very important since in terms of validity or even in software engineering, it seems harmful to switch between any of these steps. Therefore our actual effort is to be focused on this automaton architecture.

It is equivalent to characterize the phase shift between perception and action as the implementation of the problem solving loop. In particular, a context holds in step 3, cannot occur before and should not occur after in order to stay clear enough. Therefore, one should focus on the conceptual basis since contexts can be classified according to this quadryptich and the methodological sequence above.

11.3.2 Dependency of Contexts with Respect to Signal Analysis

The previous section provides qualitative remarks. In order to deal with quantitatives in contexts, one should use this main assertion: *"contexts always depend more or less on the empirical basis (contexts always depend more or less on signal analysis)"*. It is equivalent to set that the step 2 in the problem solving loop sequence in Figure 11.4 *always* holds between step 1 and step 3 in the automaton defined in the previous section.

This property can be interpreted in terms of abstract physics as the subjective nature of any definition: when an entity tries to solve a problem in step 1, using a context definition in step 3, don't forget that this attempt is implicitly curved by its signal perception on step 2. This curvature which can be called *experience* can be expressed in terms of action. It is based on two remarks: the first one is a consequence of Kant's principle: *"signal perception is the carrier of any definition and since any definition is a lemma, it definitely influences any judgment"*. In terms of formal analysis, this is not a good result because this carrier is the measure for the *subjective* nature of any conceptual definition. The definite loss of objectivity is measured by computational issues. Consequently, computation is the exact valuation of incompleteness due to step 2. That means that the signal notion is to be either used or transformed as a context but it cannot be ignored. Since that automaton is actually a loop (step 6 loops to step 1), one can conclude that this problem solving iterative activity reduces step 2 as much as possible but cannot eliminate it completely.

The second remark is a direct consequence of the previous sentence: *"knowledge representation is an iterative process which provides information at a very high level which is quasi independent from the signal (as an abstract context can be) but not totally"*. Inside the loop structure, one can interpret the automaton as follows: in order to provide a context definition according to step 3, one must skip step 2. In other words, universality or genericity has to be divided by the skip operator of step 2. Assume that signal annihilation is the actual support for any knowledge representation in step 3; one should retain the following physical principle: signal cannot be annihilated, it rather must be skipped. This skip operator is to be valuated. Its valuation means that step 3 has some kind of quasi-autonomous property; in terms of semantic that means that the subjective nature of step 3 means that objectivity is identified by pushing to the limit the autonomous property. That is objectivity generates *undecidability*.

On the other hand, *subjectivity* means that contexts are defined according to endomorphisms[6] in category theory; they form a complex structure named an endofunctor.[7] Entropy holds precisely because of the dependency of any intelligent entity to this endofunctor. Therefore, it is therefore possible to define the role of the decision process: exploiting full entropy in order to classify endomorphisms using coalgebra operators; the classification process is defined according to a Hopf algebra H representing bisequents (see next section). The classification criterion is set with respect to inner automorphisms on H.[8]

11.3.3 Knowledge Basis Indexing using Arithmetic Numbers

The automaton structure can be indexed, according to an arithmetic structure, using the abstract notion of number supporting the shift between perception and action. One can label these four knowledge bases according to quaternary numbers; they are classified using to their remainder mod 4 (see Table 11.1) according to Figure 11.3. It turns out that the importance of the action component corresponds to the residue level of the signal. The highest is the action level, the lowest is the signal level.

The outer empirical basis where signal analysis is performed from the environment is understood as the main embedding or equivalently a colimit. This outer basis is labeled by level 0 modulo 4 (noted $0 \cong 4$) those objects containing that kind of knowledge are data from the "real world". The three inner other basis are distributed in terms of knowledge according to polarities $P = \{-1, \pm 0, +1\}$. In order to have a graphical interpretation of the automaton, one should use the analogy in terms of the non-commutative version of the sine function (see Figure 11.5) which provides a harmonic representation or more simply a phase model. But in that case, a subtraction by 2 is required (see Table 11.1). From the representation modulo 4: $R = \{0, 1, 2, 3\}$, one obtains two incompatible domains:

1. the open interval $P =]-2, +2[= \{-1, \pm 0, +1\}$ characterizing the three inner bases,
2. the set $E = \{+2\} \cup \{-2\}$ characterizing the outer empirical basis (the "real world").

---

[6] In mathematics, an endomorphism is a morphism (or homomorphism) from a mathematical object to itself.

[7] A functor from a category to itself is called an *endofunctor*. Given any category C, the functor category **End**(C) = C^C is called the endofunctor category of C. The objects of **End**(C) are endofunctors $F: C \rightarrow C$ and the morphisms are natural transformation between such endofunctors.

[8] In abstract algebra an inner automorphism is an automorphism of an algebra given by the conjugation action by a fixed element, called the conjugating element. These inner automorphisms form a subgroup of the automorphism group, and the quotient of the automorphism group by this subgroup gives rise to the concept of the outer automorphism group.

The entropy level in the empirical basis is the lowest: signal analysis (harmonic analysis) may be used and Fourier algebras are common representations. *Perception* and *action* in the set $E = \{+2\} \cup \{-2\}$ are *simultaneous* components of the signal in Fourier analysis (action is mathematical action).[9] A good low entropic model in the set $E = \{+2\} \cup \{-2\}$ is also the Gaussian distribution well defined around the adherence μ, less defined around *boundaries* (see Figure 11.6(a)).

However, context is definitely not signal analysis since *perception precedes action* (from the context validity theorem) according to the AI problem solving loop and necessary leads to high entropy. In effect, when action and perception are in phase shift according to causal elasticity, high entropy occurs; that means that valid contexts are, in best cases, located at the boundaries of the delimited domain and in worst cases "totally out of bounds". Intuitively, that means, that because of that shift, the definition domain is unsafe or irregular or equivalently it contains "holes". Consequently, usable contexts require dealing with "out of bounds" algebras as in abstract differential geometry, exterior algebras[10] (see Figure 11.6(b)).

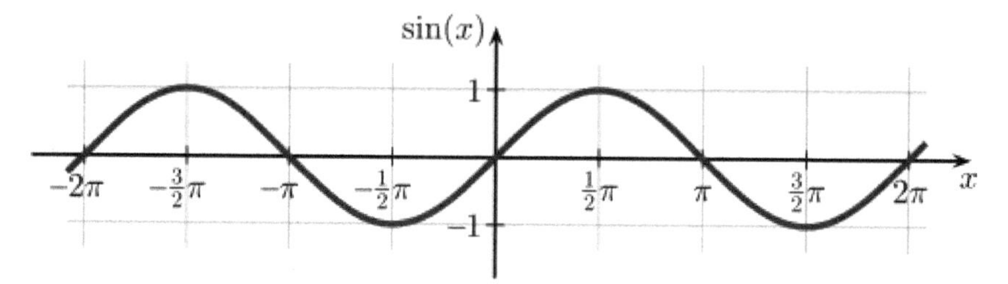

Figure 11.5. The phase model.

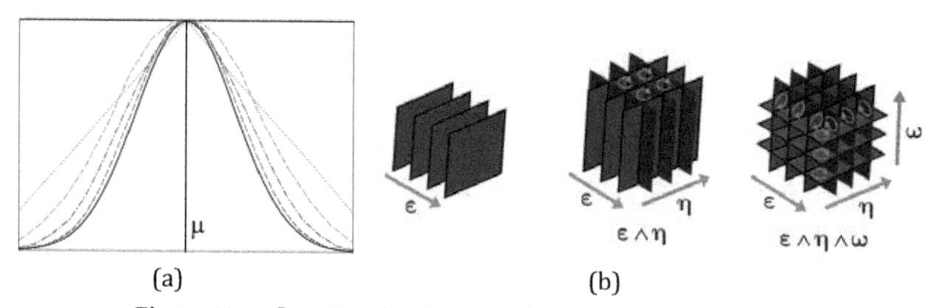

Figure 11.6. Gaussian distribution (a) and exterior product (b).

[9] Let k a field and V a k-vector space; a mathematical action α acts on vectors as a k-linear operator on V and the set of actions is a k-matrix algebra A.

[10] Let F a field and F^n the n-dimensional vector space; the set $G(k, n)$ (F) of *k-planes* in F^n is the set of all k-dimensional vector subspaces of F^n with the exterior product ∧.

11.3.4 Automata, Knowledge Bases and Contexts

The empirical basis cannot be represented by the automaton in Figure 11.7 according to the sine design. In effect, the sine design means that levels are characterized by the set of polarities $P = \{-1, \pm 0, +1\}$ corresponding respectively to the lowest value of the signal, the zero value and the highest value. That is, the conceptual basis is the upper domain around the + 1 value whereas the methodological basis is the lower domain around the – 1 value; the formal basis is in between, around ± 0. In other words, the "black hole" (or the missing link) between + 1 and – 1 is to be filled by the basis in the middle: the formal basis in level ± 0.

The automaton structure (see Figure 11.7) can be designed according to two kernel components, the rising edge ↗ from ± 0 to + 1 (the interval $]0/2\,\pi, 1/2\,\pi[$) and the falling edge ↘ from + 1 to – 1 (the interval $]1/2\,\pi, 3/2\,\pi[$). The main interest of this automaton is to reject the classical commutative notion of curve and the classical commutative notion of signal. That is, the transition from ± 0 to – 1 is not the rising edge ↗ but rather the falling edge ↙; that way – 1 is a generic point and the formal basis ± 0 is a reflector.

In other words, this formal basis seems to offer completeness: i.e., a quasi-universal representation as the harmonic model in the empirical basis but with a crucial supplementary property: *faithfulness*. Intuitively, that means that elementary components for harmonic analysis are not Fourier characters but the whole algebra. It is equivalent to assume that any element of the algebra even an element of the set of residues is a character. But faithfulness requires necessary to be able to exploit some kind of completeness. The last backward arrow ↙ suggests that the "state" – 1 is more complete that the state + 1. One can assert that when a transition is performed according the local nature of the subjective endofunctor, it is necessary to increase faithfulness. One can also assert the following result:

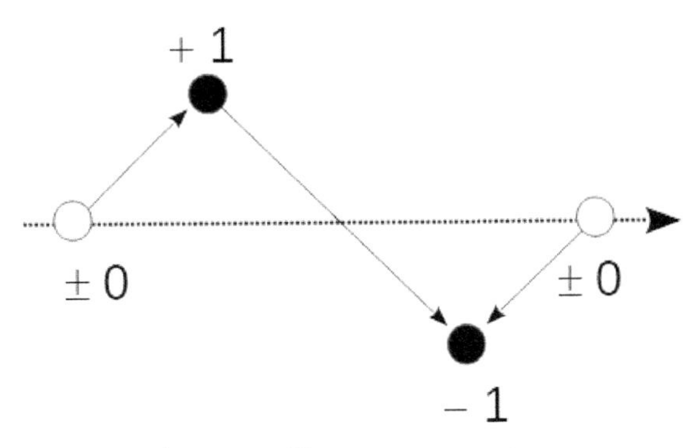

Figure 11.7. The context automaton.

1. a transition is performed,
2. faithfulness increases.

One can interpret the automaton according to ring theory and coalgebras. In terms of ring theory, the state + 1 means that the set of signal functions is a *ring*[11] whereas – 1 means that the set of signal functions is a *field*[12] (signal functions are invertible). According to this, it can be justified to locate the formal basis associated with the polarity ± 0 as some kind of kernel. That is, the formal basis is the right location to exploit *completeness*. Using coalgebras, it means that coalgebras are more faithful than algebras. In particular, co-identities are richer than identities. The topology of co-identities is continuous. In that frame, it is clear that the most difficult arrow to represent is the arrow

$$+1 \rightarrow -1 \tag{11.1}$$

Since this arrow is subjective, it corresponds exactly to endomorphisms quotiented by inner automorphisms and that way, a plan can be expected to be the quotient.

11.3.5 Integration of Contexts in the Problem Solving Loop Automaton

According to the Figure 11.1, it can be worthwhile to decompose the cognition arrow which is subjective and specific to any AI entity according to four knowledge basis used in the problem solving loop. At first sight, this arrow looks like a trivial cycle or an identity arrow but it turns out that is not the case.

One obtains the automaton corresponding to the middleware representation of a context in Figure 11.8. The main interest of this automaton is to specify the interactions of the environment and the AI entities according to the context. In order to so, one has to answer the crucial question: *What does "a problem to solve" mean?* in terms of interaction. One can propose a semantic of contexts thanks to self-orientation processes (Bartheye and Chaudron, 2015a). A typical example of self-orientation processes is a research lab: evaluated scientists are supposed to be the best experts of their own domain and as such could propose the orientations of their own discipline. A context expresses that way the mutual influence

[11] A ring R is a triple $(R, +, .)$ where R is an abelian group under +, and R is a monoid under multiplication . which is distributive with respect to +.

[12] A field k is a triple $(k, +, .)$ where k is an abelian group under + and $k-\{0\}$ acts multiplicatively on k.

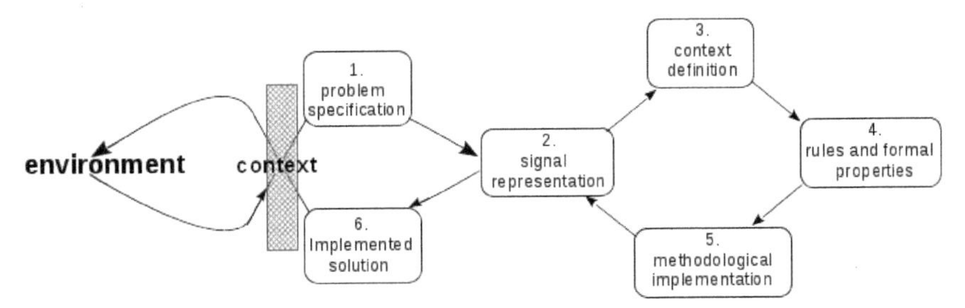

Figure 11.8. Context in the problem solving loop automaton.

between the environment and the human. Nowadays, natural environments shift more and more towards artificial environments (towns, computer networks, connected objects, smartphones, internet of things, ...); therefore the so-called *environment* feature is most of time cultural and most of time artificial. There is a fix point: cultural environments define intelligent entities and conversely, intelligent entities define cultural environments. One obtains some kind of *self-duality* and self-duality seems to be the required property to perform a context change (Chaudron et al., 2015). The problem solving loop is the expression thanks to a context to the continuous shift of the cultural environment performed by any AI entity according to the problem solving loop.

In such a self-dual world, when information coming from entities equates information coming from the environment, one can assimilate the problem solving in step 1 as shift detection. That is, the environment has to be continuously transformed according to the *valid* computed solution in step 6. Assume that shift detection can be represented by the pair (σ_i, σ_f) of states where σ_i is the *initial state* and σ_f is the *final state*. Then the solution to compute can be expected to be an *action plan*. But actually, this is really not a good idea because in that case, one is faced with the most difficult Artificial Intelligence issue (Bartheye and Chaudron, 2016): generating automatically an action plan of unbounded length according to two arbitrary distinct states whose precedence relation according to a time line matters. This is apparently a paradox, since plan generation is the favorite activity of the modern humans (Bartheye and Chaudron, 2015b).

11.4 Computational Contexts and Automata

11.4.1 Computation Contexts and AI Action Planning

Since the automatic plan generation is a computational process, one can be interested in computational contexts. The purpose of such a context is to specify a precise, operational, implementable model using affordable actions, identified situations. In practice, this notion in the most papers devoted to that topic leads to very poor models. For instance, in order to keep things simple, Reich (2011) limits them to first communicative acts, i.e., ignore sequentiality. It is equivalent to say that the validity of contexts is managed, a false statement. In the general case, the classical way to interconnect perception and actions is to use orthogonal tools:

1. *predicates* as representative tools for situations and
2. *operators* or actions as transition tools which can be fired once a certain situation holds.

For instance, PDDL (Planning Domain Definition Language) is the standard encoding language for "classical" planning tasks. Components of a PDDL planning task are:

1. *Objects*: things in the world that interest us,
2. *Predicates*: properties of objects; can be true or false,
3. *Initial state*: the state of the world that we start in,
4. *Goal specification*: things that we want to be true,
5. *Operators or actions*: ways of changing the state of the world.

Actions and transitions between the initial state and the final state compatible with the goal to achieve are expressed using PDDL. A context definition in that frame will definitely alter this duality although that this model denotes full perception as predicates and full action as operators. Introducing contexts means that the notion to be eliminated is the notion of transition between states. It is equivalent to assert that a semantic denotation of a sentence on a dynamic domain cannot be performed according to a transition on states.

Automatic computation of state transitions requires rewriting rules. This is a normalization process, i.e., a way to associate a denotational semantic and a operational semantic as for instance SL-resolution on Horn clauses in Prolog. That is, as complicated as can be the definition of a context, it has to be translated using situations and states in order to provide a clear semantic (i.e., in order to validate the behavior of any agent in a dynamic context). However, the definition of a context is very far from PDDL components; in effect, a context denotes a very complicated environment; that way, the notion of agent, autonomous entity and the notion of team as a group of autonomous entities is missing. Moreover, a context is often based on a social background. That is, taxonomies or ontologies could be used in order to classify contexts but recall that a normal form is required in order to define ultimately a decision procedure.

The reversal of the context validity theorem (i.e., the causal break) means that knowledge must be represented the other way round; that is, all is transition. Therefore a plan is a morphism from a transition to a transition; the initial state is actually a pair composed by a left transition before the initial state and the initial state $(\rightarrow, \sigma_i)$ whereas the final state is a left transition from the initial state and the final state $(\leftarrow, \sigma_f)$ (and not the right arrow $\rightarrow$ since the geometry is not commutative). Assume that transitions are dual to states; the final state is a *self-dual* pair composed by the plan and the final state. One can propose a plan as an arrow where the domain is characterized by full entropy whereas the entropy in the codomain is controlled thanks to the self-dual property.

A plan is therefore the quadruple

$$((\rightarrow, \sigma_i), (\leftarrow, \sigma_f)) \tag{11.2}$$

which be set in correspondence with the triple according to the intuitionistic semantic

$$(\sigma_i, (\rightarrow, \leftarrow), \sigma_f) \tag{11.3}$$

That is, σ_i is the set of provable formulas, σ_f is the set of refutable formulas, and the plan is entirely located in the "logical black hole" in between $(\rightarrow, \leftarrow)$ implemented according to the context automaton in Figure 11.8.

$$(\sigma_i, (^{\nearrow\searrow}, {}_{\searrow\swarrow}), \sigma_f) \tag{11.4}$$

One can propose a geometrical interpretation of a plan: σ_i is a plateau; so is σ_f. The role of the expected plan is to compute the self-dual extension in Figure 11.9(a) but the available information in the middle element of the triple corresponds to the integration of the two triangles in Figure 11.9(b).

A possible solution is to consider that this arrow $+1 \rightarrow -1$ is actually an *identity*. In a full entropy model, that means that identity is full change and consequently any non identity arrow is a deviation with respect to full change, i.e., an inertia law. One can go further: assume that full entropy

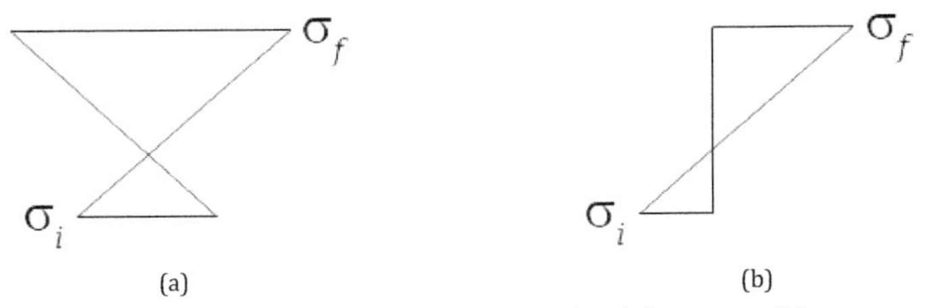

Figure 11.9. Plan extension (a) and plan information (b).

is implemented according to a pure deterministic process. That implies that any non-deterministic process is actually based on some inertia law. Assume that the context validity theorem (*perception* ***precedes*** *action*) is also built on some inertia law (that is inertia cannot be annihilated); that means that, in a full entropic world, causality is not supported. In other words, a break from the signal is required and it turns out that it is a causal break which can be managed by suppressing the context, or equivalently by reversing the context validity theorem according to the statement: *action* ***precedes*** *perception*. Once, the causal break is identified, one can go further and one can determine in which case the causal break can be managed by the decision process. In order to check whether one can jump safely from non-deterministic processes to deterministic ones, one has to use formal tools as *proof theory*.

11.4.2 Provable Computational Contexts and Torsions

Proof theory seems to be undoubtedly a very important tool. However, automatic plan generation or automatic theorem proving has to be altered in order to integrate contexts. Reversing the context validity theorem in order to manage the causal break is impossible using proof theory since proof theory fully supports causality.

In fact, every proof process depends on the topology of the logic defined by symbols, well-formed formulas and inference rules implementing logical axioms. That is, according to set theory, one can define the set of models and the set of counter-models for a given logic and proof processes cannot betray the underlying topology defined by models. Since the aim is to separate regular subdomains (models) $\top$ and singular ones (counter-models) $\bot$, one can classify using proof theory, deductive systems, or even theorem proving as qualitative processes. A natural and standard tool qualifying regular and singular domains is the Powerset functor $\boldsymbol{P}(E)$ (the set of all subsets of given set E). This functor permits to implement validity since it classifies by completeness *all* acceptable solutions (models) $\top$ and *all* unacceptable solutions (counter-models) $\bot$. That way, consistency is ensured by the incompatibility axiom

$$\top \cap \bot = \emptyset \tag{11.5}$$

That means that models and counter-models are incompatible and of equal importance. This skeptical attitude is definitely different from signal analysis. The harmonic representation is performed by models $\top$, for instance the diagonal matrix according to eigenvalues or eigenvectors. Once they are computed, no one is interested in non diagonal elements in which counter-models are included in the background noise. Equating the saliencies of the signal and the background noise automatically breaks down the harmonic representation.

Thanks to symbolic representations, logical negation can express the strength of consistency. If negation holds, then incompatibility between $\top$ and $\bot$ can be specified according to the symbolic formula A and the absurd symbol $\bot$; one can set the consistency axiom ($\supset$ means implies)

$$A \cap \neg A \supset \bot \tag{11.6}$$

In that case, strong negation localizes absurdity and localizers are proof trees. But a proof tree is not a decision provided by any AI entity, this is only a proof. If one wants the proof process to be complete, one should set the proof process as a universal equalizer U (conceptual abstraction). One can provide that way as a proof f as a "faithful" context representation of the problem to solve π (the causal break is managed, since there is no "doubt"). In doubtless proof theory, the sequent to prove, $\pi : \Gamma \Rightarrow \Delta$, has a faithful representation $f = \rho(\pi)$. We consider that this faithful representation providing *strong coherence* is definitely too costly. In effect, this best way to observe this judgment rejecting strong coherence is under the angle of sequent systems. A sequent system is a deductive system which can be depicted using coalgebras as a cocalculus from the conclusion to the axioms. In sequent systems, identity axioms (leaves of a proof tree) are co-identities

$$\frac{}{\cdots_1 \;,\; A \;\Rightarrow\; A \;,\; \cdots_2}\ \textbf{initial} \tag{11.7}$$

Inference rules are nodes of that cocyclic proof tree.

The "bad guy" in the sequent system is the cut rule since a cut-rule is a short-cut in a proof (a unsafe intuitive proof or a unsafe lemma).

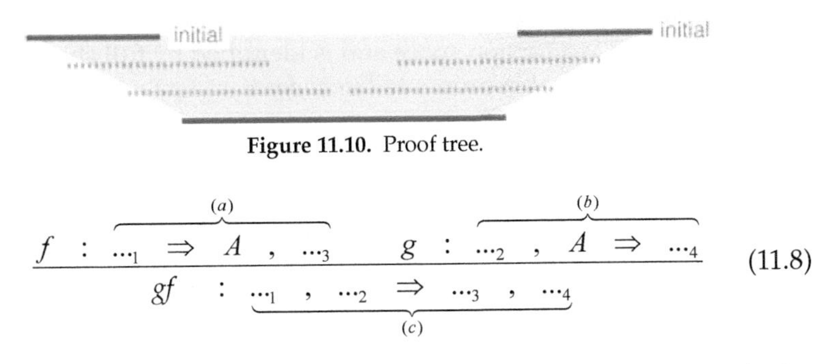

Figure 11.10. Proof tree.

$$\frac{f : \overbrace{\cdots_1 \Rightarrow A \;,\; \cdots_3}^{(a)} \qquad g : \overbrace{\cdots_2 \;,\; A \Rightarrow \cdots_4}^{(b)}}{gf : \underbrace{\cdots_1 \;,\; \cdots_2 \Rightarrow \cdots_3 \;,\; \cdots_4}_{(c)}} \tag{11.8}$$

The statement establishing the validity of any sequent system is the cut-elimination theorem:

Theorem 11.2 (Gerhard Gentzen's Hauptsatz). Completeness and correctness: *If a sequent system has a cut-elimination result, each sequent has the* *sub-formula property (the proof data required in order to decide about the validity of the sequent is self-contained) and this proof system is correct.*

But the "full proof" with as many liberty degrees as required in a sequent system with respect to a cut-elimination result is utopian since any proof is bounded. Since any proof is bounded, it contains cut-rules which mustn't be eliminated since cut-rules are precisely contexts. One can justify this using Goethe's quote

> Beware of daemons you hunt as they are often the best part of yourself.

or Leroy-Gourhan

> One can neither imagine a behavior asking for a continuous lucidity, nor a totally conditioned behavior without any conscious intervention; the former would ask for a reinvention of every part of the smallest movement; the latter would correspond to a totally preconditioned brain.

In fact, a context can also be viewed as a *lemma* (a) simplifying the proof (c) using (b) according to the "short-cut" rule. Take a cut as torsion;[13] one obtains the notion of torsion-free (or context-free) condition for proofs that we definitely do not want. Torsion is a way to encode incompleteness in order to provide weak consistency which should be defined according to weak negation (contexts are incompatible with strong consistency).

11.4.3 Decision Process and Bisequents

Dealing with weak absurdity means to classify the proof space on four areas: area 1 and area 4 is a torsion space and is identified by full absurdity; area 2 and 3 contains valuable proofs as *bisequents*:

$$\ldots_1 \;] \; \ldots_2 \;\Rightarrow\; \ldots_3 \; [\; \ldots_4 \tag{11.9}$$

A bisequent looks like the context automaton in Figure 11.7 and corresponds geometrically to a "bow tie" $\rhd\lhd$ between 2 and 3 in order to encode the finite dimension property. That is, inside that "box" which is actually an Hopf algebra H (a vector space equipped with the multiplication μ and the co-multiplication Δ; connected by the antipode S), proofs are *"proofs inside a box"* (proofs with uncontrolled inconsistent oscillations due to entropy) plus a vanishing operator in 1 and 4 handled by the torsion. Automorphisms can be classified as inner (between 2 and 3) or outer (1 or 4).

[13] If there exists a regular element r of the ring R (an element that is neither a left nor a right zero divisor) that annihilates m ($rm = 0$), an element m of a module M over a ring R, m is called a torsion element.

This weakly consistent functional box is an algebra H of (co)-dimension 4 homomorphic to the Boolean algebra B^2 up to inconsistent oscillations. Unfortunately, no proof system can manage inconsistent oscillations. In order to deal with the causal break, the reverse of the context validity theorem is preferred "*action* ***precedes*** *perception*". An anti-perception algebra provides a representation: a maximally non-commutative algebra called a simple algebra[14] attesting that the perceptive structure can be reversed. Assume that it exists a Hopf algebra H of finite dimension/codimension which is a representation of a *context automaton decision box* in which decisions can be computed by an AI entity. If a context is defined in the conceptual basis, at level + 1, "out of bounds" analysis is performed according to a subjective endofunctor defining an identity (the Descartes "*doubt*" and "*think*" state). *Doubt* requires full entropy; the aim is to qualify the reversal of Descartes law: the causal break is necessary subjective and is necessary managed by the decision process computing another identity as the transition $+1 \rightarrow -1$.

11.4.4 Decision Process and Double S-curves

One can provide an interesting geometrical representation of contexts using *S*-curves. A *S*-curve is the shape of the logistic function.[15] A *S*-curve is also known as the *Sigmoid* curve Figure 11.11 from the *sigmoid* function, a mathematical function that produces a sigmoid, or *S*-shaped, curve. *S*-curves are important project management tools. They identify a discontinuity zone during a phase transition from old to new technologies and they allow the progress of a project to be tracked visually over time, and form a historical record of what has happened to date. Analyses of *S*-curves allow project managers to quickly identify project growth, slippage, and potential problems that could adversely impact the project if no remedial action is taken.

In Hannay et al. (2015), a paper submitted in a NATO symposium, disruption using *S*-curves were studied according to double *S*-curves, Figure 11.12, where the term adaptive thinking means to change the behavior due to a context modification. This paper cites Dwight D. Eisenhower:

> In preparing for battle, I have always found that plans are useless, but planning is indispensable.

[14] An algebra A is simple if it contains no intermediate non-trivial two-sided ideals, i.e., special multiplicative subsets of the operator algebra A.

[15] A logistic function or logistic curve is a common *S*-shape (sigmoid curve), with equation: $f(x) = \frac{L}{1+e^{-k(x-x_0)}}$ where e = the natural logarithm base (also known as Euler's number), x_0 = the x-value of the sigmoid's midpoint, L = the curve's maximum value, and k = the steepness of the curve.

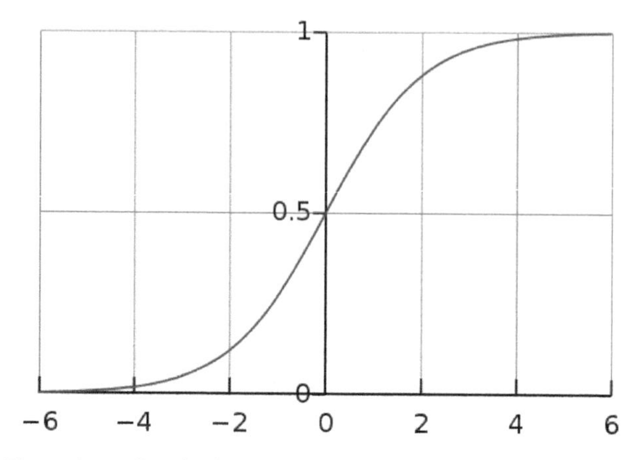

Figure 11.11. Standard logistic sigmoid function $L = 1, k = 1, x_0 = 0$.

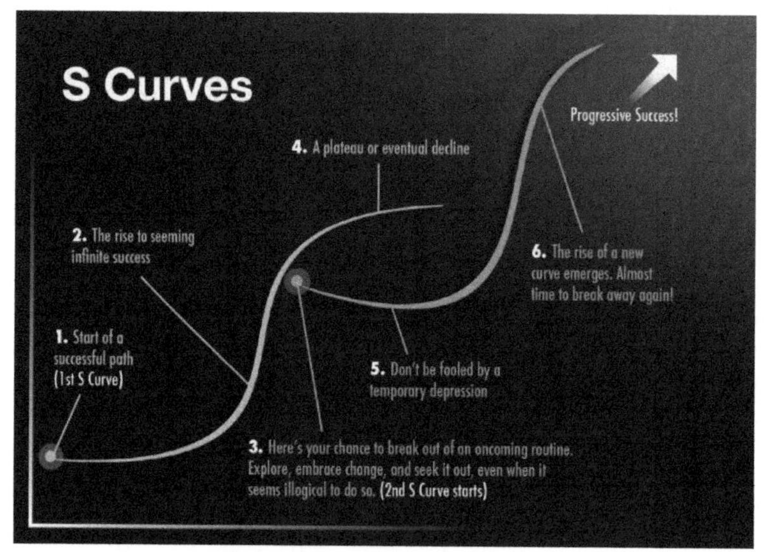

Figure 11.12. Disruption in *S*-curves.

The planning process is still perceived as useful, but

- do not plan in detail,
- do not plan too early,
- do not plan on grounds you do not have,
- do not try to predict an unpredictable future.

The intuition is to express the commander's intent, rather than detailed plans. Detailed plans at relevant levels of command are only defined by

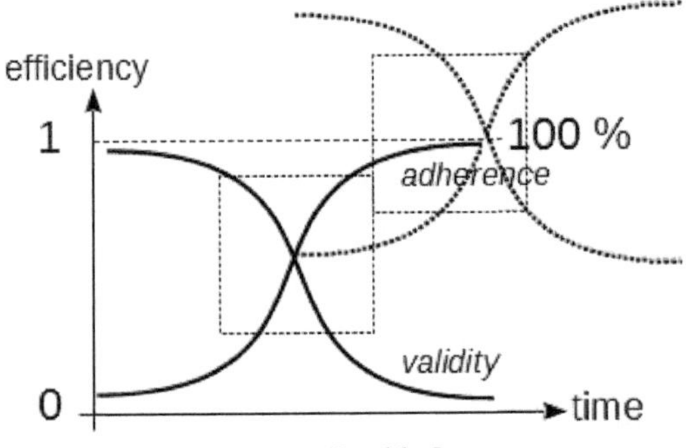

Figure 11.13. Double S-curves.

those who need the details. The idea is to "disrupt" the planning process by on-the-fly vignettes Figure 11.12 and to cue triggering. The principle of adaptive thinking is to prepare in the face of uncertainty and boost individual skills. In time, standard planning might involve adaptive thinking elements. Today's artificial elements will become realistic elements, because the actual planning process has adopted those elements. At points in time, artificial elements are necessary but they are a "necessary evil". Because you will fight as you train, artificial elements become part of a new process. In effect, human cognitive processes have evolved to make rapid and good enough (and sometimes wrong), decisions (Bartheye and Chaudron, 2015b) on marginal information. Such a decision cannot be performed outside context identification. Assume that a *S*-curve is a context; we choose to work on that double S-curves in order to associate the decision process and context change.

A *S*-curve denotes the identification and the effective adhesion of the context which becomes hazardous when the plateau is reached. A context is valid inside the square centered on the inflection point of the curve. It is equivalent to say that the context utilization curve is actually a cusp whose left hand side is context adherence and the right side is context validity in Figure 11.13. A possible transition to an alternative context is to be prepared when the situation will change. Meta rules can be used in that frame which can be fired if the truth value of a given predicate changes. A context change is to be performed as a co-cycle (pure cyclic systems cannot detect a context change). Context change may be an answer to the question: How a clear context can become suddenly unclear?

One can assume that a context change is to be performed once a reversal of the context validity theorem is required (action is preferred). From a S-curve, one can entail that we are already at a plateau level. Therefore the meta-level action is actually to perform torsion: to annihilate the context (performing a fall down) using a residue action (context is a dynamic divisor). A context change is to be performed from a cocycle (pure cyclic systems cannot detect a context change).

This leads to our last theorem:

Theorem 11.3. *Decision fills a causal break inside a decision Hopf algebra H.*

11.5 Conclusion

From the last theorem, one can conclude that the decision process compensates a causal break occurring once a context becomes invalid. We propose to use the creative decision dynamic according to torsion in the decision Hopf bialgebra H. The idea is to modify the algebraic representation of sequent systems in order to encode bisequents. In effect, the causal break attests the lack of any usable link between the current state and the states to come. Due to the validity context theorem, there is no solution using stability of contexts because this lack is necessary located inside the context definition. That means that one can retrieve as information or as data as we may, we cannot predict the context change because of the inertia of contexts. Therefore the decision process is necessary due to context change, and a future work is to show why it is sufficient. It is equivalent to show that any coalgebraic operator is necessary and sufficient when it is performed inside a decision Hopf bialgebra H.

References

Bartheye, O. and Chaudron, L. 2015a. Algebraic models of the self-orientation concept for autonomous systems. AAAI SpringSymposia 15, Standford, USA.

Bartheye, O. and Chaudron, L. 2015b. Risk management systems must provide automatic decisions according to crisis computable algebras. AAAI SpringSymposia 15, Standford, USA.

Bartheye, O. and Chaudron, L. 2016. Epistemological qualification of valid action plans: Application to UGVs or UAVs in urban areas. AAAI SpringSymposia 16, Standford, USA.

Chaudron, L. 2005. Simple Structures and Complex Knowledge. Habilitation Thesis, ONERA, Toulouse, FR.

Chaudron, L., Erceau, J., Duchon-Doris, C. and Fighiera, V. 2015. Formal approaches for autonomous systems: A generic peace engineering program. SMCE 2015, Hong-Kong.

Hannay, J., Brathen, K. and Hyndy, J.I. 2015. On how simulations can support adaptive thinking in operations planning. 2015 NMSG NATO Annual Symposium, Oct 15th–16th, Munich.

Isaksen, D.C. 2002. A cohomological viewpoint on elementary school arithmetic. The American Mathematical Monthly 109(9): 796–805.

Reich, W. 2011. Toward a computational model of context. AAAI SpringSymposia 11, Standford, USA.

A Contextual Decision-Making Framework

Eugene Santos Jr.,[1,*] *Hien Nguyen*,[2] *Keum Joo Kim*,[3] *Jacob A. Russell*,[4] *Gregory M. Hyde*,[5] *Luke J. Veenhuis*,[5] *Ramnjit S. Boparai*,[5] *Luke T. De Guelle*[5] and *Hung Vu Mac*[5]

12.1 Introduction

In Proactive Decision Support (PDS), the goal is to provide Commanders with the information that they need at the right time in order to make the right decision while dealing with a large expanse of contextual information about an uncertain and dynamic environment. Providing the Commander with too much information degrades their performance in time-critical situations; while not providing enough information can lead to uninformed and poor decisions. To compound the problem, Commanders do not all make decisions the same way, implying that identifying the right time and the right information for a Commander requires an understanding of how the individual Commander makes decisions. A Commander's decision-making process, itself, may also change over time as they learn from the consequences

[1] Professor of Engineering at the Thayer School of Engineering, Dartmouth College in Hanover, NH.
[2] Associate Professor of Computer Science, University of Wisconsin, Whitewater, WI.
[3] Research Associate, Dartmouth College in Hanover, NH.
[4] PhD graduate student, Dartmouth College in Hanover, NH.
[5] Undergraduate students, University of Wisconsin-Whitewater, Whitewater, WI.
* Corresponding author

of prior decisions. Therefore, to proactively assist a Commander, we need to understand his goals, anticipate his needs, and capture his decision-making process to correctly choose the context (relevant, critical pieces of information and actions) he needs. Unfortunately, there is a gap between the technologies involved in building such a framework to support Commanders automatically. Decision modeling techniques only focus on how decisions are made while user modeling focuses on the activities and information seeking behaviors. In addition, the use of techniques such as Cognitive Task Analysis (Heuer, 1999) do not scale well and are often useful only in well-defined and well-scoped decision-making tasks.

We develop a Double Transition Model (DTM) (Yu, 2013; Yu and Santos, 2016) to capture a Commander's decision-making process and assess the model through evaluations on synthetic and human Commanders in Naval warfare scenarios plus UAV operators in training simulations. A DTM can be used to derive a dynamic Markov Decision Process (*d*MDP) in which each state reflects a cognitive state of a decision maker that can be discovered/encountered over time and described by his context and each action represents the tasks and decisions that he makes (e.g., during a battle), while the rewards represent the effect of a decision's outcomes.

The novelty of our approach lies in the development and use of a computational model to (re-)construct, quantify, and recognize a Commander's behaviors as a decision-making process. We evaluate our approach through three assessments: The first evaluation focuses on how well our approach determines how a Commander's rewards work using synthetic Commanders. The second evaluation uses the Supervisory Control Operations User Testbed (Coyne and Sibley, 2015) to study how well DTMs can capture a real-world decision-making process. The third evaluation focuses on how well we recognize different decision-making styles by analyzing the graph structures of their DTMs, their reward functions, and the sequences of cognitive states and actions they go through with three human Commanders. Our findings show that the rewards and Double Transition Model's structure capture how a Commander makes a decision. Projecting trajectories of different decision-making styles into the same space, may help distinguish their different rewards.

In this chapter, we seek to address the following fundamental research question: How can we capture a Commander's sequential decision-making process in a computational model and evaluate it in real world situations? The heart of our approach is the idea that an individual faces a sequence of episodic tasks each with different subsequences of environmental and internal context changes caused by the actions taken by the individual. This is essential to capturing how decisions made by the Commander impact later decisions. In this way, understanding the entire process is personalized.

This chapter is organized as follows: The next section reviews related work on decision-making styles, Markov Decision Processes (MDP), and

Inverse Reinforcement Learning (IRL). The following section is a discussion of prior work on the Double Transition Model. The approach taken is discussed next, including the creation of testbeds, the proposed Inverse Reinforcement Learning algorithm, and an evaluation of the algorithm with the testbeds. The chapter ends with a discussion of what was done, what research is expected to be done next, and where the approach could potentially be useful in the future.

12.2 Background

Supporting a Commander in his decision-making process requires a sufficient understanding of how people make decisions in general. Once it is understood, a model can be built, and the learning of their decision-making style can then take place. This section describes the background knowledge that is needed to build our model: decision styles, Markov Decision Processes, and Inverse Reinforcement Learning.

12.2.1 Decision-Making Styles

A Commander's decision-making style represents how he processes data, makes sense of the received data and evaluates possible alternatives. Decision-making style has been defined as "the learned habitual response pattern exhibited by an individual when confronted with a decision situation" (Scott and Bruce, 1995). This area of research has extensively been studied in variety of domains including education management (Galotti et al., 2006), military (Thunholm, 2009), pharmaceutical training (McLaughlin et al., 2014), construction work (Esa et al., 2014), and work life balance (Michailidis and Banks, 2016). Researchers have shown that determining one's decision-making style is a crucial step to evaluating and distinguishing between good and poor decision-makers and helps improve the quality of decision-making process.

The five decision making styles proposed by Scott and Bruce (1995) have been investigated and applied in the cognitive community over the years. These are rational, intuitive, dependent, avoidant, and spontaneous decision styles. Rational decision makers consider all options and logically make a choice by reasoning over all options. Intuitive decision makers rely heavily on their own experience and choose heuristics that work best for them. Dependent decision makers rely on others' advice and decisions. Avoidant decision makers postpone the act of making decisions. Finally, spontaneous decision makers tend to make snap, quick decisions to respond to immediate needs. Similar to intuitive decision makers, in a battle setting, a naturalist decision-making style also refers to the use of past experience to make critical decisions under a lot of pressure and time constraints (Flin, 2001). In a study with 98 army captains, Thunholm (2009) has found that military leaders tend to be more spontaneous and less rational, dependent, and avoidant than their staff members. Decision-making style is a very important factor to assess the quality of a decision and is also related to a broader concept that has been studied thoroughly, called cognitive style (Syagga, 2012). Syagga explored a specific relationship between cognitive styles and decision-making style such as "*People who are more extroverted in personality are more likely to have intuitive cognitive style, while those who are more introverted in personality are more likely to have analytic cognitive style.*" Syagga's study could not confirm the relationship, however, it could not reject the relationship and lays the groundwork for further study. In this work, we aim to identify the differences of decision-making styles by using a computational framework to capture Commanders' entire decision-making processes and analyze the Commanders within this framework.

12.2.1.1 Markov Decision Processes

The Markov Decision Process (MDP) was used as early as 1957 by Bellman (Bellman, 1957) and popularized by Howard (1960). Originally used in operations research, MDPs branched out and have been used in a variety of fields, including robotics, controls, and economics. They are typically used to model decision-making in well-defined scenarios where they can aid in making optimal decisions.

A MDP M is a 5-tuple $(S, A, P_{.}(\cdot,\cdot), R_{.}(\cdot,\cdot), \gamma)$, where S is a finite set of states. A is a finite set of actions. $P_a(s, s') = Pr(\{s_{t+1} = s' \mid s_t = s, a_t = a\})$ is the probability that by taking action a from state s at time t, we will transition to state s' at time $t + 1$. $R_a(s, s')$ is the immediate reward (often relaxed as the expected immediate reward) received after action a transitions from state s to state s'. Finally, $\gamma \in [0,1]$ is the discount factor, which controls for the ratio of importance between future rewards and present rewards. The formulation makes MDPs simple to design and use when in a well-defined space.

Conventionally, MDPs have a well-defined reward function and are looking for a policy π: $S \rightarrow A$ which determines which action should be taken at any given state. More specifically, they are looking for the optimal policy π^* which maximizes the expected reward of the system. The reward function is usually hand-crafted because it is unknown. An alternative approach to hand-crafting the reward is to approximate it. In the cases where the reward function is unknown, we could perform Inverse Reinforcement Learning to approximate it.

12.2.1.2 Inverse Reinforcement Learning

Inverse Reinforcement Learning (IRL), sometimes called Inverse Optimal Control, was originally proposed by Russell (1998) in a position paper, as a means for estimating how an individual or machine could assign preferences for states when they were unknown. The first formulation was then defined by Ng and Russell (2000). IRL takes a MDP without a reward function $M \backslash R.(.,.)$, and approximates a reward function. In order to approximate the reward function, Ng and Russell's approach can take as a parameter the optimal policy π^*, but in practice the intent is to learn both the optimal policy and the reward function so they use a set Ξ_M of trajectories of the form $\vartheta = \{s_i, a_i, s_{i+1}, a_{i+1}, \ldots, s_{i+k}\}$. These trajectories are examples of the optimal behavior as demonstrated by an expert. This approach assumes that the reward function is linear and solves for the rewards via linear programming. Unfortunately, Ng and Russell's approach is under-constrained and has multiple optimal solutions, some of which are degenerate, and it is the luck of the draw as to whether you get the degenerate solutions.

Maximum Entropy (Maxent) IRL (Ziebart et al., 2008) maximizes the entropy of the system to have a single optimal solution solved via gradient descent. Unfortunately, in this work, Maxent did not converge and was fickle with respect to the hyperparameters. Maxent also happens to be the main approach upon which most other new approaches to IRL are based.

Deep IRL (Wulfmeier et al., 2015) extends the approach from Maxent but trains a fully convolutional neural network (FCNN) to learn a nonlinear reward function capturing the relationship between features on the states instead of the reward function by Ng and Russell which was linear in the features. This approach is promising, but FCNNs are still difficult to explain because of their high dimensionality of parameters and nonlinear interaction between the parameters. This poses a significant barrier to better understanding the decisions made by a Commander.

12.3 Prior Work

In this section, we introduce the DTM which is a framework that uses MDPs to model an individual's decision-making style(s) and reconstructing the way that they view the world via IRL.

12.3.1 Double Transition Model

To capture the human decision-making process, we begin by assuming that the process can be determined by a sequential choice of actions and demonstrated figuratively through connections between cognitive states. The Double Transition Model (DTM) was originally proposed in Yu (2013) as a way of describing a human opinion formation process through computational simulation and applied to multiple domains (Yu and Santos, 2016; Santos et al., 2017). Nodes are used to represent human cognitive states and edges to represent their transitions during decision-making processes. A node (representing a cognitive state) is composed of two subgraphs, a Query Transition Graph (QTG) and a Memory Transition Graph (MTG).

The QTG is specialized for the instant interest of an individual and the MTG is for the memory transition associated with the QTG. Each node in the QTG represents a single query in the individual's mind and a set of random variables associated with it can be described as $U = [X, A, B, C, ...]$. A vector $[X, ?, b_1, c_2, ...]$ denotes a single instantiation of the set of random variables and represents an instant query of the individual, while X represents the target random variable of interest, and ?, b_1 and c_2 denote the unknown/known instantiation of variables $A, B, C \in U$, respectively. The MTG represents the underlying knowledge of the individual's mind represented by a Bayesian Knowledge Base (BKB) (Santos and Santos, 1999) stored on the state. This knowledge base can also be thought of as a representation of the relevant context and how different sets of context relate to each other.

Figure 12.1 shows a portion of a sample DTM, where an individual could undergo a memory transition from one subtask to another based on the change of his perception through MTGs. During the memory transition, he would ask himself a question: "Is there any action necessary for a current subtask?", and he takes an action to speed up one of his Unmanned Aerial Vehicles (UAVs) as he believes it is the right action to take at that time.

The random variables (such as 'Action' and 'event'), their relationships in the DTM, and the associativity between QTG and MTG cannot be known *a priori*. Cognitive states and relationships must be identified and extracted from the individual's information stream to identify the relevant context and incorporate it into the MTG and QTG. Transitions between cognitive states depend on the choice of action at the current instant. The ultimate goal of our research here is to reconstruct the sequence of decisions that the individual would make and assist his future decisions by investigating potential patterns of his decisions. For assisting Commander's decision-making processes, DTMs have been applied to previous work (Yu and Santos, 2016; Santos et al., 2017).

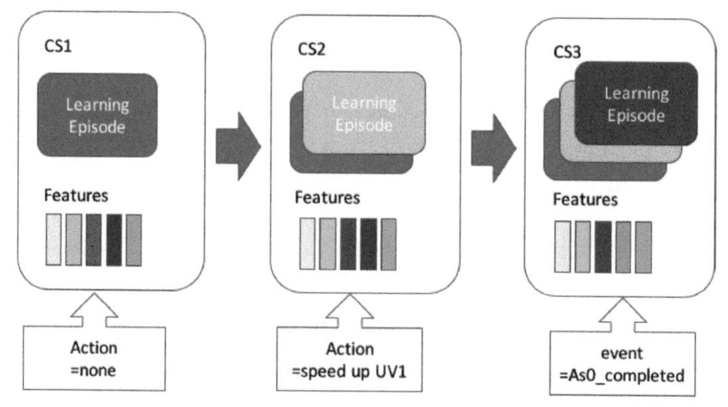

Figure 12.1. An example of DTM framework.

Here, the DTMs have been limited to just the MTG and the BKBs have been ignored in order to simplify the problem of identifying reward values for the states. A DTM at any given point in time has a mapping onto a MDP where the cognitive states and transitions in the DTM have a one-to-one mapping onto an equivalent MDP. Since the DTM is controlling and updating itself to correspond to the decision-maker being modeled the state and action spaces cannot be known *a priori*, and therefore must be dynamic. For this reason, we refer to the MDP derived from the DTM as a dynamic MPD (*d*MDP). Most of this chapter focuses on the *d*MDP derived from a snapshot of the DTM, however, it generalizes to instances where the state and action space could change.

12.4 Approach

In order to capture a Commander's decision-making process with a *d*MDP, a set of trajectories (which are representative of the decision-making process) need to be sampled from each Commander (thus forming a testbed). This section focuses on the capture and analysis of Commanders' decision-making styles and the development of a requisite, new Inverse Reinforcement Learning (IRL) algorithm. To start, it is necessary to understand limitations from previous iterations of IRL and testbeds created for evaluation. Conventionally, IRL algorithms and testbeds are used in domains such as robot motion planning which are relatively objective, but when dealing with people we do not want to regress towards the mean over all of them and lose their individuality. The new IRL algorithm tries not to make some of the assumptions common in domains such as robot motion planning. Therefore, this section begins by explaining how the testbeds were created and what types of information were included. We then discuss limitations of existing IRL algorithms and define the new IRL algorithm. We finish this section with an evaluation of the new IRL algorithm over our testbeds.

12.4.1 Testbed Construction

Due to the lack of publicly available traces of Commanders' decision-making processes or trajectories, we began our research by first synthesizing testbeds from virtual Commanders playing a simple toy game. Next, we extended to two more testbeds. The first one is the Supervisory Control Operations User Testbed (SCOUT) (Coyne and Sibley, 2015) which is under current development and analysis from the US Naval Research Laboratory. The second one is obtained from human Commanders played by students in a naval warfare simulation game.

12.4.1.1 Synthetic Commanders

Evaluation of IRL algorithms requires sufficient numbers of observed trajectories of Commanders' decisions in battlefield scenarios. Aside from the current lack of availability of such trajectories, evaluation is also not simple with human Commanders because the human Commanders' decision-making process is not fully observable and decisions about what context to include from a Commander's observations may be subjective. Furthermore, the data gathered from the human Commanders may not be repeated frequently enough to evaluate algorithm performance. In order to cope with these challenges, we designed synthetic Commanders associated with representative decision-making styles. We synthesized the Planner for a rational decision-maker and the Novice for a spontaneous decision-maker. The next few paragraphs introduce the world of actions that the Commander can take to contextualize the synthesized Commanders and then describe the design of the Commanders themselves.

The world is composed of two enemy ships, three friendly ships, and friendly scout planes. The goal is for the Commander to control the ships in a way that eliminates the enemy ships. In between actions that the Commander can take, snapshots of what the world looked like are visible and define the context of the space. These snapshots contain full instantiations of the variables in Table 12.1.

The Planner rationally chooses each action in a way that attempts to accomplish the goal of eliminating the enemy ships effectively and efficiently. When the enemy ships are concealed, they have a tactical advantage so it is advantageous to move multiple ships at a time to attack them. The design was meant to simulate an expert that knows how to traverse the domain.

The Novice makes some mistakes and can be thought of to be exploring the space at the same time as making the decisions. Generally, the novice will send one to two ships at a time to explore and will be at a tactical disadvantage (although the number of friendly ships is larger than the number of enemy ships so it is still winnable).

Table 12.1. Description of variables in synthetic Commander testbed.

Enemy ships	Variables (per ship) representing the location of the enemy ships
Fog 1-4	Variables representing explored or unexplored fog areas where the enemy ships may be hiding
Health	Variables for the health of each of the enemy ships (submarine and battleship) and each of the friendly ships (battleship 1, battleship 2, scout planes, aircraft carrier)
Friendly ships	Location of each of the friendly ships

The goal of the testbed is to evaluate the IRL algorithm under more appropriate conditions for our problem than the classic Grid World problem (Russell et al., 2003). Features in these other domains have no correlations with each other and can be treated independently, counter to what we expect to see. Since one of our goals is to address dynamism, having correlations between the features is important because we cannot make them orthogonal for factors that we have not yet seen. We use this synthetic toy problem as a means for testing before scaling up to real data such as the Supervisory Control Operations User Testbed.

12.4.1.2 Supervisory Control Operations User Testbed

This dataset was collected by the US Naval Research Laboratory from 20 volunteers executing pre-scripted Unmanned Aerial Vehicle (UAV) missions within the Supervisory Control Operations User Testbed (Coyne and Sibley, 2015). The testbed was developed to represent the tasks that operators would perform in dealing with multiple heterogeneous UAVs. Participants are encouraged to plan routes and navigate effectively by earning or losing points during twenty-minute long supervisory control missions. Each participant gets rewarded once the assigned vehicle is located within the range of the target sensor, and penalized when the vehicle enters a restricted airspace. In addition to this main task, participants are asked to respond to chat messages, which include requests to update UAV parameters (geographical locations, speed, altitude, etc.). The available context includes the location of the UAV, the score that Commanders accomplish, the type of tasks given, level of difficulty and biometric attributes of Commanders such as their pupil dilation during the mission. With the information provided, we apply our theoretical framework of DTMs into the task of building a single, unique DTM describing all Scout participants, and examine how the context embedded in the original dataset is represented in the DTM constructed. In particular, we focused on the context associated with individual characteristics such as risk-aversion, selection efficiency, and game performance.

12.4.1.3 Human Commanders

In addition to testing with synthetic users and UAV pilots, we use a team of three humans to play a Naval war game called Steel Ocean.[1] This game was selected because of its wide availability and its community of users which offer many videos with readily available examples of Naval Commanders with concrete traces. These episodic traces help us to create a platform to understand how Commanders differ in their decision-making process, why they differ, and what uniquely composes a Commander's

[1] Steel Ocean is available at http://store.steampowered.com/app/390670/Steel_Ocean/.

style over a set of tasks while still having control of how the tasks overlap. We created and recorded videos of 32 scenarios. Each scenario was played by three student volunteers who have some experience playing wargames online and involved a fleet battle between two teams: allies and enemies. One player acts as the Commander and the other two players act as fleet members. Depending on the role a person is filling, they have different ship types to choose from. Commanders use a battleship whereas the other two members use destroyers. Contextually speaking, we have three fundamental pieces in each scenario: A Commander's ship, a Support ship and an Attacker ship. A Commander's ship makes decisions predicated in part on context from the actions that the Attacker or the Support ship take. The Attacker ship is responsible for carrying out an action and the Support ship assists the Attacker ship in carrying out these actions. These pieces in each battle follow different formations in scouting the enemy and carrying out offensive and defensive actions. The formations are straight line, scattered line, triangle and D-formation as specified by United State Joint Force Command (United States Joint Forces Command, 2004). While we limit ourselves to these two specific ships, other ships do exist within the game such as submarines or aircraft carriers which enemy or ally players can use. The Commander in our simulation is the only player allowed to make any sort of decisions. The two other fleet members are allowed to make requests to the Commander which can ultimately influence the decisions a Commander makes. Along with requests, fleet members are expected to give reports to the Commander, so that the Commander has the context required to make informative decisions for their fleet. To communicate with each other in a battle, the three players use a VoIP server which allows the Commander to give verbal commands in real time to the fleet members. The fleet members are allowed to give reports and requests in real time to the Commander. We established a protocol to standardize our verbal communication. Narratives are compiled from recorded videos as we sample the movements and situations of each battle every 15 seconds and convert this data to a comma delimited file (csv). Each comma delimited file is a trace which, in turn, is used to create a portion of the DTM.

Each human Commander is also given a questionnaire (25 questions) which is used to measure their general decision-making styles (Scott and Bruce, 1995). There are five questions representative of each style and each question is rated on a scale from 1 to 5 with 1 as the lowest and 5 as the highest. We tallied the points for each group of questions for each style and determined the final style being the ones with highest scores. The results are shown in Table 12.2 below.

In summary, the testbed with Human Commanders brings several new dimensions of context to assess how the DTMs captures a Commander's decision-making process which are ground truth for Commanders' decision-

Table 12.2. Testbed of human Commanders.

Commanders	Decision Style	Number of Commanding Battles	Average Battle Length (in seconds)
Commander 1	Rational	16	643.5
Commander 2	Spontaneous and avoidant	8	542.87
Commander 3	Rational and dependent	8	518.87

making styles, and partially observable, dynamic environment with real-time actions.

12.4.2 A New Inverse Reinforcement Learning Algorithm

This work encountered limitations with existing IRL algorithms on the previously described testbeds. Here we describe the limitations briefly in order to motivate our algorithm but discuss them in more detail in the evaluation section.

The first prohibitive limitation that we encountered is that the definition of rewards in the original Ng and Russell (2000) approach yielded rewards on the states themselves. This is problematic because, when combined with the Markov Assumption, it implies that rewards are invariant to the order in which states are traversed. The assumption of the linear relationship between the features is also problematic and we will discuss this limitation in the Synthetic Commanders section.

In prior work (Santos et al., 2017), we used the Maximum Entropy Inverse Reinforcement Learning (Maxent) approach (Ziebart et al., 2008) because it yielded a, seemingly, better distribution of rewards than the Ng and Russell approach and because most of the other recent approaches build upon it. However, Maxent seems to be sensitive to the sampled trajectories being optimal, which is unlikely to be true in a dynamic environment like the ones that we are trying to work within. If the actions are not optimal, then there may not be a deterministic policy and Maxent may not converge. We also find that the linear function learned by Maxent does not hold in a dynamic environment which we take as evidence that we need a nonlinear reward function. Further analysis of Maxent appears in our Synthetic Commander testbed.

The implication of this analysis is that Commanders may have a nonlinear reward function or, in other words, the context is not defined by the full state space at every point in time. To remedy this, we could use an approach such as Deep Maximum Entropy Inverse Reinforcement Learning (Deep Maxent) (Wulfmeier et al., 2015), however, to recap, our end goal is explanation which Deep Maxent does not provide.

In order to design a new algorithm, it is necessary to avoid these common assumptions from existing IRL methods:

1. Maximum Entropy IRL does not account for the fact that the trajectory space itself is not complete, it is a sampling, and therefore it may not be representative, particularly in relatively unseen states.
2. We cannot make an assumption regarding linear combinations of states (or features) as any basis for determining rewards.
3. Trajectories may not be equally likely, there can be no assumptions on the distribution of the rewards for trajectories.
4. The frequency of state visitation, particularly with respect to features, may not correlate with reward and may correlate more with inevitability.
5. Trajectories do not have to be optimal nor do they need to be totally ordered.
6. There is an ordering (total or partial) of the importance of observed trajectories from the decision-makers.

We developed a new algorithm avoiding these assumptions while keeping any assumptions to a minimum. To recap IRL, the problem is when given a partially defined MDP $M \backslash R.(.,.)$, where only the reward function, $R.(.,.)$, is undefined, and a finite set Ξ_M of trajectories of the form $\vartheta = \{s_i, a_i, s_{i+1}, a_{i+1}, \ldots, s_{i+k}\}$ that represent our observations of the decision-maker behavior, we wish to compute a reward function $R.(.,.)$ that reflects Ξ_M. To do so conceptually, our only assumption regarding the space of all trajectories in M are that trajectories in Ξ_M should be more important than other alternative trajectories not in Ξ_M but "close" to some trajectory in Ξ_M. We now provide our specific formulation of this concept for our IRL. First, given trajectories

$$\vartheta = \{s_i, a_i, s_{i+1}, a_{i+1}, \ldots, s_{i+k}\} \tag{12.1}$$

and

$$\vartheta' = \{s_i, a'_i, s'_{i+1}, a'_{i+1}, s'_{i+2} \ldots, s'_{i+k}\} \tag{12.2}$$

from M is said to be a c-neighbor of ϑ if

$$|\{s'_j \mid s'_j \neq s_j, j = 1,.., k\} \cup \{a'_j \mid a'_j \neq a_j, j = 1, \ldots, k\}| = c \tag{12.3}$$

where c is a positive integer. Let $\Theta_c(\vartheta)$ be the set of all c-neighbors of trajectory ϑ in M. For some constant positive integer ξ, define $\overline{\Theta}_\xi(\vartheta) = \cup_{c=1}^{\xi} \Theta_c(\vartheta)$.

Given some trajectory $\vartheta = \{s_i, a_i, s_{i+1}, a_{i+1}, \ldots, s_{i+k}\}$ from M, we define the expected value of ϑ as follows:

$$E(\vartheta) = \left(\prod_{j=i}^{i+k-1} P_{a_j}(s_j, s_{j+1})\right)\left(\sum_{j=i}^{i+k-1} R_{a_j}(s_j, s_{j+1})\right). \tag{12.4}$$

We also refer to Eq. 12.4 as the Trajectory Probability*Reward Sum (TP*RS). In other contexts we use just the Trajectory Probability ($\prod_{j=i}^{i+k-1} P_{a_j}(s_j, s_{j+1})$), Reward Sum ($\sum_{j=i}^{i+k-1} R_{a_j}(s_j, s_{j+1})$), or the expected linear reward sum:

$$E(\vartheta) = \sum_{j=i}^{i+k-1} P_{a_j}(s_j, s_{j+1}) R_{a_j}(s_j, s_{j+1}) \alpha^{j-i} \tag{12.5}$$

where $\alpha \in [0,1]$ a constant discount factor.

Returning to our single assumption on the trajectory space, we have the following: For each $\vartheta \in \Xi_M$, we introduce the following constraint:

$$\left|\overline{\Theta_{\xi}}(\vartheta) - \Xi_M\right| E(\vartheta) \geq \sum_{\vartheta' \in \overline{\Theta_{\xi}}(\vartheta) - \Xi_M} E(\vartheta'). \tag{12.6}$$

These constraints are linear in terms of $R.(.,.)$. We allow this constraint to be violated but wish to penalize such violations. We rewrite this as follows:

$$\left|\overline{\Theta_{\xi}}(\vartheta) - \Xi_M\right| E(\vartheta) = \delta_{\vartheta}^{+} - \delta_{\vartheta}^{-} + \sum_{\vartheta' \in \overline{\Theta_{\xi}}(\vartheta) - \Xi_M} E(\vartheta') \tag{12.7}$$

where δ_{ϑ}^{+} and δ_{ϑ}^{-} are unique variables to each ϑ that capture the slack and possible violation of the constraint. We also guarantee that they are non-negative with the constraints $\delta_{\vartheta}^{+} \geq 0$ and $\delta_{\vartheta}^{-} \geq 0$.

Next, we introduce a positive constant ω called the peak magnitude value where for all reward instances $R_{a_j}(s_j, s_{j+1})$, we introduce the bounds

$$-\omega \leq R_{a_j}(s_j, s_{j+1}) \leq \omega. \tag{12.8}$$

Lastly, to complete our linear programming formulation for our new IRL, we define our objective function as

$$\sum_{\vartheta \in \Xi_M} \delta_{\vartheta}^{+} - \delta_{\vartheta}^{-}. \tag{12.9}$$

Our goal is to maximize this objective function. Thus, our new IRL approach can be solved as linear programming problem which admits highly efficient algorithms such as Simplex and those for convex optimization. Again, unlike the prior IRL formulations, we make no assumptions about the type, or existence, of relationships between features and can avoid degeneracy as long as the trajectories can generate neighborhoods (sets of c-neighbors, or alternatively, branching neighbors as per below) with enough constraints.

The generation of the neighborhoods for the alternative trajectory set is critical because it is a characterization of the space of potential sequences of actions that the Commander could have taken. The selection of neighborhoods, therefore is giving a semblance of what options were possibly available to take as actions at any given state. These alternative trajectories then yield different constraints and different representations of the expected possible trajectory space. In addition to the c-neighbors, we also try another method for neighborhood generation called branching.

Branching neighborhoods are formed as follows for each training trajectory $\vartheta = \{s_i, a_i, s_{i+1}, a_{i+1}, \ldots, s_{i+k}\}$: For each prefix trajectory of ϑ, say $\{s_i, a_i, s_{i+1}, a_{i+1}, \ldots, s_{i+l}\}$ where $0 \leq l < k$, we generate a random walk starting at s_{i+l} in the DTM for up to $k - l$ steps. Thus, each trajectory can have up to k neighbors through branching. Let $\Theta_b(\vartheta)$ be such a set of branching neighbors.

We can replace c-neighbors with branching neighbors, i.e., replace $\overline{\Theta}_\xi(\vartheta) = \cup_{c=1}^{\xi} \Theta_c(\vartheta)$ with $\Theta_b(\vartheta)$

$$\left|\overline{\Theta_\xi}(\vartheta) - \Xi_{\mathrm{M}}\right| E(\vartheta) = \delta_\vartheta^+ - \delta_\vartheta^- + \sum_{\vartheta' \in \overline{\Theta_\xi}(\vartheta) - \Xi_{\mathrm{M}}} E(\vartheta') \tag{12.7}$$

To

$$\left|\Theta_b(\vartheta) - \Xi_{\mathrm{M}}\right| E(\vartheta) = \delta_\vartheta^+ - \delta_\vartheta^- + \sum_{\vartheta' \in \Theta_b(\vartheta) - \Xi_{\mathrm{M}}} E(\vartheta') \qquad 12.10$$

Branching is expected to be useful in cases where there are few trajectories in order to expand the trajectory space.

Note—Given that our optimization problems are linear, the selection of the peak magnitude value for each instance of $R.(.,.)$ is theoretically irrelevant as different peak magnitudes correspond to a linear scaling of the feasible space and optimal objective value.

We still, however, have a possible problem that our state space is not separate from our feature space. While we ignore the features when learning the reward values, the cross product of the features defines an inequality of states with (possibly) irrelevant or noisy features. This means that there's an implicit dependence upon exact feature equivalency in the learning. As such, state merging is our initial attempt to abstract or generalize different states in the DTM. The idea is to merge states, resulting in drawing trajectories "closer" together and allowing for further analyses of a commander's decision-making style by interleaving trajectories.

Given a DTM, we simply define the distance between two trajectories to be the shortest path between any two states from one trajectory to the other. We further modify our distance definition as follows: Given two trajectories that share a common prefix, we remove the prefix from each trajectory. If the truncated trajectories begin with an action, delete the starting action from both truncated trajectories. If the two truncated trajectories are empty, then the distance is 0. Otherwise, the distance between the two trajectories is now defined as the distance between the two truncated trajectories. A distance of zero now indicates that there is a shared state between the two trajectories that does not occur in their common prefix. Our goal is to address situations where a common starting state for decision-making results in a delay until a different decision is made—e.g., the situation is common among "standard opening moves".

Next, we define the distance between two states s_i and s_j, denoted $d(s_i, s_j)$, in the DTM as the sum over each feature γ in the two states as follows:

- If γ is not set in both s_i and s_j, add 1.
- If γ is non-numeric but has the same value in both s_i and s_j, add 0.
- If γ is non-numeric but has different values in s_i and s_j, add 1.
- Otherwise γ is numeric, add $(\gamma_i - \gamma_j)^2$ where these are the values in s_i and s_j, respectively.

The ordered pairs of trajectories are ranked in descending order of distance for non-zero distances. The top n% (for this paper, we chose n = 90) of this ranking. For each ordered pair ϑ and ϑ' in the top n%, let s_i and s_j' be two states in ϑ and ϑ', respectively, such that $s_i \neq s_j'$ and $d(s_i, s_j')$ is smaller than any other state distance pairs.

Once each s_i and s_j' are determined for each trajectory pair in the ordered ranking, we now merge the states in the DTM as follows: Let Q be a relation on the states where sQs' if and only s and s' is one of the state pairs for some ranked trajectory pair. Q induces a partition on the states of the DTM. For each partition q in Q where $|q| > 1$, for each state $s \in q$, construct $\widetilde{P}_a(s, \hat{s}) = \sum_{s' \in q} P_a(s', \hat{s})$. Next, normalize $\widetilde{P}_a(s, \hat{s})$ so that it is probabilistic. Introduce or replace $\widetilde{P}_a(s, \hat{s})$ into the DTM. Merger thus allows for transitions across trajectories which allows for generation of additional neighbors. Merging is done before IRL is conducted.

12.4.3 Evaluation

The overall goal of this evaluation is three-fold. First, we assess whether the new IRL algorithm could address the problems the previous IRL algorithms have on the Synthetic Commander testbed. Second, we evaluate how the DTM built from the Scout dataset could reflect the original context to describe individual differences. Third, we assess how our DTMs could help us identify different decision-making styles on the human Commander testbed.

These analyses need to control for different factors in the model and align them with the testbeds individually to perform targeted analysis. For the SCOUT and Human Commander testbeds, we would like to be able to directly compare Commanders. Unfortunately, the DTMs have no information about states that they have not seen, so they cannot be directly compared. To be able to perform these analyses we build a joint DTM using the union of the trajectories from each Commander. This projects each Commander's reward values into a space in which they can be directly compared. In this projected reward space, we compare the Commanders decision-making styles and generalize over the unique characteristics of the testbeds.

12.4.3.1 Assessment with Synthetic Commanders

The Synthetic Commander testbed was designed to identify and test possible limitations of existing IRL algorithms and verify whether those limitations could be addressed by the new IRL algorithm(s). The main limitations addressed in this testbed are the linear assumption between features and the Markov Assumption in linking rewards to features on the states. The new IRL algorithm is intended to rectify these limitations by simply not using the features as a consideration, to avoid making the linear assumption, and learning the rewards on the edges instead of on the states (which we refer to as 3-tuples or triples (s, a, s')).

The most promising previous approach to us seemed to be Maxent (Ziebart et al., 2008). Maxent, like Ng and Russell, learns a linear function of rewards with coefficients $\boldsymbol{\theta} = [\theta_1, \ldots \theta_n]$, so it seems to make sense to evaluate the reward function's fit as we would evaluate an arbitrary linear regression model. Ideally, we would determine whether the fit itself is significant. However, we cannot compute an R^2 value because we have no ground truth and do not know which states the Commander would consider better or worse. The other significance test for a linear model is a test of whether the coefficients θ_i, themselves, have a significant impact on the overall model. This is performed by holding out θ_i and determining whether the reward value assigned is significantly different from the original reward. If the distribution of the rewards without the term is the same, then that term is fitting noise in the reward function and the model is unlikely to generalize if any coefficient is not significant. We held out the rewards over the Novice Commander and performed a t-test to compare the distributions, finding that 44 out of 74 coefficients were significant at a $p \leq 0.05$ level. This implies that 30 of the coefficients had no significant impact on the reward distribution and the model was likely not a very good fit (i.e., the model's fit was not significant). The t-test, however, is testing whether the value of the rewards changed by a significant amount but the magnitude of the reward values by themselves is not interpretable. Instead, we want to compare the ordering of the reward values in holding out θ_i versus the rewards with θ_i kept intact. We compared this using a Spearman Rank Order Correlation between the two distributions for each of the θ_i values held out and found that the distributions were correlated in 74/74 cases and the correlation was significant at a $p \leq 0.05$ level with a Spearman $r = 0.99$, implying that no coefficient ever had the ability to change the reward value by itself and therefore no coefficient was significant by itself. This is a strong indication that a linear model is not a good fit.

Conclusions from the analysis of Maxent are that it assumes that the features from states that are visited more frequently should have a higher reward. Therefore, Maxent increases the value of feature rewards proportionally to how often the containing state is visited. This is intuitive; however, it yields a bias towards more frequently occurring features. When the feature space is small and all the features are hand-picked to be important, this assumption may hold. Our goal is dynamic scenarios where feature relevance is not curated. In our case many features rarely change because they may not be relevant and it is ultimately a part of our task to identify which ones were relevant. As the number of features in the testbed grows, it becomes less likely that a linear model fit will do well. The number of features is expected to grow when modeling Commanders because they regularly encounter new situations, so this assumption is prohibitively expensive.

This informed the design of the IRL algorithm and biased it towards the need to accommodate nonlinear features. Explicitly making the features nonlinear would be likely to overfit so, instead, we ignore the features altogether, and nonlinear or linear features may be learned by the model. We identify some very simple structures that Maxent and the Russell and Ng approaches cannot handle as shown in Figure 12.2.

In the connections between nodes s_1 and s_2, the edges are directional because transitions between the states could have appeared either way within the trajectories. Having reward values assigned to the states implies that you can only have a symmetric relation and the weights of edges a_1 and a_2 must be equal because it is just a sum. In placing the rewards on the edges (s_1, a_1, s_2) and (s_2, a_2, s_1) we have no need to lose that information because the edges maintain the context of the prior state. The structure appearing in the left graph of Figure 12.2 appears 206 times where there are different rewards for edges a_1 and a_2 (~ 19% of all edges) in the Planner's DTM and 1420 times (~ 20% of all edges) in the Novice's DTM because the

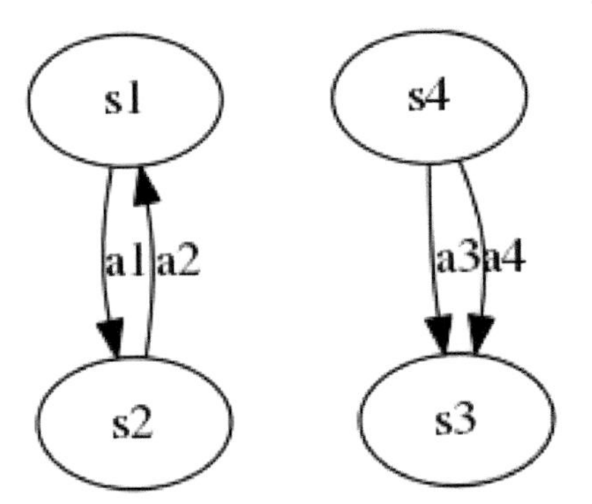

Figure 12.2. Problematic double transition model structures for conventional inverse reinforcement learning algorithms.

DTMs were constructed exhaustively. There seemed to be little difference between the c-neighbors and the branching neighbors at the node level in the graph. There may be variation amongst the relationships between the features, but it is nonlinear and analysis requires further interpretation.

In the right graph of Figure 12.2 we can see that there are 2 edges from s_4 to s_3 by taking 2 different actions a_3 and a_4. In the reward-on-state approach, the only way to differentiate these actions would be the probabilities of having taken the action.

The Synthetic Commander experiments demonstrated that we both need and have a means to learn nonlinear rewards in terms of the states and that we have potential to get a clear benefit from learning nonlinear rewards.

12.4.3.2 Supervisory Control Operations User Testbed

In this section, we gathered the context related to individual characteristics from the original Scout dataset and examined how well they were demonstrated in DTMs. To this end, we hypothesized the following based on the assumptions stated:

A1: A set of features collected from the original dataset is appropriate context to describe individuals.

A2: A trajectory in the DTM can be fully represented by the properties including trajectory probability, sum of rewards, expected linear sum and the product of probability and rewards.

H: If an appropriate DTM was built from a group of individuals' behaviors (trajectories), there should be some metric to reflect appropriate context to describe individuals in the DTM.

Each participant is involved in two consecutive sessions which correspond to two trajectories per participant. For individual characteristics, we addressed selection efficiency (Vogel et al., 2005), risk aversion (Thomas, 2016), and game performance. For selection efficiency, we collected the information regarding the number of actions, events and messages and computed the average repeated occurrence of unique instantiations in those fields during each session. For risk-aversion, we counted the number of actions violating restrictions. Therefore, the smaller the number of restricted operating zones (ROZ) violations the more risk-averse the individual was while playing the session. For game performance, we collected the maximum score an individual achieved during a session shown in Table 12.3. As stated by A1, the diverse range of attribute values in each column supports its fitness as the context to describe individuals.

For building the DTM describing Scout participants, we selected a set of 9 features in Table 12.4, which were identified as important out of 194 features through recursive partitioning and tree-based methods (Therneau et al., 2017). For the attributes having continuous values, we discretized them into multiple levels (i.e., 2, 5, 10 and 20). Due to the sparsity of the original data, we collected the observations only when the value of "ScoreChange" was meaningful (i.e., neither NA nor empty), and missing values were substituted by the observables identified at the closest time stamp.

In order to validate our hypotheses, we constructed a joint DTM from the 40 learning trajectories obtained from the 20 participants, and each trajectory was composed of difference numbers of cognitive states. We collected the derived trajectory properties from our IRL algorithm (i.e., trajectory probability, reward sum, expected linear reward with discount and trajectory probability x Reward sum) for all trajectories in Table 12.5 and confirmed the suitability of the DTM with respect to A2.

We summed up the absolute values of Pearson correlation associated with every pair and presented them in the last column in Table 12.6. By comparing the sum of correlation with respect to each trajectory property, we observed that value of RS and EXRW were higher than that of TP and TP*RW. We conducted additional correlation tests, and t-statistic (in the first row) and p-value (in the second row) were presented in Table 12.7. All p-values were larger than 0.05 and thus we could not reject the null hypothesis that each pair of attributes came from the same distribution under the correlation coefficients given in Table 12.6.

For assessing the p-values in Table 12.7, we examined the normality of each attribute with Shapiro-Wilk test (J.P. Royston, 1982) and obtained the p-values provided in Table 12.8. When the p-value is greater than 0.05, we can assert that the normality of the distribution of the attribute under consideration. Therefore, we could confirm the normality only for the five attributes: event, msgOut, score, RS and EXRW. Other attributes turned out to be significantly different from the normal distribution. In conclusion, it is hard to say whether the correlation test is significant or not from the analysis. However, this observation is pertinent to the fundamental challenge of our research in human decision making, since the distributions of some attributes describing human behaviors are significantly different from the normal distribution and is difficult to analyze under the assumption of normality.

The trajectory information in Table 12.5 was obtained from one of our implementation settings (i.e., merged with no branching) with respect to state management and search for neighbor trajectories, as explained earlier for our IRL algorithm.

We conducted further experiments to see how consistent our observations are and obtained the correlations between each individual characteristic and trajectory property under merged (M), merged-branched (MB), unmerged, and unmerged-branched (B) as shown in

Table 12.3. Individual characteristics of scout dataset.

PID	Action	event	msgIn	msgOut	ROZ	Score	PID	Action	event	msgIn	msgOut	ROZ	Score
1	2.60	7.11	1.17	3.82	0.00	9375.00	12	2.85	7.94	1.15	3.54	0.00	12875.00
1	2.52	9.63	1.13	3.12	1.00	5900.00	12	2.77	9.94	1.15	2.83	1.00	9750.00
2	2.98	7.78	1.15	4.45	0.00	9800.00	14	2.22	7.11	1.15	2.45	0.00	11075.00
2	2.34	10.06	1.13	3.00	1.00	7850.00	14	2.45	6.72	1.13	2.27	0.00	11050.00
3	2.64	8.53	1.15	3.47	0.00	12450.00	15	2.24	11.53	1.15	3.21	0.00	8075.00
3	2.81	8.39	1.15	2.78	0.00	11700.00	15	2.52	10.61	1.15	2.42	1.00	6650.00
4	2.53	9.16	1.15	4.00	0.00	11625.00	16	2.56	8.83	1.17	3.00	0.00	10525.00
4	2.58	7.79	1.13	3.79	0.00	9725.00	16	2.43	7.63	1.13	2.71	0.00	9350.00
5	2.48	9.95	1.17	4.17	0.00	10150.00	17	2.39	8.33	1.15	3.36	1.00	11850.00
5	2.72	9.79	1.13	3.19	1.00	9875.00	17	2.67	8.84	1.13	3.13	0.00	9275.00
6	2.64	10.11	1.15	4.00	1.00	12375.00	19	2.58	8.42	1.17	3.86	1.00	11675.00
6	2.76	7.53	1.13	4.00	1.00	8750.00	19	2.83	10.68	1.13	2.94	1.00	9175.00
7	3.55	8.06	1.17	4.62	0.00	11450.00	20	2.34	8.11	1.15	3.73	0.00	11250.00
7	2.90	7.65	1.13	3.69	1.00	7475.00	20	2.64	9.76	1.13	3.08	1.00	6575.00
8	1.69	8.58	1.15	2.08	1.00	8400.00	21	2.62	7.95	1.15	3.62	0.00	12850.00
8	2.25	7.79	1.13	2.53	1.00	6050.00	21	2.27	9.63	1.13	2.93	0.00	8900.00
9	2.57	8.94	1.17	3.00	0.00	11950.00	22	2.53	8.95	1.15	3.71	0.00	13600.00
9	2.75	7.84	1.13	3.06	0.00	9100.00	22	2.68	8.21	1.15	3.33	0.00	9225.00
10	2.53	7.95	1.15	4.27	0.00	12850.00	23	2.78	7.61	1.15	3.57	0.00	13525.00
10	2.35	9.17	1.13	2.40	0.00	10900.00	23	2.70	8.42	1.13	2.68	0.00	10300.00

Table 12.4. Features selected as important through recursive partitioning and tree-based methods.

Feature name	Content
MsgOut	Participant's written responses to messages in the Simulation's chat boxes
SE_EstDelay	Estimated delay across the network
RtPupilDiameter	Right pupil diameter size, not accounting for quality of the data
Asset2_NDdist	UAV's distance to the next destination
LtEyeLidOpeningQ	Quality of the left eyelid data from SmartEye
eventType	Marks when an event occurs in the simulation—both when a scripted event occurs and when the user performs an event
RtFILTPupilDiam	A quality weighted average filter for the right pupil size
ScoreChange	Shows when feedback was provided to the user about points that they have earned
Action	Shows more detailed information about what action the user took in response to an event

Table 12.9. For the case of branching, we computed the mean of 10 random runs and compared it against the results from non-branching method. The average values regarding EXRW (expected linear reward with discount) and RS (Reward sum) were 1.025 and 1.038 while those of TP (Trajectory probability) and TP*RW (trajectory probability x Reward Sum) were the same as 0.742. Therefore, we validated that our finding was consistent in different empirical settings with respect to state management and branching methods. Furthermore, we demonstrated the validity of the DTM to some degree since the sum of correlations with respect to EXRW and RW reflected the context to describe individuals found in the original data. However, we were still challenged by the low level of statistical confidence caused by the non-normality of attribute distributions observed in the Scout dataset.

12.4.3.3 Human Commanders

While assessment of the DTM model with the synthetic Commanders and SCOUT testbeds provides some insights on IRL algorithms and how features from the environment affect the decision-making process, the evaluation of human Commanders explores the relationship between decision making styles and rewards. The objectives of this evaluation are three-fold: First, we would like to assess whether we can distinguish different decision-making styles by comparing the graphical structures of different DTMs; second, we want to study how significant the reward values are by exploring the

Table 12.5. Information of Trajectories based on DTMs (TP=Trajectory probability, RS=Reward Sum, EXRW=Expected Linear Reward with Discount, TP*RW=Trajectory Probability * Reward Sum).

PID	TP	RS	EXRW	TP*RW	PID	TP	RS	EXRW	TP*RW
1	7.75E-31	-8200	-855794.5048	-6.35E-27	12	6.12E-59	-8300	-680113.0988	-5.08E-55
1	1.15E-27	-12800	-1758674.159	-1.48E-23	12	4.32E-76	-7400	-632474.3482	-3.19E-72
2	8.45E-30	-4700	-520933.5793	-3.97E-26	14	2.62E-21	-10000	-948829.9884	-2.62E-17
2	2.99E-31	-7600	-961511.1009	-2.27E-27	14	2.10E-36	-4000	-325949.996	-8.41E-33
3	1.08E-60	-7000	-900609.1025	-7.54E-57	15	7.39E-42	-5700	-458227.3474	-4.21E-38
3	1.13E-27	-4700	-513976.3881	-5.29E-24	15	4.38E-25	-11200	-1089398.327	-4.90E-21
4	1.02E-41	-7700	-889852.5266	-7.84E-38	16	1.23E-37	-10100	-1104708.871	-1.24E-33
4	3.16E-50	-4600	-419816.3757	-1.45E-46	16	3.01E-15	-12000	-1492579.663	-3.61E-11
5	4.84E-52	-7100	-884810.6541	-3.43E-48	17	1.13E-51	-10100	-1000422.76	-1.14E-47
5	1.12E-41	-10900	-1523688.639	-1.22E-37	17	1.09E-43	-2100	-198086.6633	-2.28E-40
6	2.07E-55	-9000	-1211881.894	-1.87E-51	19	3.21E-47	-11300	-1388526.463	-3.62E-43
6	7.70E-44	-6300	-801191.2841	-4.85E-40	19	1.89E-44	-8200	-1118135.543	-1.55E-40
7	1.66E-64	-5600	-508943.4366	-9.30E-61	20	6.83E-68	-5300	-448002.0003	-3.62E-64
7	1.51E-34	-6700	-763243.8041	-1.01E-30	20	3.40E-43	-5100	-565096.1046	-1.73E-39
8	4.82E-32	-7300	-861230.0344	-3.52E-28	21	1.14E-37	-5600	-745090.2646	-6.37E-34
8	5.76E-28	-8400	-830391.2891	-4.84E-24	21	4.37E-24	-9500	-1477729.987	-4.15E-20
9	5.49E-31	-8800	-1061067.041	-4.83E-27	22	6.52E-34	-9200	-1062696.482	-6.00E-30
9	6.67E-31	-9600	-1031492.496	-6.41E-27	22	3.05E-41	-6200	-822866.627	-1.89E-37
10	6.37E-69	-4000	-371733.4876	-2.55E-65	23	3.33E-37	-10800	-1325783.067	-3.60E-33
10	9.67E-41	-7200	-849345.5252	-6.96E-37	23	8.40E-32	-9900	-1290626.097	-8.31E-28

Table 12.6. Correlation between Individual characteristics and trajectory properties DTMs (TP=Trajectory probability, RS=Reward Sum, EXRW=Expected Linear Reward with Discount, TP*RW=Trajectory Probability * Reward Sum).

	Action	event	msgIn	msgOut	ROZ	Score	Sum
TP	-0.085	-0.154	-0.170	-0.153	-0.118	-0.062	0.742
RS	0.158	-0.155	-0.056	0.248	-0.292	0.072	0.982
EXRW	0.134	-0.221	0.066	0.188	-0.293	0.089	0.992
TP*RW	0.085	0.154	0.170	0.153	0.118	0.062	0.742

Table 12.7. Additional information of correlation Test (t-statistic, p-value, degree of freedom=38).

	action	event	msgIn	msgOut	ROZ	score
TP	-0.529	-0.959	-1.061	-0.955	-0.729	-0.384
	0.600	0.344	0.295	0.345	0.470	0.703
RS	0.986	-0.969	-0.349	1.579	-1.883	0.446
	0.331	0.339	0.729	0.123	0.067	0.658
EXRW	0.837	-1.397	0.406	1.181	-1.892	0.551
	0.408	0.171	0.687	0.245	0.066	0.585
TP*RS	0.529	0.959	1.061	0.955	0.729	0.384
	0.600	0.344	0.295	0.345	0.470	0.703

Table 12.8. Normality check with Shapiro test.

	action	event	msgIn	msgOut	ROZ
p	6.25E-03	1.77E-01	8.69E-06	8.62E-01	3.52E-09
	score	TP	RS	EXRW	TP*RS
p	4.07E-01	6.64E-14	9.37E-01	7.68E-01	6.64E-14

Table 12.9. Correlation analysis in different setting (M=merged, B=branched).

	action	event	msgIn	msgOut	ROZ	score	Sum
TP(M)	-0.085	-0.154	-0.170	-0.153	-0.118	-0.062	0.742
RS(M)	0.158	-0.155	-0.056	0.248	-0.292	0.072	0.982
EXRW(M)	0.134	-0.221	0.066	0.188	-0.293	0.089	0.992
TP*RS(M)	0.085	0.154	0.170	0.153	0.118	0.062	0.742
TP(MB)	0.085	0.154	0.170	0.153	0.118	0.062	0.742
RS(MB)	0.063	0.275	0.122	0.181	0.201	0.236	1.078
EXRW(MB)	0.053	0.354	0.149	0.131	0.234	0.193	1.115
TP*RS(MB)	0.085	0.154	0.170	0.153	0.118	0.062	0.742
TP	-0.085	-0.154	-0.170	-0.153	-0.118	-0.062	0.742
RS	-0.080	-0.080	0.260	0.101	-0.331	0.248	1.101
EXRW	-0.122	-0.217	0.243	0.032	-0.305	0.158	1.077
TP*RS	0.085	0.154	0.170	0.153	0.118	0.062	0.742
TP(B)	0.085	0.154	0.170	0.153	0.118	0.062	0.742
RS(B)	0.097	0.263	0.100	0.195	0.125	0.160	0.940
EXRW(B)	0.074	0.312	0.135	0.158	0.168	0.122	0.968
TP*RS(B)	0.085	0.154	0.170	0.153	0.118	0.062	0.742

reward distribution; lastly, we want to measure the effect of decision styles on rewards by combining the trajectories from Commanders with different decision-making styles.

We use the trajectories compiled from the 32 games played by the three student Commanders to create 32 DTMs (unmerged) to evaluate over the derived *d*MDP and use the number of states, the number of actions and density of the graph to measure the graph structure of each *d*MDP. The density of the graph is computed as follows: $d = \frac{|E|}{|V|*(|V|-1)}$ with E being the number of actions and V being the number of states. The Table 12.10 below shows the statistics of these values broken down by Styles (we encoded Rational style to be 1, Spontaneous-Avoidant to be 2 and Rational-Dependent to be 3). As shown in Table 12.10, there are differences of means between states, actions and density among three Commanders. However, one-way ANOVA analysis confirmed that these differences are not statistically significant (as shown in Table 12.11).

To understand whether we could detect the differences of decision-making styles based on their rewards, we decided to use one training set which consists of all the trajectories from all three Commanders. For each Commander, we construct a test set by using only his own trajectories. We create a DTM for each test set and run IRL on each corresponding *d*MDP to determine his reward function and expected linear reward discount. The intuition behind this set-up is that different Commanders with different decision-making styles may have different ways of rewarding decisions. In order to be able to compare them, we should have the same reward space for all decision-makers.

As shown in the Table 12.12, there are differences between average reward sum and expected linear reward discount among three Commanders. Specifically, the first Commander with Rational decision-making style tends to achieve the highest expected linear reward discount compared to the other two Commanders with Spontaneous-Avoidant and Rational-Dependent styles. However, we performed one-way ANOVA analysis and found these differences are not statistically significant, as shown in Table 12.13.

Lastly, we are going to see how the DTMs constructed by using each Commander's own trajectories compared to the joint DTM created by combining his own trajectories with other Commander's trajectories would affect the rewards that he puts on each action from a specific state. For each Commander i, we generate a set of DTM which includes a DTM using his own trajectories, a DTM using his trajectories and the other Commander j($j \neq i$) unmerged and merged (as defined earlier in IRL section), respectively. For each DTM, we determined a set of attributes that is most significantly related to the reward distribution and used the k-Means clustering algorithm

Table 12.10. Statistics of graph structures of DTMs for three human Commanders.

Descriptives

		N	Mean	Std. Deviation	Std. Error	95% Confidence Interval for Mean		Minimum	Maximum
						Lower Bound	Upper Bound		
States	1	16	20.56	8.422	2.105	16.07	25.05	11	38
	2	8	19.00	6.141	2.171	13.87	24.13	12	31
	3	8	15.25	4.743	1.677	11.28	19.22	9	23
	Total	32	18.84	7.265	1.284	16.22	21.46	9	38
Actions	1	16	15.75	7.197	1.799	11.91	19.59	7	29
	2	8	13.00	4.567	1.615	9.18	16.82	6	19
	3	8	11.00	4.342	1.535	7.37	14.63	4	18
	Total	32	13.88	6.179	1.092	11.65	16.10	4	29
Density	1	16	.045035	.0177797	.0044449	.035561	.054509	.0206	.0705
	2	8	.042585	.0219408	.0077573	.024242	.060928	.0204	.0934
	3	8	.054629	.0208191	.0073607	.037223	.072034	.0292	.0972
	Total	32	.046821	.0195332	.0034530	.039778	.053863	.0204	.0972

Table 12.11. One-way ANOVA on graph structures of DTMs for three Commanders.

ANOVA

		Sum of Squares	df	Mean Square	F	Sig.
States	Between Groups	150.781	2	75.391	1.472	.246
	Within Groups	1485.438	29	51.222		
	Total	1636.219	31			
Actions	Between Groups	128.500	2	64.250	1.766	.189
	Within Groups	1055.000	29	36.379		
	Total	1183.500	31			
Density	Between Groups	.001	2	.000	.888	.423
	Within Groups	.011	29	.000		
	Total	.012	31			

(MacQueen, 1967) to cluster its states using the reduced set of attributes. The main idea of clustering task is to (i) find out the values of attributes describing the centroid of each cluster corresponding with reward being zero and reward being negative; and (ii) evaluate how the states assigned to a single cluster.

The results are summarized in Table 12.14 below. As we can see, the DTM created through state merging between Commander 1 and Commander 2's trajectories provided the lowest errors within clusters and percentage of incorrectly clustered instances for Commander 1. The reward being zero is given to states of low damage for Commander 1 while reward being negative is given to no damage given. For Commander 2, the DTM created by using unmerged Commander 1's and Commander 2's trajectories generated the lowest errors and lowest percentage of incorrectly clustered instances. The reward being negative for movement to a specific location. For Commander 3, the DTM created by using unmerged Commander 2's and Commander 3's trajectories provided the lowest errors and lowest percentage of incorrectly clustered instances. Negative rewards are given to the actions of moving east of the Commander 3. Recall that Commander 2 has a spontaneous decision style while Commander 1 has a rational decision style and Commander 3 is a rational and dependent decision maker. The states in the DTMs created by using the combination and merged process of trajectories of a rational and spontaneous decision maker are better segmented than the states of the DTMs created by unmerged between Rational and Rational dependent decision styles or Rational alone. The rational decision makers often make decision based on thorough search and evaluation of alternatives (Scott and Bruce, 1995) while the spontaneous decision makers make react to the situations quickly. This gives some insights on how the combination of two highly contrasted decision-making styles would help a Commander to form his own style by distinguishing between actions with negative rewards from actions with zero rewards.

Table 12.12. Descriptive table of rewards broken down by decision making styles.

Descriptives

						95% Confidence Interval for Mean			
		N	Mean	Std. Deviation	Std. Error	Lower Bound	Upper Bound	Minimum	Maximum
RewardSum	1	16	506.250000	766.3495721	191.5873930	97.891138	914.608862	.0000	2400.0000
	2	8	512.500000	372.0119046	131.5260702	201.490265	823.509735	.0000	1000.0000
	3	8	462.500000	726.9063606	257.0002084	-145.208926	1070.208926	.0000	1900.0000
	Total	32	496.875000	659.6599197	116.6125006	259.042237	734.707763	.0000	2400.0000
ExpectedLinearRewardWDiscount	1	16	6533.910894	9349.1443907	2337.2860977	1552.103503	11515.718285	.0000	28544.1248
	2	8	5035.241277	4076.4086944	1441.2281153	1627.278324	8443.204231	.0000	11076.5884
	3	8	3337.142067	5554.3166676	1963.7474903	-1306.382872	7980.667007	.0000	14183.0809
	Total	32	5360.051283	7403.1352361	1308.7017819	2690.936402	8029.166164	.0000	28544.1248

Table 12.13. One-way ANOVA analysis of rewards and decision-making styles.

ANOVA

		Sum of Squares	df	Mean Square	F	Sig.
RewardSum	Between Groups	12812.500	2	6406.250	.014	.986
	Within Groups	13476875.000	29	464719.828		
	Total	13489687.500	31			
ExpectedLinearRewardWDiscount	Between Groups	55628448.057	2	27814224.029	.491	.617
	Within Groups	1643370302.989	29	56667941.482		
	Total	1698998751.047	31			

Table 12.14. Effect of rewards on clustering of features using combinations of Commanders' trajectories.

Commanders	DTM trajectories	Description of centroid for cluster 0 (reward = 0)	Description of centroid for cluster 1 (reward = –100)	Sum of SQR error	Incorrectly clustered instances (%)
Commander 1	Commander 1	Reporting = Yes	Reporting = No	32	45.31%
	Commander 1 & 2 Unmerged	Damage_Given_high=Yes	Damage_Given=No	13	16%
	Commander 1& 3 unmerged	Movement_stationary=No	Movement_stationary=Yes	60	30.6548%
	Merged 1 & 2	Damage_Given_low=yes	Damage_Given_low=No	12	9.0395%
	Merged 1 & 3	Movement_stationary=No	Movement_stationary=Yes	24	28.2738%
Commander 2	Commander 1 & 2 Unmerged	Command__move_to_E4:_ship1_ ship2 = No	Command__move_to_E4:_ship1_ ship2 = Yes	4.0	1.9774%
	Commander 2	Movement__north=Yes	Movement__north=no	91.0	20.6897%
	Commander 3 & 2 Unmerged	Report__ship1:_Ship1_sank_ ShipEnemy=No	Report__ship1:_Ship1_sank_ ShipEnemy=Yes	35	4.1026%
	Merged 1 & 2	Command__move_to_E4:_shipYes,_ ship2=No	Command__move_to_E4:_ shipYes,_ship2=Yes	23	1.9774%
	Merged 2 & 3	Report__ship2_ShipEnemy_sank_ Ship2=No	Report__ship2_ShipEnemy_sank_ Ship2=yes	103.00	12.8205%
Commander 3	Commander 1 & 3 Unmerged	Movement__south =No	Movement__south =yes	25.00	24.4048%
	Commander 2 & 3 Unmerged	Movement__east =No	Movement__east =Yes	17.00	4.1026%
	Commander 3	Movement__east =Yes	Movement__east = No	91.0	48.9362%
	Merged 1 & 3	Command__engage_ShipEnemy_ at_F1:_ship1_ship2 =No	Command__engage_ShipEnemy_ at_F1:_ship1_ship2 =Yes	76.0	5.6548%
	Merged 2 & 3	Report_Is_Set =No Report__ship2:_ShipEnemy_sank_ Ship2=No	Report_Is_Set =Yes Report__ship2:_ShipEnemy_sank_ Ship2=Yes	171.0	12.8205%

For the spontaneous decision maker, using the trajectories from a rational decision maker and a spontaneous decision maker but not merging the states would help him learn separately how to evaluate the choices and segment the states more clearly. Similarly, a rational but dependent decision maker could learn from a spontaneous decision maker to help him to form his own unique style.

12.5 Summary of Evaluation

Our evaluation with three testbeds have assessed the DTM, the IRL algorithm used in this model, and the applications of this model in three different settings. As we could see from the Table 12.15, the complexity of our testbed is increasing from the Synthetic to Human Commanders testbed. The main results that we have achieved so far are:

- With the Synthetic Commander dataset, we identified limitations with existing IRL algorithms and demonstrated that learning rewards on (s, a, s') triples decreases the amount of information lost versus just having the rewards on the states.
- With the Scout dataset, we confirmed that the DTM constructed is valid, where EXRW and RW of trajectories could reflect the context to describe individuals, but the level of confidence is limited due to the non-normality of attribute distributions.
- With Human Commanders, there is a difference of graphical structures of the DTMs and expected linear reward discount for various decision-making styles. Unfortunately, due to the small sample size, the change is not statistically significant. Secondly, combining the trajectories from Commanders with different decision-making styles leads to a better segmentation of the actions of Commanders based on rewards.

Table 12.15. Attributes of three testbeds.

	Partially observable	**Dynamic**	**Complete control of environment**	**Real time**	**Ground truth availability**
Synthetic	Semi	No	Yes	No	Complete
SCOUT	No	No	Yes	Yes	No
Human Commander	Yes	Yes	Yes	Yes	Partially

12.6 Discussion and Future Work

This chapter has introduced a framework for modeling Commanders decision making styles, described Inverse Reinforcement Learning as a means for describing a Commander's decision-making process, and evaluated it over three testbeds: synthetic Commanders, Supervisory

Control Operations User Testbed (SCOUT), and human Commanders. Here, we discuss the difficulties we encountered in this work followed by future directions.

The Synthetic Commander testbed showed that it is useful to have rewards on the (s, a, s') triples. The exhaustiveness of the generation of neighborhoods for the Synthetic Commanders meant that many of the possible c-neighbors were already existing within the trajectories themselves. This is not expected to happen frequently with real-world cases but if it does occur in more real-world cases then an alternative method for generating neighbors might be required.

While investigating the records of the 20 participants in the SCOUT testbeds, we realized that the cognitive processes of individuals are difficult to classify into any "typical" cognitive style, since they are composed of multiple layers and facets that would be observed differently according to the situations given. We focused on specific tendencies in human behavior such as the risk-aversion, selection efficiency and game performance, which are known to be critical in affecting behaviors. From our empirical study, we observed some limited associations between the individual characteristics and trajectory properties observed in DTMs. There are likely to be other individual characteristics that would impact a Commander's decision and we will continue to work on further understanding them.

Lastly, there are differences in values from the following three categories: graph structure, expected linear rewards with discount and trajectories for our three human Commanders. However, these differences are not statistically significant (with p-value < 0.05). Differences are caused in part due to the small sample size of battles ($n = 32$) and the average length of the battles (around 10 minutes). Interestingly, we found that Commander 1 has achieved highest expected linear reward with discount and also achieved the highest number of wins among the three Commanders (6 wins versus 2 wins for Commanders 2 and 3). The wins and losses are automatically defined by the Steel Ocean game. His games are also longer in length (642.5 seconds versus 542.9 seconds for Commander 2 and 518.9 seconds for Commander 3). Interestingly, in our data, games that took longer more often resulted in wins. For instance, games that lasted longer than 10 minutes had a 66.67% chance of winning and games that were shorter than 10 minutes had a 6.25% chance of winning. This is also backed up by Commander 1's personality test coming up strongly rational. One could infer that a rational commander would be more willing to focus on future events rather than focusing on the instant gratification of the reward at hand.

There are a number of directions to pursue for our future work. First, we would like to study how the DTMs change over time, especially when we have a junior Commander working to get experience and get better at Commanding. How a junior Commander's rewards and sequences of actions change gives us a potential to understand how people learn and when they use different decision styles. This means that, in a battle, a Commander may have a sequence of actions where he is acting spontaneously but also has a sequence of actions where he is acting more rationally. A DTM could capture these changes and that could be used to distinguish dynamic decision styles over time. Second, we also want to strengthen our evaluations with hypothetical and human Commanders by adding more data points and finding how using a more experience Commander's DTM could help train less experience Commander. Lastly, we would like to verify whether the reward distribution plays a significant role in reinforcing good decisions, enabling us to learn from bad decisions.

Our new IRL approach with its basis as a linear programming formulation enables us to extend and scale with updated computations as new trajectories with yet unseen states and/or actions are introduced. A large, well-understood literature of tools and techniques from linear and convex optimization are readily applicable to our new IRL algorithm including standard sensitivity analysis, variable impact analysis, and gap-analysis to explain changes in reward values reflecting our dMDPs.

Acknowledgements

This work supported in part by ONR Grant No. N00014-1-2154, ONR/NPS Grant No. N00024-15-1-0046, AFOSR Grant No. FA9550-15-1-0383, DURIP Grant No. N00014-15-1-2514, and University of Wisconsin - Whitewater undergraduate research grants summer 2017.

References

Bellman, Richard. 1957. A Markovian decision process. Journal of Mathematics and Mechanics: 679–684. http://www.iumj.indiana.edu/IUMJ/FULLTEXT/1957/6/56038.

Coyne, Joseph T. and Ciara, M. Sibley. 2015. Impact of task load and gaze on situation awareness in unmanned aerial vehicle control. Naval Research Laboratory Washington United States. https://www.nrl.navy.mil/itd/imda/sites/www.nrl.navy.mil.itd.imda/files/pdfs/079%20Coyne,%20Sibley.pdf.

Esa, Muneera, Anuar Alias and Zulkiflee Abdul Samad. 2014. Project managers' cognitive style in decision making: a perspective from construction industry. International Journal of Psychological Studies 6(2): 65.

Flin, Rhona. 2001. Decision making in crises: the Piper Alpha disaster. pp. 103–116. In MANAGING CRISES: Threats, Dilemmas, Opportunities.

Galotti, Kathleen M., Elizabeth Ciner, Hope E. Altenbaumer, Heather J. Geerts, Allison Rupp and Julie Woulfe. 2006. Decision-making styles in a real-life decision: Choosing a college major. *In*: Personality and Individual Differences 41(4): 629–639.

Heuer Jr., Richards J. 1999. Psychology of Intelligence Analysis. Central Intelligence Agency Washington DC Center for Study of Intelligence.

Howard, Ronald A. 1960. Dynamic Programming and Markov Processes.

McLaughlin, Jacqueline E., Wendy C. Cox, Charlene R. Williams and Greene Shepherd. 2014. Rational and experiential decision-making preferences of third-year student pharmacists. American Journal of Pharmaceutical Education 78(6): 120.

Michailidis, Evie and Adrian P. Banks. 2016. The relationship between burnout and risk-taking in workplace decision-making and decision-making style. Work & Stress 30(3): 278–292.
MacQueen, James B. 1967. Some Methods for classification and analysis of multivariate observations. Proceedings of 5-th Berkeley Symposium on Mathematical Statistics and Probability. Berkeley, University of California Press 1: 281–297.
Ng, Andrew Y. and Stuart J. Russell. 2000. Algorithms for inverse reinforcement learning. In Icml, pp. 663–670.
Royston, J.P. 1982. An extension of Shapiro and Wilk's W test for normality to large samples. Applied Statistics 31(2): 115–124.
Russell, Stuart. 1998. Learning agents for uncertain environments. pp. 101–103. *In*: Proceedings of the Eleventh annual conference on Computational learning theory, ACM.
Russell, Stuart J., Peter Norvig, John F. Canny, Jitendra M. Malik and Douglas D. Edwards. 2003. Artificial Intelligence: A Modern Approach. Vol. 2, no. 9. Upper Saddle River: Prentice Hall.
Santos Jr., Eugene and Eugene S. Santos. 1999. A framework for building knowledge-bases under uncertainty. Journal of Experimental & Theoretical Artificial Intelligence 11(2): 265–286.
Santos Jr., Eugene, Hien Nguyen, Jacob Russell, Keumjoo Kim, Luke Veenhuis, Ramnjit Boparai and Thomas Kristoffer Stautland. 2017. Capturing a Commander's decision making style. *In* Sensors, and Command, Control, Communications, and Intelligence (C3I) Technologies for Homeland Security, Defense, and Law Enforcement Applications XVI, vol. 10184, p. 1018412. International Society for Optics and Photonics.
Scott, Susanne G and Bruce, Reginald A. 1995. Decision-making style: The development and assessment of a new measure. Educational and Psychological Measurement 55(5): 818–883.
Syagga, Laura. 2012. Intuitive Cognitive Style and Biases in Decision Making. PhD diss., Eastern Mediterranean University (EMU).
Therneau, Terry, Beth Atkinson, and Brian Ripley. "R Rpart Manual." Accessed April 12, 2017. https://cran.r-project.org/web/packages/rpart/rpart.pdf.
Thomas, P.J. 2016. Measuring risk-aversion: The challenge. Measurement 79: 285–301.
Thunholm, Peter. 2009. Military leaders and followers—do they have different decision styles? Scandinavian Journal of Psychology 50(4): 317–324.
United States Joint Forces Command. 2004. Commander's Handbook for Joint Battle Damage Assessment. Retrieved from http://www.dtic.mil/doctrine/doctrine/jwfc/hbk_jbda.pdf.
Vogel, Edward K., Andrew W. McCollough and Maro G. Machizawa. 2005. Neural measures reveal individual differences in controlling access to working memory. Nature 438(7067): 500–503.
Wulfmeier, Markus, Peter Ondruska and Ingmar Posner. 2015. Deep inverse reinforcement learning. CoRR, abs/1507.04888.
Yu, Fei. 2013. A Framework of Computational Opinions. Dartmouth College.
Yu, Fei and Eugene Santos Jr. 2016. On modeling the interplay between opinion change and formation. pp. 140–145. In FLAIRS Conference. Proceedings of the 29th International FLAIRS Conference, Key Largo, FL 140–145.
Ziebart, Brian D., Andrew L. Maas, J. Andrew Bagnell and Anind K. Dey. 2008. Maximum entropy inverse reinforcement learning. In AAAI 8: 1433–1438.

Cyber-(in)Security, Context and Theory

Proactive Cyber-Defenses

W.F. Lawless,[1,*] *R. Mittu,*[2] *I.S. Moskowitz,*[3] *D.A. Sofge*[4] and *S. Russell*[5]

13.1 Overview

In his fifth annual Presidential message, George Washington (1793) warned that when a nation cannot defend its boundaries, its "reputation of weakness" encourages foreign adversaries. From Sun Tzu (513), "You can ensure the safety of your defense if you only hold positions that cannot

[1] Departments of Mathematics and Psychology, Paine College, Augusta, GA.

[2] Branch Head, Information Management & Decision Architectures Branch, Naval Research Laboratory, Information Technology Division, Washington, DC.
Email: ranjeev.mittu@nrl.navy.mil

[3] Mathematician, Information Management & Decision Architectures Branch, Information Technology Division, Naval Research Laboratory, Information Technology Division, Washington, DC.
Email: ira.moskowitz@nrl.navy.mil

[4] Computer scientist, Distributed Autonomous Systems Group, Navy Center for Applied Research in Artificial Intelligence, Naval Research Laboratory, Washington, DC.
Email: donald.sofge@nrl.navy.mil

[5] Branch Chief, Battlefield Information Processing Branch, Army Research Laboratory, Adelphi, MD.
Email: stephen.m.russell8.civ@mail.mil

* Corresponding author: w.lawless@icloud.com

be attacked." But our nation's cyber-boundaries have been attacked and penetrated. Admiral Michael Rogers, former head of the National Security Agency (NSA) and United States Cyber-Command (USCC), told a Senate committee that the "US must improve its ability to engage in cyber attacks, as deterrent against enemy nations" (Sanger, 2015a).

13.1.1 Overview and Background. Cyber-risks for Organizations and National Security

Based on self-reports, no more than 10% of the variance in human behavior can be predicted (Zell and Krizan, 2014). Rejecting self-reports, based on a narrow range of the physical activities of users (e.g., mobile phone calls, etc.), Barabasi (2012) found that human behavior is stable. Our overarching goal is to review the context of proactive cyber-defenses. We believe the best defense hinges on improved theory that includes both subjective and physical human behavior. Further, the focus on independent individuals has led to overly simple models; instead, introducing interdependence shifts the focus to teams, how they disarm illusions to understand the context, and to how teams produce superior decisions, including cyber-defenses. We begin with subjective data followed by physical data.

Interviews. Cyber-security has become a major factor in the context of protecting organizations. Pew (Rainie et al., 2014) surveyed experts for their cyber-security perspective:

- Joel Brenner, former counsel to the National Security Agency: "The Internet was not built for security, yet we have made it the backbone of virtually all private-sector and government operations, as well as communications. Pervasive connectivity has brought dramatic gains in productivity and pleasure but has created equally dramatic vulnerabilities."
- Dave Kissoondoyal, CEO for KMP Global Ltd., added: "I would not say that a major cyber-attack will have caused widespread harm to a nation's security and capacity to defend itself and its people, but the risks will be there."
- Lee McKnight, a professor of innovation at Syracuse University's School of Information Studies: "... it is easy to ... imagine significant harm done to individual users and institutions given the black hats' upper hand in attacking systemic vulnerabilities, to the extent of tens of billions in financial losses; and in loss of life. But security systems are progressing as well; the white hat good guys will not stop either. ... Imagining bad scenarios ... is worrisome, but I remain optimistic the good guys will keep winning ...."

- The Ponemon Institute (2014) reported: 43% of firms in the United States had experienced a data breach in the past year. ... One of the most chilling breaches was discovered in July at JPMorgan Chase & Co., where information from 76 million households and 7 million small businesses was compromised.

Surveys. From the Ponemon Institute (2015) survey of senior-level information technology and information technology security leaders in the US, UK/Europe and Middle East/North Africa:

- Insider negligence risks are decreasing (but see "Insurance Perspective"). Due to investments in technologies, organizations will gain better control over employees' insecure devices and apps. Training programs will increase awareness of cyber-security practices. (p. 3)
- There will be significant increases in the risk of nation state attackers and Advanced Persistent Threats, cyber-warfare or terrorism, data breaches involving high value information and the stealth and sophistication of cyber-attackers, along with slight improvements in mitigating the risk of hacktivism and malicious or criminal insiders. (p. 6)
- Of respondents, a majority believe their organization needs more knowledgeable and experienced cyber-security practitioners, and that cyber-intelligence activities are necessary to protect information assets and IT infrastructure. Less than half believe their organization has adequate security technologies. Finally, under one-third percent believe their organization is prepared to deal with cyber-security risks or issues in the Internet of Things. (p. 16)
- ... respondents rated the top five cyber-threats over the next 3 years: zero day attacks, data leakage in the cloud, mobile malware/targeted attacks, SQL injection and phishing attacks ... insider threats would not be a major concern. (p. 18)
- Big data analytics will have both a negative and positive impact on organizations. Vast amounts of sensitive and confidential data need to be protected, but analytics will be helpful in detecting and blocking cyber-attacks. (p. 19)

Physical examples. Cyber threats target denial of service, surreptitious control of physical devices and assets, reconnaissance and monitoring, and theft of intellectual property and identity ... An example is the cyber-defense activities in aviation today (Pasztor, 2015):

> U.S. aviation regulators and industry officials have begun developing comprehensive cyber security protections for ... the largest commercial jetliners to small private planes ... as major design criteria. ... [e.g.,]

> Boeing Co.'s decision to pay outside experts dubbed "red hat testers"—essentially authorized hackers—to see if built-in protections ... can be defeated. ... certification of ... [its] 787 Dreamliner required Boeing to purposely allow such teams inside the first layer of protection to demonstrate resilience.

A poor example comes from the Federal Aviation Administration (FAA), which oversees the air traffic control system in the USA. In 2009, hackers stole personal information for 48,000 agency employees, prompting an investigation ... [identified] high-risk vulnerabilities that could give attackers access to FAA's computers. An audit of FAA found (Shear and Perlroth, 2015):

> [Many and] significant security control weaknesses ... placing the safe and uninterrupted operation of the nation's air traffic control system at increased and unnecessary risk.

A third example illustrates that cyber-threats are rapidly mutating: A moving vehicle has been hacked and taken in an unapproved demonstration (Yadron and Spector, 2015):

> Two computer-security researchers demonstrated they could take control of a moving Jeep Cherokee using the vehicle's wireless communications system, raising new questions about the safety of Internet-connected cars. ... [they] show in a video that they can ... disengage a car's transmission or, when it is moving at slower speeds, its brakes.

This hack alarmed the public, leading to an editorial in the *New York Times* (EB, 2015):

> As car technology advances, driving will become safer and more enjoyable. One system being developed jointly by industry and government would let cars communicate with one another using wireless frequencies. The Department of Transportation says this could reduce accidents by as much as 80 percent by warning drivers when they are getting close to other cars. ... [But the] revelation that hackers have found ways to remotely take control of cars made ... is alarming. ... The National Highway Traffic Safety Administration ... should ... require automakers to test the software and make sure a car's wireless system cannot be used to control the engine and brakes.

In other examples, hackers have apparently tricked many small-businesses into compromising bank accounts (Simon, 2015), rail signals (Westcott, 2015) and medical records (Armour, 2015):

> crooks are using personal data stolen from millions of Americans to get health care, prescriptions and medical equipment. ... Ponemon ...

found 65% of victims ... spent an average of $13,500 to restore credit, pay health-care providers for fraudulent claims and correct inaccuracies in their health records.

For the "Internet of things", the lure of free services creates new cyber-vulnerabilities (Goodman, 2015); e.g., Internet-enabled artificial limbs, webcams, refrigerators.

13.2 Introduction. Outline of the Chapter

In this section, we review the problem in recent events and in perspective. We discuss boundary-level defense postures. We review how information is gathered across boundaries. To prevent future attacks, we review strategies and vulnerabilities. We briefly review existing mathematical models. After a discussion, we draw conclusions and offer a path forward.

We address the context of cyber-security from the perspective of an attack that penetrates a firm's physical boundary. Boundaries (Lamont and Molnar, 2002) have been studied traditionally by their effect on cognition, for example, identity and culture. Physical boundaries have been less well-studied as a bound and influence of social effects, especially on the limits of meaning.

Simon (1972) addressed cognitive limitations: "... rationality can be bounded by assuming that the actor has only incomplete information about alternatives." (p. 163) We expand Simon's bounded rationality in attempting to capture the actions a human executes (Lawless et al., 2015), making all interpretations of physical and social reality incomplete, illustrated by the poor correlations derived from questionnaires collecting subjective information about behavior (e.g., Zell and Krizan, 2014); e.g., the correlation between self-esteem and academic or work performance is negligible (Baumeister et al., 2005). The discipline of network science, based on physical behavioral data from cell-phone use, dismisses subjective information (Barabasi, 2012).

13.2.1 The Problem

In 1996, Alan Greenspan, the Chair of the U.S. Federal Reserve Board, spoke about how difficult it was at that time to measure the increases in productivity from computers in American society, industry and business. Today, there is little doubt that the changes wrought in productivity by computers have transformed modern society (Hsu et al., 2015). At the national level, CPS systems have become the backbone of modern economies and security. These benefits have come with a cost—cyber-crime and cyber-attacks.

In an editorial, Voas (2015), at the National Institute of Standards and Technology (NIST), concluded that our computer systems and software are highly reliable, uncompromising systems, yet, because software security is related, trust, a subjective belief, is hurt by uncertainty from intruders and attackers. Engineers are countering with simulated attacks, stronger firewalls and encryption. Because data is invaluable for years versus the short shelf-life of defensive and malicious software, personal details are worth much more than software, making the supreme "engineering challenge ... to be in secure and private data engineering." (p. 538)

But when and where the boundary of a system is vulnerable to a cyber-attack and what can be done to reduce the likelihood of an attack is unknown (Colbaugh and Glass, 2012):

> ... even fundamental issues associated with how the "arms race" between attackers and defenders actually leads to predictability in attacker activity, or how to effectively and scalably detect this predictability in the relational/temporal data streams generated by attacker/defender adaptation, haven't been resolved.

We address how we as a society have progressed in securing our cyber-boundaries. Have boundaries for communities and businesses become secure and cyber-insecure?

13.2.2 The Problem. Recent Events

Recent events. In previous research (Marble et al., 2015), we concluded that the Advanced Persistent Threats (APTs) by other nations posed a long-term threat to users and national security. Jasper (2015, p. 63) reported that the APT attacks by China against Google and 30 other companies in 2010 were undertaken for competitive advantage.

Recent events amplify the persistent vulnerability in our nation's cyber-systems: the theft of private data for federal employees from the Office of Personnel Management (OPM) not only included employee privacy data but also security clearances. In this "unprecedented hacking by China of confidential databases" (Crovitz, 2015):

> "... hackers obtained the records of more than four million federal employees, which include listings of "close or continuous contacts." That tells Beijing which of its citizens are in contact with American officials. Another hacking incident affected many millions more—apparently nearly everyone who has applied for a security clearance ... with personal details such as mental-health conditions, police records, drug use and bankruptcy. Chinese intelligence could use this information to blackmail federal employees. The form ... includes ... "people who

know you well," enabling China to piece together networks of people, including Chinese citizens, linked to federal employees."

While the full context of the theft by China may not be fully known, the theft may encompass military personnel (Larter and Tilghman, 2015), members of Congress (Taranto, 2015), and be for sale on black markets (Pagagini, 2015). The breech, discovered accidentally during a demonstration to OPM (Nakashima, 2015a), gave the "Chinese government intruders access to sensitive data for a year"; once alerted, "OPM identified traffic moving to suspicious Web sites or domains that were not known." Sanger and colleagues (2015) reported that the lack of sales on black markets indicated that China was behind the thefts, adding:

> For more than five years, American intelligence agencies followed several groups of Chinese hackers ... systematically draining information from defense contractors, energy firms and electronics makers ... to fit Beijing's latest economic priorities. But last summer, officials lost the trail as ... hackers changed focus again, burrowing deep into United States government computer systems that contain vast troves of personnel data .... Undetected for nearly a year, the Chinese intruders executed a sophisticated attack that gave them "administrator privileges" into the computer networks at the Office of Personnel Management, mimicking the credentials of people who run the agency's systems.... Much of the personnel data had been stored in the lightly protected systems of the Department of the Interior, because it had cheap, available space.... The hackers' ultimate target: the one million or so federal employees and contractors ... details personal, financial and medical histories for anyone seeking a security clearance.

Baker (2015), the editor-in-chief of the *Wall Street Journal*, interviewed General M. Hayden, the former CIA and NSA chief, about the records stolen from OPM.

> GEN. HAYDEN: The current story is this was done by the Ministry of State Security—very roughly the [Chinese] equivalent of the CIA. Those records are a legitimate foreign intelligence target. If I, as director of the CIA or NSA, would have had the opportunity to grab the equivalent in the Chinese system, I would not have thought twice, I would not have asked permission. So this is not shame on China. This is shame on us for not protecting that kind of information. ... my deepest emotion is embarrassment.

Gen. Hayden clarified earlier accounts that the value of what the Chinese had stolen in the past, until the OPM event, was primarily for the commercial advantages that have eluded China:

> I've met with PLA 3 [the People's Liberation Army, Third Department], the Chinese cyber stealing thing. I never had this conversation with PLA 3, but I can picture it as: "You know, we're both professionals. You steal stuff, I steal stuff, but you know.... You can't get your game to the next level by just stealing our stuff. You're going to have to innovate. And as soon as you start to innovate, you're going to be as interested as we are in people not stealing your innovation."

Discussing the theft of OPM's records, the Editorial Board at the *Washington Post* (2015) wrote:

> The director of national intelligence, James R. Clapper, said China is the "leading suspect" in the breach. The FBI has issued a "flash" alert that ... identified some malware ... previously been associated with Chinese cyber attacks ... on the mammoth health insurer Anthem this year. ... the perpetrators did not seem to be the usual Chinese outfits that try to steal military and industrial secrets through espionage, but another group affiliated with China's Ministry of State Security. This is a worrisome prospect. The Chinese security service may be attempting to use the stolen personal data from Anthem and from OPM to build a directory of Americans who work in sensitive government positions and who can be targeted for further espionage.

In an update of the cyber-theft of OPM records, Nakashima (2015b) reported:

> Two major breaches last year of U.S. government databases holding personnel records and security-clearance files exposed sensitive information about at least 22.1 million people, including not only federal employees and contractors but their families and friends, U.S. officials said Thursday. ... The vast majority ... were included in an OPM repository of security clearance files ...

Reportedly, the Chinese team of hackers also hacked into insurance and medical companies and United Airlines (Riley and Roberson, 2015),

> That data could be cross-referenced with stolen medical and financial records, revealing possible avenues for blackmailing or recruiting people who have security clearances. ... [including] at least 10 companies and organizations ...

Supporting the claim that the Chinese are not necessarily stealing for commercial benefit but maybe preparing for conflict (Jasper, 2015):

> ... the director of the Federal Bureau of Investigation, James Cook, said China has hacked every big US company looking for useful information; however, the cases investigated by the US Senate related to Chinese hackers breaking into computer networks of private transportation

companies working for the US military point more to preparing the digital battlefield for a potential conflict.

13.2.3 Industry

From Line and colleagues (2014), the "power industry is an attractive target (p. 13) … not well prepared for … targeted attacks" (p. 21). Nicholson and colleagues (2012) note that for railways, power generation, electrical grids, nuclear reactors, etc., cyber-security has been an afterthought, meaning catastrophes and fatalities could occur (p. 418; 421):

> Supervisory Control and Data Acquisition (SCADA) systems are deployed worldwide in many critical infrastructures ranging from power generation ... [to] public transport to industrial manufacturing systems ... [where] the key problems seem to be the increased connectivity and the loss of separation between SCADA and other ... IT infrastructures.

Nicholson and colleagues add that the most common cyber-security threats have been by insiders (p. 422) and operator mistakes (e.g., Chernobyl, p. 424), but that cyber-attacks have exploited SCADA systems (e.g., the first weapon used against a SCADA system was Stuxnet, p. 424). Most SCADA attacks are made in the energy sector, industrial sector second, then transport and health sectors equally (p. 425). Based on their survey, a third of SCADA attacks were by insiders; a third by accidents and security vulnerabilities; and a third by APTs and hackers (p. 425). They recommended best practices (p. 432) and training similar to war games (p. 431).

Other malware espionage codes have been built with Stuxnet (Fidler, 2012). Overall, cyber-espionage creates a formidable problem, possibly a destabilizing cyber-arms race.

13.3 The Problem in Perspective

13.3.1 National Rankings

Hathaway (2013) published the first index ranking the cyber-security of 125 nations; she calculated that along with Australia, Canada, Netherlands, and the United Kingdom, the U.S. is one of the top countries fighting cyber-crime and cyber-terrorism (p. 6).

13.3.2 Economic Harm

Cyber-threats harm the GDP of the USA and other countries, but by how much is unknown. From the *Cyber-Readiness Index 2.0* (Hathaway et al., 2015):

> Equally important is bringing transparency to the GDP erosion from illicit and illegal activities that is ... threatening national security and our economic prosperity ... (p. 18).

Bremmer (2015) at *Time magazine* estimated these costs for cyber-crime:

> Hacking costs the U.S. some $300 billion per year ... Worldwide that figure is closer to $445 billion, or ... 1 percent of global income. The research firm Gartner projects that the world will spend ... $101 billion in 2018—and that still won't be enough.

To keep the problem in perspective, with a GDP of $17.7 trillion, the annual cost of cyber-crime in the U.S. is less than 2% of GDP. Regardless, consumers are mostly indifferent (Popper, 2015):

> At Target, 40 million customers had their credit card information exposed to hackers. At JPMorgan Chase, personal details associated with 80 million accounts were leaked. Last month, a hacker gained access to 4.5 million records from the University of California, Los Angeles, health system. Enormous numbers like these can make it feel as if we're living through an epidemic of data breaches, in which no one's bank account or credit card is safe. But the actual effect on consumers is quite different from what the headlines suggest. Only a tiny number of people exposed by leaks end up paying any costs, and for the rare victims who do, the average cost has actually been falling steadily.

13.3.3 Insurance Perspective—From Biener and Colleagues (2015)

> ... human behavior is the main source of cyber risk, while ... external [cyber-events] are rare ... [and yet] average losses [across all groups] ... are ... similar.

From the insurance perspective, most attacks are in North America (51.9%) compared to Europe (23%) and Asia (18%), and primarily in the financial industry (78.6%, p. 140). Contradicting the Ponemon Institute (2015), these authors ranked as the highest insurance risk the actions of people (e.g., human errors; malicious events by insiders); second were systems and technical failures; and third were failed internal processes; least of all was cyber-risk. (p. 157)

This perspective indicates not only the need for best practices and the need for reporting to reduce the uncertainty about the cyber-attacks actually occurring (Carter and Zheng, 2015):

> Cyber attacks and data breach activity at financial institutions have increased significantly in recent years ... the FBI estimates that over

> 500 million financial records were stolen in 2014, the vast majority by cyber means. (p. 11)

13.3.4 SCADA in Perspective

To place SCADA in perspective, Kenney (2015, p. 112) states that despite the warnings about major disruptions to industry, for the most part, they have failed to materialize. He pointed out that Stuxnet was successful by creating an illusion that the SCADA systems were operating normally even as they were spinning out of control (p. 116). The USA has experienced hundreds of thousands of cyber-attacks, mainly disruptive, not destructive (p. 112). An exception is that in response to Stuxnet, Iran counterattacked against Aramco in the U.S., but with a modest impact (Axelrod and Iliev, 2014). Also, a disgruntled employee in Australia took control of a SCADA system in 2000 to release 800,000 gallons of raw sewage into the environment (p. 124).

In addition, an important exception was made by Admiral Rogers over the hack of Sony (Sanger, 2015b; also Noto, 2014, below):

> In the Sony attack, the theft of emails was secondary to the destruction of much of the company's computer systems ... to intimidate the studio ... from releasing a comedy that portrayed the assassination of Kim Jong-un, the North Korean leader.

13.3.5 U.S. Reported Skills

The U.S. is deeply involved in cyber-intelligence. Reports indicate that the U.S. has the skills and capability to infiltrate networks (Bell, 2015),

> ... the two intelligence agencies [National Security Agency, or NSA; and the UK Government Communications HQ, or GCHQ] hacked into the popular software packages to "track users and infiltrate networks", monitoring email and web traffic to discreetly obtain intelligence. ... repeatedly singled out ... Kaspersky ... adding that GCHQ aimed to subvert Kaspersky anti-virus ... NSA studied Kaspersky's software for weaknesses ...

13.3.6 China's Perspective

China's Vice Premier in 2015, Yang (2015), deflected the problem:

> Over the past six years, direct investment from Chinese companies to the U.S. has increased fivefold, creating more than 80,000 jobs ... A recent study by the National Committee on U.S.-China Relations ... found that Chinese investments in the U.S. now total nearly $50 billion and

will increase to between $100 billion and $200 billion by 2020, creating between 200,000 and 400,000 jobs for U.S. workers.

13.4 Defense Postures

Two types of defense contexts exist: passive and active. Passive defenses, dependent on human factors, have not stopped nor slowed cyber-attacks (Marble et al., 2015). The reason is that passive defenses protect static targets against dynamic attackers (Ge et al., 2014). In contrast, active defenses move "in multiple dimensions to foil the attacker and increase resilience ... [by] migrating a mission critical application from one platform to another ...[or] changing the platform at randomly chosen time intervals" (Okhravi et al., 2012). We focus on the context of active defenses.

13.4.1 Active Defense

An example of active defense comes from Perlroth and Sanger (2015):

> Russian cyber security firm Kaspersky Lab says [the] United States ... [can] permanently embed surveillance and sabotage tools in computers in countries like Iran, Russia, Pakistan and China; ... many of [these] tools ... run on computers ... not connected to Internet, and lets US intelligence agencies unlock scrambled contents unnoticed.

After the OPM breach, active defensive steps were taken (from Shear and Perlroth, 2015):

> At some agencies, 100 percent of users are, for the first time, logging in with two-factor authentication ... Security holes that have lingered for years despite obvious fixes are being patched. And thousands of low-level employees and contractors with access to the nation's most sensitive secrets have been cut off.

From Jasper (2015), active defense "relies on forensic intelligence and automated countermeasures":

> Deterrence instills a belief that a credible threat of unacceptable counteraction exists, that a contemplated action cannot succeed ...

Regarding deterrence, the Editorial Board (2015) from the *Washington Post* added:

> ... the United States should begin preparations for retaliation aimed specifically at the alleged Chinese attackers. ... the thieves must feel the heat ... to deter future attacks.

According to the *New York Times* (Sanger, 2015b),

> The Obama administration has determined that it must retaliate against China for the theft of the personal information of more than 20 million Americans from ... [OPM.] One of the ... actions ... involves finding a way to breach the so-called great firewall, the complex network of censorship and control that the Chinese government keeps in place to suppress dissent inside the country ... to demonstrate to the Chinese leadership that the one thing they value most—keeping absolute control over the country's political dialogue—could be at risk if they do not moderate attacks on the United States.

Innovations provide more active defensives.

13.4.2 New Technology as an Innovative Defense: HTTPS

Instead of HyperText Transfer Protocol (HTTP), secure websites use HyperText Transfer Protocol Secure (HTTPS). HTTPS (SSL, 2015) uses a "code" that scrambles messages so no hackers can read them. The "code" uses a Secure Sockets Layer (SSL), sometimes called Transport Layer Security (TLS), to transmit information safely.

13.4.3 Honeypots

Some systems use decoys to permit intruders to enter, logging the entry of intruders or attackers, giving them safe or false information, and tracing their activities (Even, 2000; Juels, 2015).

Honeypot defenses have lost value in the face of APTs. Honeypots were used to establish a legal case against thiefs. However, for APTs, the thieves may be in a country inaccessible to prosecution (Kotlarchuk, 2015).

13.5 Gathering Information in the Context of Illusions and Deceptions

13.5.1 Deception as an Active Defense

13.5.1.1 Deceiving Convergent Systems Created by Deep Neural Networks with Bistable Illusions

Machine learning systems with deep neural networks (DNNs) have achieved near-human-level performance on a variety of pattern-recognition tasks; e.g., visual classification problems (Goertzel, 2015). But supervised and unsupervised machine learners are vulnerable to the information to train them. For example, changing an image (e.g., of a lion) in a way imperceptible to humans can fool a DNN into labeling the image as, say, a library, indicating brittleness. It is possible to generate data that can shape results by machine learners. Nguyen and colleagues (2015) have shown images unrecognizable to humans, yet DNNs label the data incorrectly with 99.99% confidence (e.g., classifying white noise as a lion). They used convolutional neural networks trained with ImageNet or MNIST datasets and tested images with evolutionary algorithms or gradient ascent to find that DNNs can mislabel with high confidence on these datasets. Their results raise questions about the generality of DNN computer vision.

Another technique that fools machine learners is to have them attempt to solve an illusion (LaFrance, 2015). For example, messages on freedom of the press transmitted in China might be hidden in an image of the Communist party flag (i.e., a steganograph). Further, regardless of training (Fawzi et al., 2015, Sect. 6), all classifiers have been found to be unstable, susceptible to small perturbations that afford multiple interpretations.

The challenge for any system that uncovers deception is to identify the intent of an adversary. In this sense, challenging the relevancy of information, either that used in machine learning or by humans, might uncover the hidden intent and goals of an adversary (Kovach et al., 2015).

13.5.1.2 Polling and Social Networks

The problem of mis-classification also occurs with polling. Nate Silver (Byers, 2015), who rose to fame after successfully predicting two presidential contests, has declared a crisis with polling after failing to predict the outcome of five national and international contests.

> Nate Silver fared terribly in Thursday's UK election ... Silver claimed ... "It's becoming increasingly clear that pre-election polls underestimated how well Conservatives would do and overestimated Labour's result," ... But the problem went beyond the UK. ... Silver went on to cite four examples where the polls had failed to provide an accurate forecast of the election outcome: the Scottish independence referendum, the 2014 U.S. midterms, the Israeli legislative elections, and even the 2012 U.S. presidential election ...

13.5.1.3 Two Associated Problems: Cyber-Stealth and the Majority Illusion

There are two other problems associated with cyber-security. First, Axelrod and Iliev (2014) used gaming theory contexts to address when a cyber-conflict could be imminent based on the stealthiness of a cyber-weapon available; the persistence of a cyber-weapon until it is discovered; and the stakes for the relevant teams and organizations at the launch of a cyber-attack:

> What is most needed is a sophisticated understanding of how to estimate the potential Gains (and losses) from actually using ... [a cyber-weapon in a particular setting. (p. 1302)

Second is the majority illusion discovered in Social Network Analysis (SNA). SNA scientists discovered that social networks can create an illusion that something is common when it is not (MIT-TR, 2015), possibly why some postings "go viral" (Lerman et al., 2015). With this phenomenon, an individual generalizes information after observing the network with a small subset of friends, even though it is rare in the larger network. The majority illusion occurs when the local belief is believed to be more common than the truth, motivating a contagion.

13.5.1.4 Section Summary

To summarize Section 13.5.1 on deception and illusion on deception and illusion, illusions with machine learning, polling and viral beliefs can occur. Lawless and colleagues (2015) have theorized and found that to determine the context, debates (e.g., in politics, courtrooms, and science) can uncover an illusion. This approach of debate to determine the context is general and can be used with teams.

13.5.2 Teams & Firms

Previously we reviewed how individuals contribute to cyber-security or cyber-vulnerabilities (Marble et al., 2015). We expand on our earlier research by advancing the theory of teams. Until now, the difficulty with promoting team solutions has been the intractable theory for teams. We begin by proposing that constructing the best cyber-teams to respond to cyber-attacks is critical, questioning what it means to have a good team; to be competitive; and when are team solutions superior to individual ones. The latter question suggests that a team offers value unavailable to an equivalent set of individuals. We also want to know whether physical training or education is the better path forward.

13.5.2.1 Review of Teams

Briefly, here is what is known about teams regarding leadership; team size; diversity; competition; dysfunction; and training. Afterwards, we will review our theory of teams.

Leadership. From Hackman and Coutu (2009), leaders become less important once a team is performing. In their view, a leader's key functions are to set team boundaries; to determine team members; to ruthlessly minimize team members; to set a team's direction; to harness its internal criticisms; and to coach the team as a team. In addition, the clash from the differences in opinions and attitudes for diverse teams makes them unstable (Harrison and Klein, 2007). Thus, leadership is needed to hold diverse teams together.

Team size. Regarding a team's correct size, Hackman and Coutu (2009) concluded that large teams underperform despite having the resources afforded from having extra personnel,

> "Big teams usually just wind up wasting everybody's time." Better to use the smallest possible team that can get the job done.

But the exact size of teams remains unsolved; a solution is proposed next.

Diversity. From Karlgaard and Malone (2015), teams should be constructed with maximum diversity. A team with members of different viewpoints is not only less likely to err because of groupthink, but also more likely to come up with novel solutions to problems.... This doesn't mean finding people of different races ... but rather individuals with different life experiences, talents, cognitive skills and personalities.

Page (2010) has explored the contributions of diversity within complex systems. Unlike systems near equilibrium, where diversity produces a variation around the mean, in a complex adaptive system, diversity increases robustness; allows for multiple responses to external and internal perturbations; and drives novelty and innovation.

However, is diversity open-ended? A cyber-threat can often pose as an ill-defined problem. In that case, estimating the configuration of opposing forces is critical to team performance; e.g., scientific research. On the productivity of scientists in interdisciplinary teams, Cummings (2015) found that research by teams composed of different disciplines (diversity) adversely impacted their performance; Cummings also found that the best performing teams were the most interdependent, indicating teams well-fitted to the problem being addressed, a possible answer to the question of size.

Competition. Teams are formed to multitask (Lawless et al., 2015), increasingly important as the problems a team is tasked to solve increase in complexity. Teamwork productivity is important in competitive environments, the winner driving social evolution (Wilson, 2012) by forcing the losing team to adjust for its next round of competition (Lawless et al., 2015). When the best two teams compete, they produce Maximum Entropy (MEP), where MEP represents the fullest exploration for possible solutions to a problem (Martyushev, 2013), interestingly, attracting an audience of observers. Wissner-Gross and Freer (2013) have invoked MEP where "the maximization of entropy production ... [is] a thermodynamic proxy for [an] intelligent observer".

Dysfunction. Functional teams serve as a social resource (Lawless et al., 2015), but dysfunctional teams can end ambitions (Karlgaard and Malone, 2015); e.g.,

> If history is any guide, many ... [Presidential] contenders will fall to earth ... dragged down by dysfunctional organizations ...

Physical training. The activity that improves teamwork best is to physically train teams with the best teammates available; in contrast, education does not appear to improve teams (e.g., air combat maneuvering; in Lawless et al., 2010). For the best Cyber-security training, to generate MEP, red "adversary" teams opposing blue "friendly" teams should be roughly equal in skill, creating a Nash equilibrium when two competitors independently make a choice that cannot be mutually improved (Blasch et al., 2015). In contrast, officials of Homeland Security charged with protecting cyber-security are faced by "bureaucratic obstacles [to hiring and firing that] make it difficult to compete in the cutthroat war for talented security specialists" (Shear and Perlroth, 2015).

13.5.2.1.1 Theory. Under Theory, We Cover Multitasking and Uncertainty (e.g., action-observation uncertainty)

Multitasking. Individuals are poor at multitasking (Wickens, 1992). In contrast to individuals, the function of teams is to multitask (Lawless et al., 2015). Characterizing teamwork and serving as a metric, teamwork interdependently defends the physical and social boundaries around the assets that the same individuals cannot defend as independent individuals. A team does this by converting a group into a configuration that multitasks to solve a well-defined or ill-defined problem better than the same individuals working alone; this result occurs by reducing the degrees of freedom among the members of a team (Kenny et al., 1998). This reduction serves several functions: *Its degree of interdependence* converts a group of independent individuals into a team; the reduction in its degrees of freedom measures team effectiveness; its reduction underscores that interdependence is a team and social resource. It provides an objective means to know when a team is effective against competition and threats, like cyber-attacks.

Action-observation uncertainty. Unraveling the puzzle of what makes teams and organizations work effectively is unclear. The traditional approach to teams has been the cognitive approach (e.g., Bell et al., 2012). But what people cognitively (self-report) about behavior is often not how people actually behave, known as mind-body dualism (i.e., uncertainty from subjective-objective complementarity). To develop self-aware machines and robots, Jack (2014) concludes:

> ... how can we relate our knowledge of the physical to our knowledge of the mental? ... is [dualism] merely an illusion ... Why have philosophers been more concerned by the distinction between minds and machines than by the distinction between self-awareness and external awareness?

Bohr (1955) first recognized that action-observation duality among humans was an example of complementarity: Uncertainty in action at the human level maps to uncertainty in momentum at the atomic level; uncertainty in observation at the human level maps to uncertainty in position at the atomic level. We propose that the information about action-observation uncertainty collected from an interaction not only causes turn-taking and belief incompleteness, the uncertainty during each turn as an actor speaks complements a listener as an observer listens (Lawless et al., 2015). Complementarity implies that a single, simultaneous ground truth for action and observation is impossible. As an observation of an action converges to a single belief, incompleteness precludes observers from fully grasping the essence of the action, both supported by experimental results (Zell and Krizan, 2014).

13.5.2.2 Energy-time Uncertainty

Mindful that the sum of squared frequencies is power, we have generalized Bohr to include energy-time uncertainty partly based on Cohen's (1995) conclusion about the transformation between Fourier pairs in signal detection, where a

> narrow [time] waveform yields a wide [frequency] spectrum, and a wide waveform yields a narrow spectrum and that both the time waveform and frequency spectrum cannot be made arbitrarily small simultaneously. (p. 45)

Cohen provides a Heisenberg energy uncertainty principle (Lawless et al., 2010). All organisms, including humans and their teams, need energy to survive (Nicolis and Prigogine, 1989). Survival demands that the energy be collected effectively, and possibly, efficiently. Individual humans multitask poorly (Wickens, 1992) but the purpose of teams is to multitask, an uncertainty tradeoff. Independent of the cognitions and uncertainties involved, enough energy must be collected to offset entropy losses to keep a team's structure intact, disarm attacks against it, and to allow a team to fully address the problem(s) for which it was constructed (i.e., MEP). By reducing its degrees of freedom, a team converts a group into a configuration that multitasks to solve a dedicated problem better than the same individuals working independently (Lawless et al., 2015). This result allows a multitasking team to apply more energy to solving a problem than a collection of the same individuals acting independently.

Productivity is the effectiveness of applying energy; it also offers an example of team size and productivity, Apple's productivity is about $2.1 Million per employee compared to the industry average of less than $500,000 per employee (Elmer-DeWitt, 2013): "Roughly 1,000 companies make smartphones. Just one reaps nearly all the profits." From Ovide and Wakabayashi (2015):

> Apple Inc. recorded 92% of the total operating income from the world's eight top smartphone makers in the first quarter, up from 65% a year earlier, estimates Canaccord Genuity managing director Mike Walkley. Samsung Electronics Co. took 15% ...

Apple's results suggest that a Hilbert space based on counting the productivity versus the number of agents is possible. In addition to physical counts to determine productivity, by combining cognitive and physical data, the agreement about a team's goals among its team members and those of an opposing teams's goals can be determined (where *cos* 0 deg equals 1 for agreement, and *cos* 90 deg equals 0 for no agreement; from van Rijsbergen, 1979).

Energy considerations are exemplified by mergers that seek to gain control of a market beset by rapid advances in technology or government regulation, as with the recent agreement by Aetna and Humana (Mathews et al., 2015). The destabilization of an industry across the marketplace from cyber-theft has also motivated mergers; e.g., from Noto (2014):

> Sony Pictures ... stands to lose millions of dollars from the recent attack by the hacker group known as Guardians of Peace, which posted illegal copies of upcoming Sony movies on file-sharing hubs.... Sony's recent experience serves as a cyber security wake-up call for companies big and small, but it also spells opportunities for acquirers.

With cyber-security as the driving force, Noto (2014) adds, "The pace of acquisitions isn't expected to slow down as the urgency to fend against cyber-threats continues to grow."

Mergers are also important in the context of battle-management plans (Ridder, 2015). Suppose that a human pilot leading a wave of unmanned aircraft (UAVs) is in a sector under-represented with targets alongside of another sector overloaded with targets; in this context, a merger might make both sectors of blue forces more effective. Well-defined problems like playing football, building a car, or operating cyber-security in a low risk environment, imply that an optimum team configuration exists as the minimum number of teammates necessary to solve the targeted problem (Hackman and Coutu, 2009). Not all mergers succeed, likely because the problem to be solved is ill-posed, making the estimation of team size with intelligence difficult.

13.5.2.3 A proposed Mathematical Solution using Artificial Intelligence (AI)

Given that China is involved in cyber-theft due to its inability to innovate, an inability likely due to its inferior, centralized command-decision approach to making social decisions compared to the West's open distributed approach; i.e., decisions regarding political, scientific, and legal matters (Rothstein, 2015). Regularly devalued in the literature (e.g., Layman et al., 2006; Kitto and Widdows, 2015), Western democracies mine their vast stores of information and knowledge by pitting two polarized groups against each other in front of witnesses (e.g., audiences; citizens; juries), where the witnesses commonly help to determine the context of the decisions proposed by polarized opponents (e.g., the polarization in the US Senate; in Moody and Mucha, 2013).

Polarization in a context arises only with "free speech" (Holmes, 1919), found across the political spectrum and often associated with intractable problems (e.g., climate change, from Ginsberg, 2011). In this view, an audience decides the winner of a debate. We speculate that the polarized, orthogonal beliefs (characterized by *cos* 90 deg = 0), not only helps an audience to make better decisions to improve social well-being, but also they reduce the likelihood of tragic mistakes as proposed for the US Navy by Smallman (2012), especially for the use of advanced technology in complex situations. We believe that this process can be modeled with AI.

13.5.2.4 Boundary Maintenance Trade-offs: LEP and MEP

We have concluded that the best performing teams have the strongest boundaries (Lawless et al., 2015). Strong boundaries serve multiple functions: they allow unfettered communication and coordination among a team's members; they reduce external interference; and they help to identify a team as the in-group from those not in their group (Tajfel, 1970). However, strong boundaries require a defense. For optimum MEP, we calculate that the entropy production to protect a boundary in the limit goes to zero; i.e., least entropy production (LEP; see Nicolis and Prigogine, 1989), suggesting a tradeoff balances boundary defense (LEP) and problem-solving tasks (MEP). Often, the tendency is to do the opposite for cyber-defense, i.e., to reduce MEP to expend the maximum effort sealing a cyber-border as does China (Areddy, 2015):

> [China] envisions a future in which governments patrol online discourse like border-control agents, rather than let the U.S., long the world's digital leader, dictate the rules. President Xi Jinping ... is moving to exert influence over virtually every part of the digital world in China.... Mr. Xi is trying to fracture the international system that makes the Internet basically the same everywhere, and is pressuring foreign companies to help.

China's focus on suppression and censorship reduces the information that drives innovation, exactly what China has experienced. Suppression is manpower intensive (Dou, 2015):

> The physical police units at Web companies are part of Beijing's broader efforts to exert greater control over China's Internet. China tightened regulation of social networks ... banning usernames that could harm national security or promote illegal services. The country has long required Web companies to delete accounts that it believes are spreading rumors, criticizing the Chinese Communist Party or ... contain illegal content.

But in promoting cyber-thefts, because it is centrally controlled and suppresses the citizens who could help it to solve the social problems it faces, China can make a mistake with cyber-thefts as it is has by failing to clean up its environment (Qiu, 2011), or by interfering with its stock market (Frangos, 2015), the latter interference troubling to the World Bank (Magnier, 2015).

13.5.2.5 Cross-entropy Methods: Brazil and Exxon

We address size in this section. Recently in the *Washington Post* (DeYoung, 2015):

> When Susan E. Rice took over as President Obama's national security adviser two years ago, she was struck by how the White House had grown. Since she had last served on the National Security Council, during the Clinton administration, its staff had nearly quadrupled in size, to about 400 people. ... Rice embarked on an effort to trim that number, hoping to make the policymaking process more agile.

Although intuitively suggestive, what is a satisfactory team size rationally? We addressed this question with nine oil companies in different countries (the data in Figure 13.1 are from: Petrobras-Brazil; Sinopec-China; Gazprom-Russia; Pemex-Mexico; Rosneft-Russia; BP-UK; Exxon-USA; Royal Dutch Shell-UK and Netherlands; Chevron-USA).*

Statistically, in two-tailed tests, the daily productivity from these oil companies strongly correlated with their number of employees (r=0.54, p=n.s.) but weakly correlated with its country's freedom index (r=.21, p=n.s.). The correlation between an oil company's number of employees and its country's freedom index was nearly significant at the .05 level (r=.64, p=n.s.). These results suggests if an oil company is controlled by an autocratic government, the less likely that its size will be appropriate for the problem that is being addressed.

* Oil production in 2012 from http://www.forbes.com/pictures/em45gmmg/not-just-the-usual-suspects/; oil employees in 2013 from http://www.oilgaspost.com/2013/04/25/top-20-largest-oil-gas-employers/; and country of origin Freedom Index in 2015 from http://www.heritage.org/index/ranking.

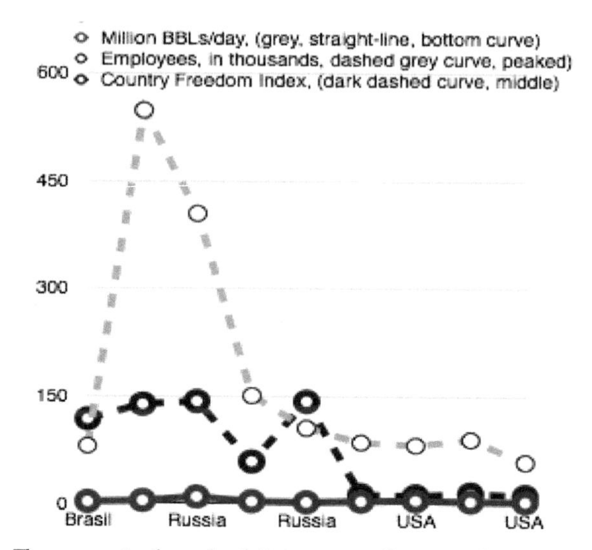

Figure 13.1. The amount of productivity among oil companies is within an order of magnitude, but wide variance exists for the number of employees working at a company based on the company's country of origin.

13.5.2.6 Section Summary

In this section, we reviewed the theory of teams, until now, an intractable problem. We began by proposing that the best cyber-teams respond most effectively to cyber-attacks. We found that a good team reduces the degrees of freedom of its members; that this reduction only occurs with competition among teams to determine the best team; and that physical training for teams using physical skills is superior to education (e.g., flying jets in formation; air-to-air combat; team sports; and cyber-securing physical boundaries). Finally, multi-tasking team solutions to problems are superior to individual ones for multitasking contexts. A team of multitasking members is more likely to reach MEP, obtaining a maximum performance far superior to an equivalent set of individuals.

In response to the burgeoning interest in cyber-security, Colbaugh and Glass (2012) wrote there is an increasing:

> … interest in developing predictive capabilities for social diffusion processes, for instance, to permit early identification of emerging contentious situations, rapid detection of disease outbreaks, or accurate forecasting of the ultimate reach of potentially "viral" ideas or behaviors.

With a machine learning algorithm based on the cross entropy* in data from WikiLeaks versus the Japan earthquake-tsunami in 2011, Colbaugh and Glass (2012b) predicted accurately that:

> … in the case of the WikiLeaks DDoS, the blog entropy ... experiences a dramatic increase several days before the event, while in the case of the Japan earthquake blog entropy is small for the entire collection period. Similar social media behavior is observed for all events in the case study, suggesting that network dynamics-based features, such as dispersion of discussions across blog network communities, may be a useful early indicator for large mobilization events.

What about cyber-hackers? Aggressive actions of deterrence may cause collateral damage or an arms race, but what about the value of alternative interpretations in the case of cyber-hacking?

We conclude that individuals acting alone are unable to determine the context under uncertain conditions (Lawless et al., 2015). To best determine the context under uncertainty and to resolve the questions raised by Nguyen and colleagues, Silver, and others, Lawless and colleagues have argued that it takes two teams of competitors to pose as orthogonal operators while each struggles independently to assemble the best evidence available for the interpretation of the context that best supports its team's self-interest (based on Smallman, 2012); also needed is a third team of superposed neutrals free to join with or leave either team as they judge who has made the most persuasive argument (similar to quorum sensing among bees, a product of disagreement and contest rather that consensus or compromise; see Seeley and Visscher, 2004). Interestingly, although they have not yet offered a solution as detailed as ours, Nguyen and colleagues (2015b) propose informally that searching over multiple objectives approximates a divergent search, more or less what we have formally proposed with our orthogonal approach.

13.6 Mathematical Models & Predictive Analyses

13.6.1 From Lampson (1973), historical boundary controls on multi-level computer systems are concerned with information leaking from a high level (High) to a lower level (Low).

Paraphrasing Lampson, "… these covert channels are communication channels from High to Low that should not exist by system design, yet they still exist …". Covert channels are an unintended consequence of both

* On average, the cross-entropy is the entropy that identifies an event compared to another event when taken from two probability distributions; it measures how close the newly identified event is to the compared event.

hardware and software complexity, coupled with the need for pragmatic engineering solutions. Even though a countermeasure may be put in place, such as the NRL Network Pump® (Kang et al., 1996), the only way for total security in a system that allows High and Low classes of users is to use a "secure brick". This result is due to the concept of the "small message criterion" (Moskowitz and Kang, 1994).

Traditionally covert channel insecurity has been measured by Shannon's information theory. However, for these types of measurements, other than the exceptions provided by Allwein (2004) and Moskowitz and Kang (1994), very little other work discusses the semantic issues of how dangerous a covert channel may be. This is a serious lack in the literature. The essence of the small message criterion means it is possible that even one bit (as in the one bit from a lantern of Paul Revere's fame) may convey a very important message, and the speed at which that bit, or bits are transmitted may not be as important as the message conveyed.

The issue of physical controls against these small messages is also weak when it comes to the practice of steganography. Greek for covered writing, steganography is the art and science of sending a message so that the existence of the message is only known to the sender and intended receiver. It differs from cryptography, which hides the meaning of a message, but not, necessarily, its existence. The weaknesses of qualitative-only measures of steganographic communication are discussed by Moskowitz and colleagues (2000). Predictive analysis concerned with the inability to detect all steganographic payloads is discussed by Moskowitz and colleagues (2007). In their paper, a proactive approach of removing, rather than detecting steganography, is put forward as a pragmatic solution to the seemingly intractable problem.

The essence of the above models and approaches is that mathematical models and metrics must be viewed with a skeptical eye. Additionally, especially with respect to the stego-scrubber, it is sometimes necessary to think completely outside the box.

We conclude by reminding the reader that the hugely popular methods of public key cryptography are seriously compromised (Shor, 1995) if quantum computation becomes mainstream. This again reminds us that mathematical measures of assurance and security may only have a short shelf life.

13.7 Discussion and Conclusions

The cyber-security context is a critical one (Shear and Perlroth, 2015):

> The dangers are accelerating as hackers repeatedly target computer networks used to collect taxes, secure ports and airports, run air traffic control systems, process student loans, oversee the nation's nuclear

> stockpile, monitor the Federal Reserve and support the armed services. Last year, officials say, there were more than 67,000 computer-related incidents at federal agencies, up from about 5,000 in 2006.

Several context challenges exist. First, testimony before the U.S. Senate in 1998 indicated that a cyber-security programming problem has been unsolved for almost two decades (Timberg, 2015),

> "There's this tremendous push to get code out the door, and we'll fix it later." ... Hackers—the black-hat kind—have consistently outrun efforts to impose security.

A second challenge in any context is to prevent users from becoming insider threats.

A third challenge is to move from reacting to cyber-attacks to managing cyber-threats before the context exists of an attack. After the OPM data thefts (Shear and Perlroth, 2015), digital swat teams descended on government agencies to close many of the worst security vulnerabilities.

A fourth context challenge is to build defenses based on individual psychology (e.g., Marble et al., 2015), which has not worked well, or based on teams. Comparing individuals and teams, we have long known that individuals multitask poorly (Wickens, 1992). But, the function of a team is to convert a group into a configuration that multitasks to solve a dedicated problem better than can the same individuals working alone (Lawless et al., 2015); e.g., teamwork interdependently better defends physical and social assets than the same independent individuals.

The physical boundary of a team is important. However, too much emphasis on its cyber-security by increasing LEP reduces a team's MEP. Thus, a fifth challenge is to find the context where the tradeoff between boundary defense and problem-solving, the primary function of a team, is balanced. By exchanging something of value and information, all (social) interactions determine the context, whether clear, uncertain or illusory. Debate clarifies uncertain contexts and challenges illusory ones. But all (human) team structures are dissipative (Nicolis and Prigogine, 1989), requiring that teams be shaped with intelligence to maximize the energy used to address the problems a team faces, forcing a tradeoff between a team's structure and MEP. The stronger a structure, the lower the MEP available for a problem's solution.

Traditional perspectives understate the value of an organization's physical presence, notably its operational and resource boundaries. Undervaluing physical skills to address a problem also overstates the value of cognitive solutions. This conclusion means that traditional approaches focus more on how an organizational boundary impacts its interpretation of context, but not on the tools that physically protect its boundaries. Boundary maintenance is a natural, physical activity based on an organizations's ability to interdependently coordinate its teams engaged in multitasking as its teams attempt to solve the problems that the same individuals acting independently cannot solve alone, like cyber-security. The cognitive approach is likely the reason why the self-reported observations that feed into group theories have been unsuccessful at unraveling the physical properties of teams, like context; why an organization may trust its employee pledges to safeguard its cyber-boundaries; and why that trust can be misplaced.

Regarding training, we conclude that cyber-war games should be held and structured to train "blue" (friendly) teams by playing against "red" (hacker) teams to learn how to best detect hacking, deception and counter them (Heckman and Stech, 2015; Marble et al., 2015).

To multitask optimally, a team must be maximally interdependent (Cummings, 2015). The degree of interdependence among a team's independent teammates informs the effectiveness of a team in the context of competition and threats (cyber-attacks). The value an interdependent team of mates offers is measured by the reduction in a team's degrees of freedom (Kenny et al., 1998), making interdependence a team and a social resource (Lawless et al., 2015).

We have postulated and found evidence suggesting that the best teams are not only highly interdependent (e.g., Cummings, 2015), but minimally fitted to better address the problem(s) they are tasked to solve, easily determined for well-defined problems (e.g., team sports), more problematic for ill-defined problems (e.g., military campaigns; scientific research; innovation). Interestingly, the size of the best teams are dramatically different as a function of whether the government for its country of origin is free or not (e.g., a free market or an autocracy; in Lawless, 2017).

The distribution of MEP among teams across a society may indicate its overall level of intelligence (Wissner-Gross and Freer, 2013). Our results support this speculation in the context that a society's overall intelligence increases as its labor and money are freer to solve the problems a society deems important. We propose to study this idea in future research. Moreover, we suspect that managing a hybrid team (i.e., teams arbitrarily composed of humans, machines and robots) will be similarly affected.

Cyber-threats will always exist. In the context of a tradeoff between the maintenance of boundaries and the function of teams, the inability of the Chinese to innovate is instructive. Putting too much emphasis on defensive barriers impedes the primary task of a team to solve the problems it was designed to solve.

Finally, when deception and illusion affect context, the increased competitiveness suggested by Lawless and colleagues (2015) offers a ready means with AI to disarm them as well as the machine learning illusions discovered by Nguyen and colleagues; the polling problems raised by Silver; and the majority illusions discovered by Lerman and colleagues.

To advance theory for the context of improved cyber-defenses, we conclude that cyber-risks are increasing, as are cyber-defenses, but that a combined new theory can point the way to further improve cognitive and physical cyber-defenses.

References

Allwein, G. 2004. A Qualitative Framework for Shannon Information Theories. Proceedings NSPW 2004.

Areddy, J.T. 2015, 7/28. China Pushes to Rewrite Rules of Global Internet. Officials aim to control online discourse and reduce U.S. influence. Wall Street Journal, from http://www.wsj.com/articles/china-pushes-to-rewrite-rules-of-global-internet-1438112980.

Armour, S. 2015, 8/8. The Doctor Bill From Identity Thieves. Crooks use stolen personal data for medical care, drugs; victims get the tab. Wall Street Journal, from http://www.wsj.com/articles/SB20130211234592774869404581082383327076792.

Axelrod, R. and Iliev, R. 2014. Timing of cyber-conflict, Proceedings of the National Academy of Scientists (PNAS) 111(4): 1298–1303.

Baker, G. 2015, 6/21. Michael Hayden Says U.S. Is Easy Prey for Hackers. Former CIA and NSA chief says 'shame on us' for not protecting critical information better, interview by the Wall Street Journal's editor in chief, from http://www.wsj.com/articles/michael-hayden-says-u-s-is-easy-prey-for-hackers-1434924058.

Barabasi, A.L. 2012. Network science: Understanding the internal organization of complex systems. Invited talk. AAAI Publications, 2012 AAAI Spring Symposium Series, Stanford.

Baumeister, R.F., Campbell, J.D., Krueger, J.I. and Vohs, K.D. 2005, Jan. Exploding the self-esteem myth. Scientific American.

Bell, J. 2015, 6/23. Joint programme designed to 'track users and infiltrate networks'. The Inquirer, from http://www.theinquirer.net/inquirer/news/2414415/nsa-and-gchq-worked-together-to-hack-into-kaspersky-lab-software.

Bell, B.S., Kozlowski, S.W.J. and Blawath, S. 2012. Team learning: A theoretical integration and review. The Oxford Handbook of Organizational Psychology. Steve W. J. Kozlowski (Ed.). New York, Oxford Library of Psychology. Volume 1.

Biener, C., Eling, M. and Wirfs, J.H. 2015. Insurability of Cyber Risk: An Empirical Analysis, The Geneva Papers 40: 131–158.

Blasch, E., Shen, D., Pham, K.D. and Chen, G. 2015. Review of game theory applications for situation awareness. Proceedings, SPIE, 9469: 94690I-1, AFRL, Rome, NY.

Bohr, N. 1955. Science and the unity of knowledge. pp. 44–62. *In*: L. Leary (ed.). The Unity of Knowledge. New York: Doubleday.

Bremmer, I. 2015, 6/19. These 5 Facts Explain the Threat of Cyber Warfare. Time Magazine, from http://time.com/3928086/these-5-facts-explain-the-threat-of-cyber-warfare/.

Byers, D. 2015, 5/8. Nate Silver: Polls are failing us. Politico, from http://www.politico.com/blogs/media/2015/05/nate-silver-polls-are-failing-us-206799.html.

Carter, W.A. and Zheng, D.E. 2015, July. The Evolution of Cyber security Requirements for the U.S. Financial Industry. A Report of the CSIS Strategic Technologies Program, Washington, DC, from https://csis.org/files/publication/150717_Carter_CybersecurityRequirements_Web.pdf.

Cheng, J. 2024, 8/12. Air Force looks to get proactive on cyber defense. DefenseSystems, from http://defensesystems.com/articles/2014/08/12/air-force-cyber-resilience.aspx?m=2.

Cohen, L. 1995. Time-frequency analysis: theory and applications. Prentice Hall Signal Processing Series.

Colbaugh, R. and Glass, K. 2012. Proactive Defense for Evolving Cyber Threats, SANDIA REPORT SAND2012-10177, Sandia National Laboratories, Albuquerque, NM.

Colbaugh, R. and Glass, K. 2012b. Early warning analysis for social diffusion events. pp. 1–22. *In*: Colbaugh, R. and Glass, K. (eds.). Proactive Defense for Evolving Cyber Threats, SANDIA REPORT SAND2012-10177, Sandia National Laboratories, Albuquerque, NM.

Crovitz, G. 2015, 6/14. Blessed by Alibaba, Cursed by Beijing. A giant online market for American goods. A huge cyber raid on U.S. government files. Wall Street Journal, from http://www.wsj.com/articles/blessed-by-alibaba-cursed-by-beijing-1434319492.

Cummings, J. 2015. Team Science Successes and Challenges. National Science Foundation Sponsored Workshop on Fundamentals of Team Science and the Science of Team Science (June 2), Bethesda MD.

DeYoung, K. 2015, 8/4. How the Obama White House runs foreign policy. Washington Post, from https://www.washingtonpost.com/world/national-security/how-the-obama-white-house-runs-foreign-policy/2015/08/04/2befb960-2fd7-11e5-8353-1215475949f4_story.html.

Dou, E. 2015, 8/5. China to Embed Internet Police in Tech Firms. Cyber cops are Beijing's latest bid to reduce online freedoms and prevent 'spreading of rumors'. Wall Street Journal, from http://www.wsj.com/articles/china-to-embed-internet-police-in-tech-firms-1438755985.

Editorial Board. 2015, 7/5. The OPM cyber attack was a breach too far. Washington Post, from http://www.washingtonpost.com/opinions/the-opm-cyber-attack-was-a-breach-too-far/2015/07/05/de2b98b2-20e9-11e5-aeb9-a411a84c9d55_story.html.

EB. 2015, 8/8. Regulators Should Develop Rules to Protect Cars From Hackers. New York Times Editorial Board, from http://www.nytimes.com/2015/08/09/opinion/sunday/regulators-should-develop-rules-to-protect-cars-from-hackers.html.

Elmer-DeWitt, P. 2013, 10/31. Analyst: Apple's revenue per employee is 'off the charts'. At $2.13 million per full-time equivalent, Apple towers over other tech companies. Fortune, from http://fortune.com/2013/10/31/analyst-apples-revenue-per-employee-is-off-the-charts/.

Even, L.R. 2000, 7/12. Honey pots explained. SANS, http://www.sans.org/security-resources/idfaq/honeypot3.php.

Fawzi, A., Fawzi, O. and Frossard, P. 2015. Fundamental limits on adversarial robustness. Proceedings of the 31st International Conference on Machine Learning, JMLR: Workshop and Conference Proceedings, Vol 37, Lille, France, from http://lts4.epfl.ch/files/content/sites/lts4/files/frossard/publications/pdfs/icml2015a.pdf.

Fidler, D.P. 2012. Tinker, Tailor, Soldier, Duqu: Why cyber-espionage is more dangerous than you think. Internal Journal of Critical Infrastructure Protection 5: 28–29.

Frangos, A. 2015, 7/3. Troubling Lessons in China's Crumbling Stock Market. Wall Street Journal, from http://www.wsj.com/articles/troubling-lessons-in-chinas-crumbling-stock-market-1435913235.

Greenspan, A. 1996, 10/16. Fed Chair in 2000. Remarks by Chairman Alan Greenspan. Technological advances and productivity At the 80th Anniversary Awards Dinner of The Conference Board, New York, NY, from http://www.federalreserve.gov/boarddocs/speeches/1996/19961016.htm.

Ge, L., Yu, W., Shen, D., Chen, G., Pham, K., Blasch, E. and Lu, C. 2014. Toward Effectiveness and Agility of Network Security Situational Awareness Using Moving Target Defense (MTD), SPIE Proceedings, 9085, in Sensors and Systems for Space Applications VII, Khanh D. Pham & Joseph L. Cox (Eds).

Ginsburg, R.B. 2011. American Electric Power Co., Inc., Et Al. V. Connecticut Et Al., 10-174, http://www.supremecourt.gov/opinions/10pdf/10-174.pdf.

Goertzel, B. 2015. Are there Deep Reasons Underlying the Pathologies of Today's Deep Learning Algorithms? OpenCog Foundation, from http://goertzel.org/DeepLearning_v1.pdf.

Goodman, M. 2015. Future crimes. Everything is connected, everyone is vulnerable, and what can be done about it. Doubleday.

Hackman, J.R. and Coutu, D. 2009, 5/1. Why Teams Don't Work. Harvard Business Review, from https://hbr.org/product/why-teams-dont-work/R0905H-PDF-ENG; See also Coutu's interview of Hackman: Coutu, D. 2009, May. Why Teams Don't Work. Harvard Business Review, from https://hbr.org/2009/05/why-teams-dont-work.

Harrison, D.A. and Klein, K.J. 2007. What's the difference? diversity constructs as separation, variety, or disparity in organizations, Academic Management Review 32(4): 1199–1228.

Hathaway, M.E. 2013. Cyber Readiness Index 1.0. Hathaway Global Strategies LLC, from http://belfercenter.hks.harvard.edu/files/cyber-readiness-index-1point0.pdf.

Hathaway, M., Demchak, C., Kerben, J., McArdle, J. and Spidaleri, F. 2015, February, Draft. Cyber-readiness index 2.0. A plan for cyber readiness: A baseline and an index. Potomac Institute for Policy Studies, Arlington, VA; from www.potomacinstitute.org.

Heckman, K.E. and Stech, F.J. 2015. Cyber counterdeception: How to detect denial & deception (D&D). *In*: Jajodia, S., Shakarian, P., Subrahmanian, V., Swarup, V. and Wang, C. (eds.). Cyber Warfare. Building the Scientific Foundation (Describes the latest research on the attribution problem to help identify culprits of cyber attacks), . Springer, Ch. 9 (http://www.springer.com/us/book/9783319140384).

Holmes, O. W. 1919. Dissent: Abrams v. United States.

Hsu, D.F., Marinucci, D. and Voas, J.M. 2015. Cyber security: Toward a secure and sustainable cyber ecosystem, Computer, from.

Jack, A. 2014. Finding the Mind in the Brain: The opposing domains hypothesis. Proceedings, International meeting on: Self-awareness–An emerging field in neurobiology, 17–19 September 2014, Royal Danish Academy for Sciences and Letters, Copenhagen.

Jasper, S. 2015, Spring. Deterring malicious behavior in cyber space. Strategic Studies Quarterly, pp. 60–85, from http://www.au.af.mil/au/ssq/digital/pdf/Spring_2015/jasper.pdf.

Juels, A. 2015, 7/28. A bodyguard of lies: The use of honey objects in information security. Proceedings forthcoming from Springer; presented at the ARO Invitational Workshop on Cyber Deception, George Mason University, Fairfax, VA, July 28–29.

Kang, M.H., Moskowitz, I.S. and Lee, D.C. 1996. A Network Pump. IEEE Transactions on Software Engineering Archive 22(5): 329–338.

Karlgaard, R. and Malone, M.S. 2015, 7/9. Building a Winning Political Team. Ronald Reagan knew how to do it. So did Bill Clinton. Their secret? They ignored the conventional wisdom. Wall Street Journal, from http://www.wsj.com/articles/building-a-winning-political-team-1436482423.

Kenny, D. A., Kashy, D.A., and Bolger, N. 1998. Data analyses in social psychology. Handbook of Social Psychology. D.T. Gilbert, Fiske, S.T. and Lindzey, G. Boston, MA, McGraw-Hill. 4th Ed. 1: 233–65.

Kotlarchuk, Ihor O.E. 2015. retired trial lawyer with the Justice Department's Internal Security Section; personal communication.

Kovach, J., Sadler, L., Suri, N. and Winkler, R. 2015, 5/27. Addressing information management and dissemination challenges for the next generation analyst, Proc. SPIE 9499, Next-Generation Analyst III, 94990C, Barbara D. Broome, Timothy P. Hanratty, David L. Hall and James Llinas, (Eds.). from http://spie.org/Publications/Proceedings/Paper/10.1117/12.2184176?origin_id=x4323.

Kenney, M. 2015, Winter. Cyber-terrorism in a post-stuxnet world, Elsevier's Foreign Policy Research Institute, 111–128.

Kitto, K. and Widdows, D. 2015, 7/16. Ideologies and their points of view. Presented at Quantum Interaction 2015, Filzbach, Switzerland; July 14–17, 2015 (Proceedings forthcoming).

LaFrance, A. 2015. How to Fool a Computer With Optical Illusions. New research highlights the distinction between how artificial intelligence sees and how it knows what it's looking at. The Atlantic, from http://www.theatlantic.com/technology/archive/2014/12/how-to-fool-a-computer-with-optical-illusions/383779/.

Lamont, M. and Molnar, V. 2002. the study of boundaries in the social sciences. Annual Review of Sociology 28: 167–95.

Lampson, B.W. 1973. A Note on the Confinement Problem. Communications of the ACM 16(10): 613–615.

Larter, D. and Tilghman, A. 2015, 6/18. Military clearance OPM data breach 'absolute calamity'. NavyTimes, from http://www.navytimes.com/story/military/2015/06/17/sf-86-security-clearance-breach-troops-affected-opm/28866125/.

Lawless, W.F., Rifkin, S., Sofge, D.A., Hobbs, S.H., Angjellari-Dajci., F., Chaudron, L. and Wood, J. 2010. Conservation of Information: Reverse engineering dark social systems. Structure and Dynamics 4(2).

Lawless, W.F., Moskowitz, I.S., Mittu, R. and Sofge, D.A. 2015, 3/24. Thermodynamics of Teams: Towards a Robust Computational Model of Autonomous Teams. Proceedings AAAI Spring 2015, Stanford University.

Lawless, W.F. 2017. The entangled nature of interdependence. Bistability, irreproducibility and uncertainty. Journal of Mathematical Psychology 78: 51–64.

Layman, G.C., Carsey, T.M. and Horiwitz, J.M. 2006. Party polarization in American politics: Characteristics, causes, and consequences, Annual Review of Political Science 9(1): 83–110.

Lerman, K., Yan, X. and Wu, X.-Z. 2015. The Majority Illusion in Social Networks, arXiv, from http://arxiv.org/abs/1506.03022.

Line, M.B., Zand, A., Stringhini, G. and Kemmerer, R. 2014. Targeted attacks against industrial control systems: Is the power industry prepared? SEGS'14 Proceedings of the 2nd wWorkshop on Smart Energy Grid Security, ACM, pp. 13–22.

Magnier, M. 2015, 7/3. World Bank Deletes Critical Passage on China. Institution removes from report its call on Beijing to reduce government influence on financial system. Wall Street Journal, from http://www.wsj.com/articles/world-bank-deletes-critical-passage-on-china-1435940676.

Marble, J.L., Lawless, W.F., Mittu, R., Coyne, J., Abramson, M. and Sibley, C. 2015. The human factor in cyber security: Robust and intelligent defense. *In:* Jajodia, S., Shakarian, P., Subrahmanian, V., Swarup, V. and Wang, C. (eds.). Cyber Warfare. Building the Scientific Foundation (Describes the latest research on the attribution problem to help identify culprits of cyber attacks). Springer, Ch. 9 (http://www.springer.com/us/book/9783319140384).

Martyushev, L.M. 2013. Entropy and entropy production: Old misconceptions and new breakthroughs, Entropy 15: 1152–70.

Mathews, A.W., Hoffman, L. and Mattiolo, D. 2015, 7/3. Aetna-Humana Merger Marks Sway of Health-Care Law. Companies race to bulk up in a market reshaped by the recent health-care overhaul. Wall Street Journal, from http://www.wsj.com/articles/aetna-humana-merger-marks-sway-of-health-care-law-1435947049.

MIT-TR. 2015, 6/30. The Social-Network Illusion That Tricks Your Mind. Technology Review, from http://www.technologyreview.com/view/538866/the-social-network-illusion-that-tricks-your-mind/.

Moody, J. and Mucha, P.J. 2013. Portrait of Political Party Polarization. Network Science 1(1): 119–121.

Moskowitz, I.S. and Kang, M.H. 1994. Covert Channels – Here to Stay? Proceedings COMPASS 94: 235–243.

Moskowitz, I.S., Longdon, G.E. and Chang L.W. 2000. A New Paradigm Hidden in Steganography. Proceedings NSPW, pp. 41–50.

Moskowitz, I.S., Longdon, G.E. and Chang, LiWu. 2007. A new paradigm hidden in steganography. Proceedings, NSPW '00 Proceedings of the 2000 workshop on New security paradigms, pp. 41–50.

Moskowitz, Ira S., Lawless, W.F., Hyden, Paul, Mittu, Ranjeev and Russell, Stephen. 2015. A Network Science Approach to Entropy and Training, Technical Report AAAI Spring Symposium, Stanford.

Nakashima, E. 2015a, 6/18. Officials: Chinese had access to U.S. security clearance data for one year, Washington Post, from http://www.washingtonpost.com/blogs/federal-eye/

wp/2015/06/18/officials-chinese-had-access-to-u-s-security-clearance-data-for-one-year.

Nakashima, E. 2015b, 7/9. Hacks of OPM databases compromised 22.1 million people, federal authorities say, Washington Post, from http://www.washingtonpost.com/blogs/federal-eye/wp/2015/07/09/hack-of-security-clearance-system-affected-21-5-million-people-federal-authorities-say/.

Nicholson, A., Webber, S., Dyer, S. Patel, T. and Janicke, H. 2012. SCADA security in the light of Cyber-Warfare. Elsevier's Computers and Security 31(4): 418–436.

Nguyen, A., Yosinski, J. and Clune, J. 2015. Deep Neural Networks are Easily Fooled: High Confidence Predictions for Unrecognizable Images. In Computer Vision and Pattern Recognition (CVPR '15), IEEE.

Nguyen, A., Yosinski, J. and Clune, J. 2015b. Innovation Engines: Automated Creativity and Improved Stochastic Optimization via Deep Learning. Proceedings of the Genetic and Evolutionary Computation Conference, from http://www.evolvingai.org/InnovationEngine.

Nicolis, G., and Prigogine, I. 1989. Exploring complexity, New York: Freeman.

Noto, A. 2014, 12/2. Hacked Sony Movies Highlight Opportunities for Buyers of Cyber security Firms. Mergers&Acquisitions, from http://www.themiddlemarket.com/news/tech_telecom/hacked-sony-movies-highlight-opportunities-buyers-cyber-security-firms-253643-1.html.

Okhravi, H., Comella, A., Robinson, E. and Haines, J. 2012. creating a cyber moving target for critical infrastructure applications using platform diversity, International Journal of Critical Infrastructure Protection 5: 30–39.

Ovide, S. and Wakabayashi, D. 2015, 7/12. Apple's Share of Smartphone Industry's Profits Soars to 92%. Apple's share of profits is remarkable given that it sells less than 20% of smartphones. Wall Street Journal, from http://www.wsj.com/articles/apples-share-of-smartphone-industrys-profits-soars-to-92-1436727458.

Paganini, P. 2015, 6/15. Government records stolen in the recent data breach at the US OPM (Office of Personnel Management) are surfacing from the Dark Web. Security Affairs, from http://securityaffairs.co/wordpress/37803/cyber-crime/opm-data-dark-web.html

Page, S.E. 2010. Diversity and Complexity, Princeton University Press.

Pasztor, A. 2015, 6/28. U.S. Panel Aims to Shield Planes From Cyber attack. FAA advisory committee was scheduled to meet this month amid rising concern over vulnerability to computer hackers. Wall Street Journal, from http://www.wsj.com/articles/u-s-panel-aims-to-shield-planes-from-cyber-attack-1435537440.

Perlroth, N. and Sanger, D.E. 2015, 2/16. U.S. Embedded Spyware Overseas, Report Claims. New York Times, from http://www.nytimes.com/2015/02/17/technology/spyware-embedded-by-us-in-foreign-networks-security-firm-says.html?ref=topics&_r=0.

Ponemon Institute (2014, 10/30), 2014 Global Report on the Cost of Cyber Crime. From https://www.ponemon.org/blog/2014-global-report-on-the-cost-of-cyber-crime.

Ponemon. 2015, February. 2015 Global Megatrends in Cyber security. Ponemon Institute LLC. Sponsored by Raytheon, http://www.raytheon.com/news/rtnwcm/groups/gallery/documents/content/rtn_233811.pdf.

Popper, N. 2015, 7/31. Stolen Consumer Data Is a Smaller Problem Than It Seems. New York Times, from http://www.nytimes.com/2015/08/02/business/stolen-consumer-data-is-a-smaller-problem-than-it-seems.html?abt=0002&abg=1.

Qiu, J. 2011. China to Spend Billions Cleaning Up Groundwater, Science, News, 334(6057): 45; from http://www.sciencemag.org/content/334/6057/745.summary.

Rainie, L., Anderson, J. and Connolly, J. 2014, 10/29. Cyber Attacks Likely to Increase. Pew Research Center, from http://www.pewinternet.org/2014/10/29/cyber-attacks-likely-to-increase/.

Ridder, J. 2015, 6/30. personal communication.

Riley, M. and Roberson, J. 2015, 7/29). China-Tied Hackers That Hit U.S. Said to Breach United Airlines. Bloomberg, from http://www.bloomberg.com/news/articles/2015-07-29/china-tied-hackers-that-hit-u-s-said-to-breach-united-airlines.

Rothstein, E. 2015, 7/17. Practice Makes Perfect. Today leaders promise peace and prosperity. In early modern Europe, their job was to wage war. Wall Street Journal, from http://www.wsj.com/articles/practice-makes-perfect-1437166276.

Sanger, D.E. 2015a, 3/18. U.S. Must Step Up Capacity for Cyber attacks, Chief Argues. New York Times, from http://www.nytimes.com/2015/03/20/us/us-must-step-up-capacity-for-cyber-attacks-chief-argues.html?ref=topics.

Sanger, D. 2015b, 7/31. Decides to Retaliate Against China's Hacking. New York Times, from http://www.nytimes.com/2015/08/01/world/asia/us-decides-to-retaliate-against-chinas-hacking.html?hp&action=click&pgtype=Homepage&module=first-column-region®ion=top-news&WT.nav=top-news&_r=0.

Sanger, D.E., Perlroth, N. and Shear, M.D. 2015, 6/20. Attack Gave Chinese Hackers Privileged Access to U.S. Systems, New York Times, from http://www.nytimes.com/2015/06/21/us/attack-gave-chinese-hackers-privileged-access-to-us-systems.html?ref=todayspaper&_r=0.

Seeley, T.D. and Visscher, P.K. 2004. Quorum sensing during nest-site selection by honeybee swarms. Behavioral Ecololgy Sociobiology 56: 594–601.

Shear, M.D. and Perlroth, N. 2015, 7/18. U.S. vs. Hackers: Still Lopsided Despite Years of Warnings and a Recent Push, New York Times, from http://www.nytimes.com/2015/07/19/us/us-vs-hackers-still-lopsided-despite-years-of-warnings-and-a-recent-push.html.

Shor, P. W. 1995. Algorithms for quantum computation: Discrete logarithms and factoring, in Shafi Goldwasser (ed.). Proc. 35nd Annual Symposium on Foundations of Computer Science, IEEE Computer Society Press (1994), 124–134.

Simon, H. 1972. Theories of bounded rationality, in C.B. McGuire and R. Radner (eds.). Decision and Organization. North-Holland.

Simon, R. 2015, 7/29. Hackers Trick Email Systems Into Wiring Them Large Sums. Scrap processor thought it paid $100,000 to its vendor: 'We in fact had sent a wire to who knows where'. Wall Street Journal, from http://www.wsj.com/articles/hackers-trick-email-systems-into-wiring-them-large-sums-1438209816.

Smallman, H.S. 2012. TAG (Team Assessment Grid): A Coordinating Representation for submarine contact management. SBIR Phase II Contract #: N00014-12-C-0389, ONR Command Decision Making 6.1-6.2 Program Review.

SSL. 2015. HTTPS and HTTP Difference. Comodo, CA Ltd., from https://www.instantssl.com/https-tutorials/what-is-https.html.

Tajfel, H. 1970. Experiments in intergroup discrimination. Scientific American 223(2): 96–102.

Taranto, J. 2015, 6/18. Hacking Government Apart. The administration that just can't administer. Wall Street Journal, from http://www.wsj.com/articles/hacking-government-apart-1434650217.

Timberg, C. 2015, 6/22. NET OF INSECURITY. A disaster foretold—and ignored. LOpht's warnings about the Internet drew notice but little action", Washington Post, from http://www.washingtonpost.com/sf/business/2015/06/22/net-of-insecurity-part-3/.

Tzu, S. 513 BC, estimated. Sun Tzu's Art of War, Ch. 5, Energy, Book 6. Weak Points and Strong, 7th Section, John Watson, LLC; from http://suntzusaid.com/book/6.

van Rijsbergen, C.J. 1979. Information retrieval (2nd Ed.). Information Retrieval Group, University of Glasgow.

Voas, J. 2015. So where are we? A guest opinion editorial. IEEE Transactions on Reliability, 64(2): 538.

Washington, G. 1793, 12/3. Fifth Annual Message, from http://westillholdthesetruths.org/quotes/283/there-is-a-rank-due-to.

Westcott, R. 2015, 4/24. Rail signal upgrade 'could be hacked to cause crashes', BBC, from http://www.bbc.com/news/technology-32402481.

Wickens, C.D. 1992. Engineering psychology and human performance (second edition). Columbus, OH, Merrill.

Wilson, E.O. 2012. The social conquest of earth, New York: Liveright Publishing/W. W. Norton & Company.

Wissner-Gross, A.D. and Freer, C.E. 2013. Causal Entropic Forces, Physical Review Letters, 110, 168702: 1–5.

Yadron, D. and Spector, M. 2015, 7/21. Hackers Show They Can Take Control of Moving Jeep Cherokee. Using a wireless communications system, researchers manipulate the SUV's electronics. Wall Street Journal, from http://www.wsj.com/articles/hackers-show-they-can-take-control-of-moving-jeep-cherokee-1437522078.

Yang, W., China's Vice Premier. 2015, 6/21. U.S.-China Dialogue Pays Dividends. Chinese companies' direct U.S. investment since 2009 has increased fivefold, adding more than 80,000 jobs. Wall Street Journal, from http://www.wsj.com/articles/u-s-china-dialogue-pays-dividends-1434922739.

Zell, E. and Krizan, Z. 2014. Do People Have Insight Into Their Abilities? A Metasynthesis? Perspectives on Psychological Science 9(2): 111–125.

Index

K

M

N

O

P

R

S

T

U

W

9781138320642